Biomass: Remote Sensing Theory

Biomass: Remote Sensing Theory

Edited by **Brad Hill**

NY RESEARCH
P R E S S

New York

Published by NY Research Press,
23 West, 55th Street, Suite 816,
New York, NY 10019, USA
www.nyresearchpress.com

Biomass: Remote Sensing Theory
Edited by Brad Hill

International Standard Book Number: 978-1-63238-062-3 (Hardback)

Printed in the United States of America.

Contents

Preface

The main aim of this book is to educate learners and enhance their research focus by presenting diverse topics covering this vast field. This is an advanced book which compiles significant studies by distinguished experts in the area of analysis. This book addresses successive solutions to the challenges arising in the area of application, along with it; the book provides scope for future developments.

This book primarily discusses the remote sensing theory of biomass. Usually, the term biomass refers to all those materials that have originated from photosynthesis. Nevertheless, it can also apply to animals. It is important to conserve and manage biomass. There are several methods and techniques for its evaluation. One such technique is remote sensing which supplies data about biomass and also includes the appraisal of biodiversity and ecological factors over a large area. The immense capability of remote sensing has received huge attention over the last few decades in several different areas in biological sciences like nutrient status assessment, weed abundance, deforestation, glacial features in Arctic and Antarctic regions, density mapping and depth sounding of coastal and ocean depths. The book covers the following significant issues: various aspects of biomass study and survey, use of remote sensing for evaluation of biomass, evaluation of primary productivity through case studies and evaluating carbon storage in ecosystems.

It was a great honour to edit this book, though there were challenges, as it involved a lot of communication and networking between me and the editorial team. However, the end result was this all-inclusive book covering diverse themes in the field.

Finally, it is important to acknowledge the efforts of the contributors for their excellent chapters, through which a wide variety of issues have been addressed. I would also like to thank my colleagues for their valuable feedback during the making of this book.

Editor

Part 1

Forests

Lidar Remote Sensing for Biomass Assessment

Jacqueline Rosette[1,3,4], Juan Suárez[2,4], Ross Nelson[3],
Sietse Los[1], Bruce Cook[3] and Peter North[1]
[1]Swansea University,
[2]Forest Research
[3]NASA Goddard Space Flight Center,
[4]University of Maryland College Park
[1,2]United Kingdom
[3,4]USA

1. Introduction

Optical remote sensing provides us with a two dimensional representation of land-surface vegetation and its reflectance properties which can be indirectly related to biophysical parameters (e.g. NDVI, LAI, fAPAR, and vegetation cover fraction). However, in our interpretation of the world around us, we use a three-dimensional perspective. The addition of a vertical dimension allows us to gain information to help understand and interpret our surroundings by considering features in the context of their size, volume and spatial relation to each other.

In contrast to estimates of vegetation parameters which can be obtained from passive optical data, active lidar remote sensing offers a unique means of directly estimating biophysical parameters using physical interactions of the emitted laser pulse with the vegetation structure being illuminated. This enables the vertical profile of the vegetation canopy to be represented, not only permitting canopy height, metrics and cover to be calculated but also enabling these to be related to other biophysical parameters such as biomass.

This chapter provides an overview of this technology, giving examples of how lidar data have been applied for forest biomass assessment at different scales from the perspective of satellite, airborne and terrestrial platforms. The chapter concludes with a discussion of further applications of lidar data and a look to the future towards emerging lidar developments.

1.1 Context

Aside from destructive sampling, traditional methods of calculating biomass for forest inventory, monitoring and management often rely on taking field measurements within sample plots, such as diameter at breast height (DBH) or Top/Lorey's height. This effort can be time, cost and labour intensive. Extrapolation of field measurements to larger areas relies on representative sampling of trees within a land-cover type and correct classification of land cover over large areas; both of which have inherent uncertainties.

Lidar remote sensing complements traditional field methods through data analysis which enables the extraction of vegetation parameters that are commonly measured in the field.

Additionally, establishing allometric relationships between lidar and field measurements enables estimates to be extrapolated to stand, forest or national scales, which would not be feasible or very costly using field survey methods alone. Key aspects of biomass estimation from satellite, airborne and terrestrial lidar systems are outlined below.

1.2 Principles of lidar remote sensing

When walking through a woodland on a sunny day, some of the sunlight reaches the ground through gaps between the foliage, woody branches and stems; some produces more diffuse light at the ground by transmittance though the foliage or reflection between different vegetation components and the ground, and some is absorbed by the intercepted surfaces. A proportion of the energy is reflected from these surfaces back towards the source. The same principles apply to lidar.

Lidar (Light Detection and Ranging) is an active remote sensing technology, which involves the emission of laser pulses from the instrument positioned on a platform, towards a target (e.g. woodland). Here, it interacts with the different surfaces it intercepts as outlined above (Figure 1). Features further from the sensor will intercept and reflect the laser energy back to the sensor later than those closer to it.

The area which is illuminated by the laser pulse is known as the lidar 'footprint'. The size of the footprint is determined by the laser divergence and the altitude/distance from the target of the lidar instrument. Whether the footprint is of large dimensions in the region of tens of metres from the altitude of a satellite sensor, tens of centimetres as generally produced from airborne platforms or several millimetres in the case of terrestrial laser scanners, the principles remain the same.

Interactions of the laser pulse with the vegetation depend on the wavelength of the emitted pulse and its reflectance, transmittance and absorption rates for each foliage, bark and background type (e.g. bare soil, litter, snow, etc). At wavelengths of 1064 nm (in the near-infrared region of the spectrum and typical of many lidar systems used for vegetation analysis), reflectance and transmittance values may each be commonly ~45%.

The time for the reflected pulse echoes to be returned to the sensor is measured and, using the fact that the laser pulse travels at the speed of light, the total return distance travelled between the sensor and the intercepted surfaces can be calculated. The distance between the altimeter and the intercepted object is therefore half of this value (Baltsavias, 1999; Wehr *et al.*, 1999). This permits the three-dimensional reproduction of the Earth surface relief and above-surface object structures (e.g. vegetation, ice cover, atmospheric aerosols and cloud structure).

Very accurate timing is necessary to obtain fine vertical resolutions. Lidar time units are generally recorded in nanoseconds (ns), each being equal to approximately 15cm in one-way distance between the sensor and target. Time is measured by a time interval counter, initiated on emission of the pulse and triggered at a specific point on the leading edge of the returned pulse. This position is not immediately evident and therefore is set to occur where the signal voltage reaches a pre-determined threshold value. The steepness of the received pulse (rise time of the pulse) is a principal contributory factor to range accuracy and depends on the combination of numerous factors such as incident light wavelength, reflectivity of targets at that wavelength, spatial distribution of laser energy across the footprint and atmospheric attenuation (Baltsavias, 1999). The return pulse leading edge rise time is therefore formed by the strength of the return signal from the highest intercepted surfaces within the footprint. This will vary with the nature of the surface; flat ice sheets producing abrupt returns with fast leading edge rises and multilayered, complex vegetation creating broad returns (Harding *et al.*, 1998; Ni-Meister *et al.*, 2001).

Fig. 1. Representation of the interception of foliage, bark or ground surfaces by an emitted laser pulse. At each surface, some energy is reflected, transmitted (in the case of foliage) or absorbed.

The location of every returned signal to a known coordinate system is achieved by precise kinematic positioning using differential GPS and orientation parameters by the Inertial Measurement Unit (IMU). The IMU captures orientation parameters of the instrument platform such as pitch, roll and yaw angles. Therefore, the GPS provides the coordinates of the laser source and the IMU indicates the direction of the pulse. With the ranging data accurately measured and time-tagged by the clock, the position of the returned signal can be calculated.

1.3 Full waveform and discrete return systems

A waveform is the signal that is returned to the lidar sensor after having been scattered from surfaces that the laser pulse intercepts. Full waveform lidar systems record the entire returned signal within an elevation range window above a background energy noise threshold. An example of this from NASA's Geoscience Laser Altimeter System (GLAS; Section 2) is shown in Figure 2 (left). The scene shows a two-storey Douglas Fir canopy (*Pseudotsuga menziesii*) on a gentle slope of 4.9°. Typically, for vegetated surfaces on relatively flat ground, a bimodal waveform is produced.

The beginning and end of the waveform signal above the background noise threshold are represented by the upper and lower horizontal blue lines respectively (mean noise + 4.5σ in the case of GLAS). Amplitude of the waveform (x axis) represents both intercepted surface area at each elevation plus the reflectivity of the surfaces at the emitted wavelength (1064nm).

The gradient at the beginning of the signal increases slowly initially due to the relatively small surface area of foliage and branch elements at the uppermost canopy. As the energy penetrates down through the canopy, the waveform amplitude increases as more features are intercepted, before decreasing towards the base of the tree crowns. A small peak, which corresponds to a shorter tree, can be observed above the abrupt, narrow peak, which is returned from the ground. Below the ground surface, the signal can be seen to trail off gradually. This relates to both a gentle slope found at this site plus the effect of multiple scattering between features within the scene, which serves to delay part of the signal that is returned to the sensor.

Due to the complex waveform signal which is produced, this is often simplified using Gaussian decomposition of the waveform (Figure 2, left). Representing the waveform as the sum of the Gaussians, smoothes the signal yet allows a means of retaining and identifying the dominant characteristics of the signal for easier interpretation.

Small footprint lidar systems can produce dense sampling of the target surface. The returned signal is also in the form of a waveform, however with discrete return systems, only designated echoes within the waveform are recorded. These can be the first and last returns, or at times, also a number of intermediate points. The amalgamation of these returns from multiple emitted lidar pulses allows the scene to be reconstructed as a 'point cloud' of geolocated intercepted surfaces. This is seen in Figure 2, right, which illustrates the same location as seen within the GLAS waveform. The small footprint lidar point cloud can be interpreted more intuitively as a dominant upper storey of approximately uniform height and a single tree of lower height at the centre of the scene. Points are coloured with respect to their elevation.

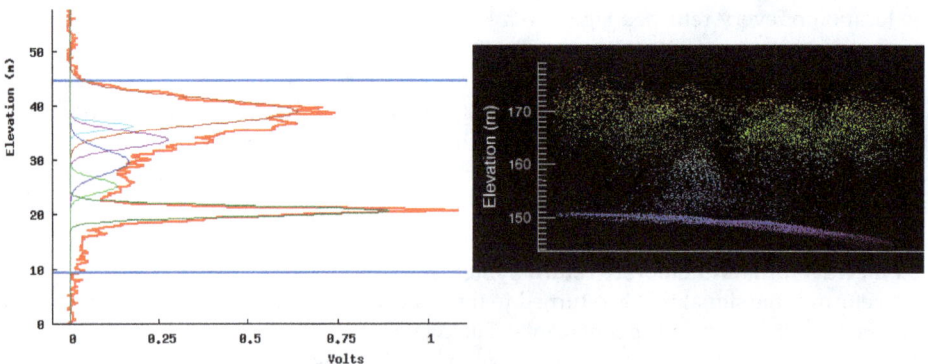

Fig. 2. Example of a waveform produced by a large footprint lidar system (left) and a discrete return lidar point cloud (right) for a coincident area. Location: Forest of Dean, Gloucestershire, UK

1.4 Lidar footprint distribution patterns

Distribution patterns of footprints differ between lidar systems. Lidar profiling involves the systematic location of footprints at intervals along the sensor's path on the ground. These may be contiguous such as the Portable Airborne Laser System of Nelson *et al.*, 2003 (PALS), or placed at discontinuous distances along the ground track in the case of NASA's Geoscience Laser Altimeter System, GLAS (Schutz *et al.*, 2005). This generally permits the sampling of extensive areas, however requires a means to extrapolate biophysical parameter estimates for areas where data were not acquired.

Laser scanning, obtained from an airborne platform, occurs perpendicular to the direction of travel and generally produces a dense distribution of small footprints. Swath width and footprint density are dependent on the altitude and speed of the aircraft plus the scan angle applied. A scanning mirror directs laser pulses back and forth across the flightline causing data to be captured typically in a sawtooth arrangement. The maximum off-nadir scan angle for the instrument can be customised according to the needs of each campaign. Narrower scan angles improve the chances of each shot penetrating dense vegetation canopies and of the sensor receiving a returned pulse from the ground as there is greater likelihood of a clear path through the canopy to the ground. The usual practice is to create an overlap of flightlines similar to photogrammetric surveys of ~60%. Multiple flightlines can then be combined to provide full coverage of the desired area by means of specialised software. Small footprint laser scanning is generally operated at the forest scale, largely due to cost implications.

1.5 Lidar system configurations

As discussed above, lidar sensors can be operated at different scales from different altitudes and different viewing perspectives in relation to the target surface; from above in the case of satellite and airborne systems and from below or to the side for terrestrial laser scanners. Lidar instrument specifications therefore vary considerably, combining different sampling patterns on the ground, density and size of individual laser footprints.

Nelson *et al*'s (2008; 2003; 2004) PALS is an example of a small footprint, discrete return, lidar profiler which is operated from an aircraft. Its innovative and portable design has permitted sampling and vegetation parameter estimation at regional scales throughout the world and may be considered a predecessor to the satellite lidar profiling sensor discussed in Section 2.

The Laser Vegetation Imaging Sensor (LVIS) is an experimental lidar instrument developed at NASA Goddard Space Flight Center (GSFC, 2010). It is a full waveform, scanning lidar that emits a 1064 nm laser beam at a pulse repetition rate of 100-500Hz. LVIS can operate at an altitude in excess of 10 km and this offers the capability of producing swaths up to two km wide and medium-sized footprints of 1-80 m diameter (Blair *et al.*, 1999; Dubayah *et al.*, 2010; GSFC, 2010).

Until relatively recently, small footprint lidar were almost exclusively restricted to discrete return systems within the commercial and operational sector whilst full waveform instruments remained a research and development tool. It should be noted that recent advances in data storage capacity are beginning to open opportunities for small footprint, full waveform scanning systems, however to date, software to process such data is not readily available.

By necessity, this chapter cannot attempt to fully present all combinations of lidar specifications. Readers should note that the multiple vegetation applications of lidar data lead to wide-ranging variations in sensor design and characteristics as outlined above. In the descriptions within sections 2-4, an example of a satellite sensor is used to demonstrate the principles of large footprint, full waveform profiling data, whilst airborne and terrestrial lidar instruments are used to provide examples of small footprint, discrete return laser scanning.

1.6 Key concepts for biomass assessment

Lidar remote sensing provides a direct estimation of the elevation of intercepted features. In the context of vegetation, if signals from the ground and vegetation can be distinguished, the relative heights above ground of forest canopies can be calculated. Since an adequate stem diameter and canopy structure are needed to support tree dimensions, vegetation height is closely related to volume and therefore biomass. The sections below provide examples of applications of lidar systems for biomass assessment.

2. Satellite lidar profiling

NASA's Geoscience Laser Altimeter System (GLAS) aboard the Ice, Cloud and land Elevation Satellite (ICESat) is currently the only satellite lidar system to have provided near global sampling coverage over an extended period of time. It therefore offers a unique dataset of vertical profiles of the Earth's surface.

ICESat was launched in January 2003 and the mission continued until the final laser ceased firing in the Autumn of 2009. During this period, the laser was operated for approximately month-long periods annually during Spring and Autumn, and additionally during the Summer earlier in the mission lifetime.

This is a full waveform, lidar profiling system which operated at an altitude of 600km, travelling at 26,000 km h[-1] and emitting 1064nm laser pulses at 40Hz. This caused the Earth's surface to be sampled at intervals with footprint centres positioned 172m apart. Footprint diameter and eccentricity have varied considerably between laser campaigns from a major axis of 148.6±9.8m to 51.2±1.7m (Figure 3). Comprehensive information regarding the sensor are available from Abshire *et al.*, 2005; Brenner *et al.*, 2003; NSIDC, 2010a; Schutz *et al.*, 2005; Schutz, 2002. Data plus tools to process them are available free of charge from NSIDC, 2010b.

2.1 Characteristics

The systematic sampling pattern and representation of the vegetation profile within the returned lidar waveform signal (Figure 2, left) enables the spatial distribution of vegetation parameters to be mapped for large areas. The seasonal coverage allows near repeat measurements at a frequency which would not be feasible using conventional survey methods.

However the system was designed primarily for cryospheric applications and therefore the configuration is not considered optimal for vegetation analysis. This poses some challenges for forestry applications. Upon sloped, vegetated terrain, vegetation and ground surfaces may occur at the same elevations. This causes the signals from ground and vegetation to be combined within the waveform and, where it is not possible to distinguish a ground peak, this prevents the signal returned from the vegetation from being reasonably identified. Furthermore, dense cloud cover prevents a valid return signal causing gaps in footprint sampling (Figure 3). For regions with high cloud cover such as the tropics, this can serve to worsen the already-sparse sampling density near the equator, produced by the polar orbit.

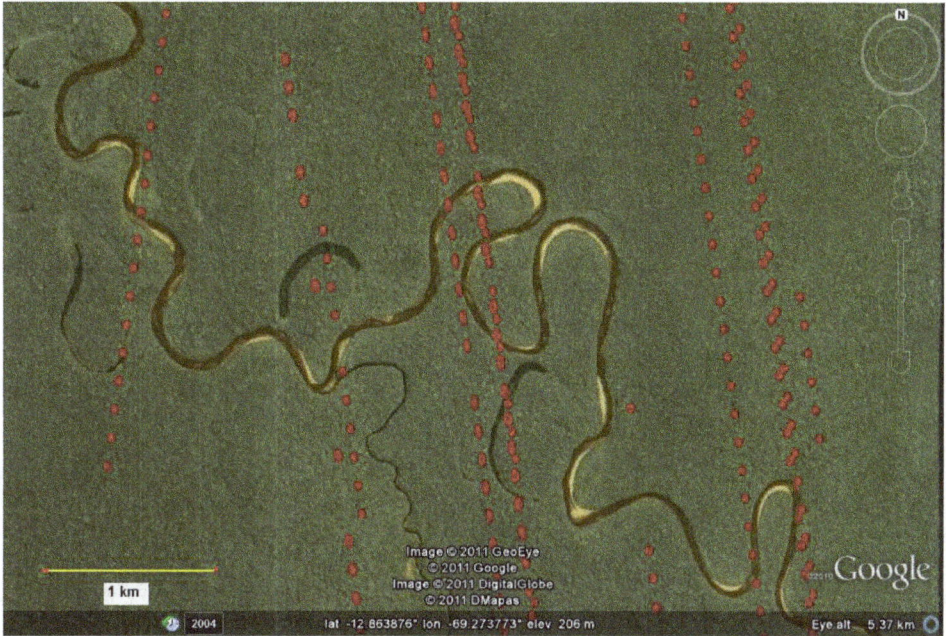

Fig. 3. Multiple GLAS laser campaigns sampling overlaid on a GoogleEarth image of Tambopata, Peru. Missing data are found where dense cloud prevents sufficient energy from penetrating to the Earth's surface and returning to the sensor.

Within each footprint, if the top of the canopy is assumed to be the start of the waveform signal (upper horizontal blue line, figure 2 left), the accuracy with which the signal returned from the vegetation can be identified depends on the ability to identify a representative ground surface within the waveform. Methods to achieve this have included the use of an independent DTM to account for terrain slope within lidar footprints (Lefsky et al., 2005; Rosette et al., 2008) or using Gaussian decomposition of the waveform to locate a peak corresponding to the ground surface (Rosette et al., 2008; Sun et al., 2008a; Sun et al., 2008b). Vegetation height can therefore be estimated as the elevation difference between the start of the waveform signal and the identified ground surface estimated within the waveform. The studies above report RMSE as 3+ metres.

2.2 Applications for biomass estimation

Most commonly, waveform indices of Height of Median Energy ($HOME$) or relative height percentiles above ground (RH_i) are calculated using the cumulative energy distribution within this region of the waveform returned from vegetation. More recently, Lefsky et al., 2007, devised an alternative method of estimating a vegetation height parameter which accounts for terrain and vegetation roughness using the waveform leading and trailing edges rather than isolating the signal returned from vegetation.

The sampling measurements produced within the satellite lidar footprints are typically combined with coincident field measurements of biomass. This enables regression equations to be developed using waveform metrics to estimate biomass for the areas sampled by the lidar footprints. The continuous spatial coverage of optical or radar data permit these estimates to be extrapolated.

Similar data fusion techniques have been used to determine biomass distribution by several authors and encompass a wide breadth of vegetation types and have been applied from regional to continental scales, including a focus on Africa's mangrove forest (Fatoyinbo & Simard, in press 2011), Siberia (Nelson *et al.*, 2009), Quebec (Boudreau *et al.*, 2008), and for mapping throughout Africa (Baccini *et al.*, 2008). Work is currently underway by the latter research group at Woods Hole Research Center, led by Josef Kellndorfer, to estimate tropical forest biomass globally (WHRC, 2011). Additionally, research utilising GLAS is in progress as part of NASA's Carbon Monitoring System initiative (NASA, 2010) to determine biomass distribution within the US (as well as to produce higher resolution biomass maps at a county level using airborne lidar data).

Global vegetation height products derived from GLAS and optical data (Lefsky, 2010; Los *et al.* 2011) or combining GLAS and radar data (Simard, 2011) open possibilities to improve our understanding of global processes (Los *et al.*, 2008) as well as allowing applications for biomass assessment.

However, comparability of data and methods must be taken into consideration. These methods rely on the application of regression equations to extend vegetation parameter estimates across large areas. Nelson, 2010 demonstrates how the calculation of biomass is often sensitive to the equation applied and lidar sensor characteristics. These inconsistencies have implications for repeat analysis and monitoring of change due to the effect of model selection and lidar system evolution on the outcome of biomass assessment.

3. Airborne lidar systems

3.1 Characteristics

The use of airborne laser scanning data in forest applications has attracted increasing interest over the last decade. Nowadays, lidar is perceived to provide a cost-effective and precise assessment of the vertical and horizontal structure of woodland areas and, therefore, a valid alternative or complementary approach to current field methods for inventory (Wulder *et al.*, 2008). Vertical location of points are often reported to 1ns (~15cm) precision whilst horizontal gelocation accuracy may be expected in the region of 20-30cm.

In this section, an overview is given of small footprint airborne laser scanning applications for both stand-level assessment and the estimation of vegetation parameters at an individual tree level. For a comprehensive description of the use of airborne lidar for forestry purposes including forest community structure, growth assessment, tree stability and timber quality, please refer to Suárez, 2010.

Airborne lidar systems provide relatively dense sampling coverage with footprint size in the region of tens of centimetres. However, they are commonly reported to underestimate canopy height as a result of point distribution and density. This is due to some degree of canopy penetration of the signal that varies according to species (Næsset, 2004). In general, this underestimation is less pronounced for cone-shaped trees like spruce or Douglas fir (Pseudotsuga menziesi) than for spherical-shaped trees like many broadleaves or even Scots pine (Pinus sylvestris L.). Conifers normally create more compact shapes with less energy penetration through the canopy than broadleaves. So, energy returns tend to be produced from the outer layers of the canopy. However, the degree of penetration is ultimately related to a combination of factors such as the sampling density, beam divergence and scanning angle (Suárez *et al.*, 2005).

Possible scenarios may be as illustrated in Figure 4 below. With higher density of laser pulses, these difficulties are reduced but at the cost of higher operating expense and flight duration restrictions limiting spatial coverage.

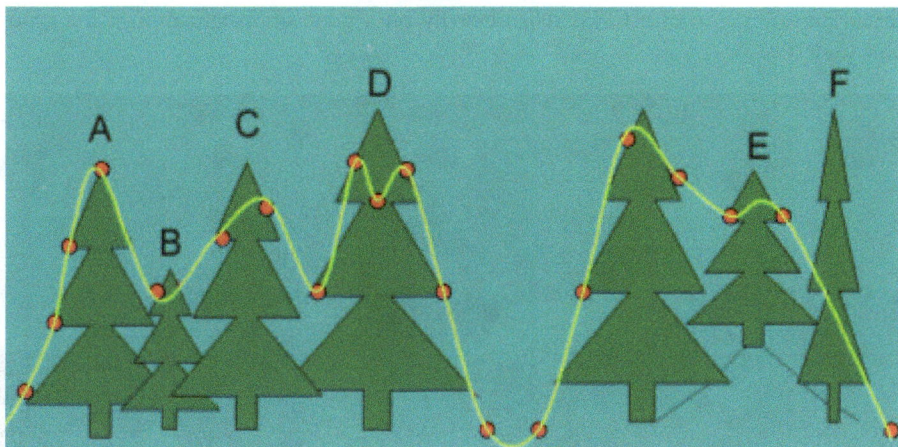

Fig. 4. Possible scenarios of lidar point cloud interception of the forest canopy. Source: Suárez *et al.*, 2005.

a. Laser hits the true top of the canopy,
b. Small trees close to larger neighbours are ignored,
c. The most likely situation: laser returns do not hit the true top of the tree,
d. One of the pulses is intercepted at a lower height due to canopy penetration wrongly suggesting two tree tops,
e. Trees on a mound can be assigned a greater height in the absence of a good model of the ground surface beneath,
f. In a situation of sparse density of returns some trees can be ignored completely

Despite common perception, lidar does not create tomographic images and, therefore, they cannot be considered as 3-D representations in the strict sense. Only the gaps in canopy cover and transmittance through leaves will allow laser energy to be returned from the ground. As with full waveform data, the critical step for the calculation of vegetation height metrics is to distinguish between those points returned from ground and non-ground surfaces.

Since lidar energy penetration through the vegetation canopy will vary with forest structure, density and laser scanning angle in particular, the last return of an emitted lidar pulse may not necessarily be returned from the ground surface. Therefore a means of filtering points is necessary in order to differentiate those returns reaching the ground from those being intercepted at different heights within the canopy (e.g. Kraus & Pfeiffer, 1998; Zhang *et al.*, 2003). A thorough comparison of different approaches and a complete description of filters can be found in Sithole & Vosselman, 2004. This allows the classification of points into ground and vegetation classes (Figure 5). Other algorithms can refine the classification further to additionally identify features such as buildings, electricity cables, etc.

The use of airborne lidar data in forestry was originally focused on the construction of two cartographic products: Digital Terrain Models (DTMs) and Digital Surface Models (DSMs), which are used to describe the underlying terrain and top of forest surfaces respectively. These products are used to generate canopy height models (CHM) that subsequently provide accurate estimates of important forest parameters such as canopy heights, stand volume, and the vertical structure of the forest canopy. The estimation of canopy heights is performed by the subtraction of bare ground values (DTM) from the canopy layer (DSM). An accurate estimation of a CHM relies heavily on a good approximation of the ground surface underneath.

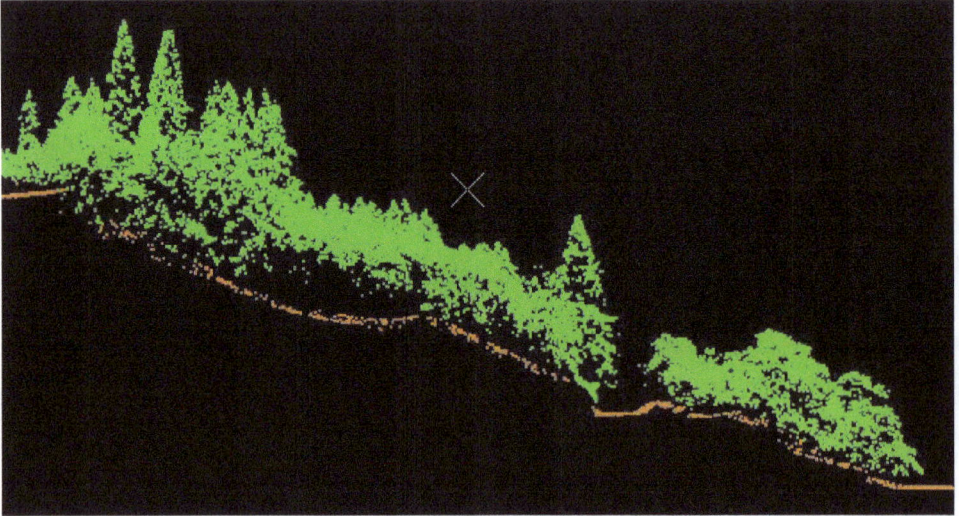

Fig. 5. Vegetation distribution and structure shown by classification of an airborne lidar point cloud. Data illustrated were provided courtesy of Forest Research, UK. Location: Scottish Highlands, UK.

More recently, increasingly sophisticated means of analysis are being applied for the estimation of important parameters at both stand (number of trees, volume, basal area, top height, percentage of canopy cover and crown layers) and individual tree level (individual tree heights, stem diameters and crown metrics). The potential areas of application span from timber production to biological diversity, carbon sequestration or general environmental protection. Section 3.2 below considers both approaches.

3.2 Applications for biomass estimation
3.2.1 Stand level analysis
Stand-level inventory from airborne lidar follows a method originally devised by Næsset, 1997a, b. These two studies are particularly relevant because both have encouraged further work based on the notion that lidar data can be used in large-scale forest inventories, provided that georeferenced data from field plots could be used in a first phase to develop empirical relationships between lidar metrics such as percentiles of relative height above ground and the main parameters for forest management. Such relationships are used to

estimate, in a second phase, forest stand parameters for all the test plots in the study area, known as two–stage procedure for stand inventory (Næsset, 2002).

In this way, the stand-level approach provides a useful estimation of key stand parameters such as top height, canopy cover, tree density, basal area and volume. The established relationships allow forest parameters such as biomass to be directly inferred from the use of lidar metrics and for this assessment to be implemented across large forest areas where there is lidar coverage. Lidar systems provide point-wise anisotropic sampling unlike the full area coverage common in optical systems. As a result, laser data are interpolated in order to convert the same coverage to a continuous surface working-image and allow the distribution of forest parameters to be mapped.

This method does of course have some limitations. It is heavily dependent on abundant field data collection to train empirical relationships between field and lidar data that often are not easily transferable to other study areas. Different relationships may be present for morphologically similar species (Norway spruce, Sitka spruce or larch) and additionally, stand structure is a significant determinant factor. The effects derived from the spatial distribution of gaps, their size and the spatial distribution of standing trees (whether perfectly aligned, growing in a natural stand or thinned at different intensities) on the vertical interception of laser hits are not parameterised in this approach.

However, the real value of this method is demonstrated in its application in regional studies, where the combination of lidar and field measurements can optimise traditional surveys, particularly as part of large area forest inventories (Hollaus et al., 2009).

3.2.2 Individual tree based inventories

Variability in characteristics is a feature of natural systems. Even in systems designed to be as uniform as possible, such as planted forests in monocultures, growing differences are inevitable. Understanding the factors controlling variability is important in developing our knowledge of how ecosystems operate and behave such as the way that trees grow and accumulate biomass in response to their environment and close interaction with their neighbours (e.g. competition for space, light and nutrients). Identifying and locating trees using airborne lidar data permits the spatial variability of biomass distribution within a forest to be considered.

Canopy delineation algorithms have been developed by several authors and most of these detect single trees using an interpolated canopy height model (section 3.1) by the detection of local maxima for tree location and watershed or pouring algorithms (as well as derivations of these algorithms) for the delineation of single tree crowns (e.g. Popescu et al., 2003). However, latterly, approaches are becoming increasingly based on raw data using clustering or blob detection methods (e.g. Morsdorf et al., 2004).

Figure 6 illustrates the delineation of individual tree crowns shown outlined by blue polygons and overlaid on a lidar canopy height model. Lighter grey shades indicate taller heights which are used to identify tree tops to infer height of individual trees.

Using known allometric relationships between tree height and crown width (derived from crown area) with stem diameter at breast height, volume can be calculated on an individual tree basis by approximating tree stems as a cone or by using more sophisticated taper functions or a crop form parameter which defines mean taper characteristics in a stand (Edwards & Christie, 1981; Matthews & Mackie, 2006). Individual tree volume can then be converted to biomass (by accounting for specific density) allowing the spatial distribution and variability of biomass to be mapped.

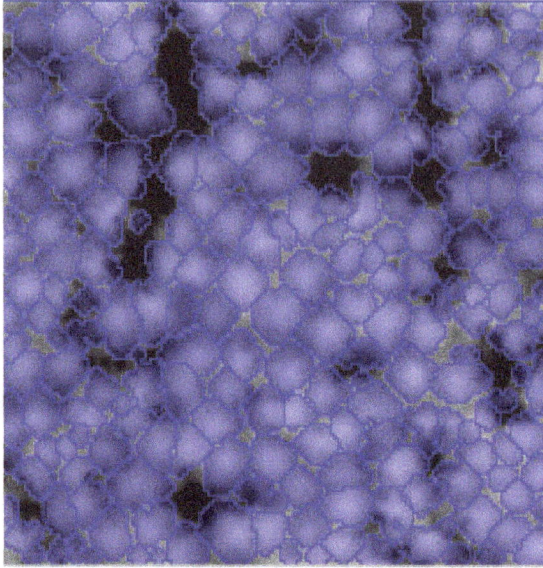

Fig. 6. An example of tree crown delineation with airborne lidar data using object oriented analysis in Definiens Developer (Source: Suárez, 2010).

Results show that lidar can detect most of the crowns of dominant and co-dominant trees in mature stands dominated by coniferous species, but finds difficulties for suppressed or sub-dominant trees, young stands, groups of trees growing in close proximity and for deciduous species. This will have an effect on biomass estimation when aggregating individual tree estimates to a stand level.

Small individuals will contribute relatively less to the overall stand volume than larger trees. However in many situations they may form a significant contribution to total biomass which should be accounted for. Authors such Maltamo *et al.*, 2004 have shown it to be possible to predict the height of small trees not detected by lidar using Weibull distributions.

The detection and estimation of individual tree heights for deciduous species is more difficult because of their more spherical crown shape compared to conifers (meaning that the tree top and crown boundaries can be located with less certainty using a CHM) and the higher probability for the occurrence of more than one apex in each tree. Moreover, crown delineation in deciduous stands becomes more difficult because the crowns of neighbouring trees often overlap (this is a common situation in conifer stands too).

The single tree approach allows individual tree counts, crown volume calculations, canopy closure or single tree height estimates. These are important inputs in order to derive estimates of diameter at breast height (DBH) distributions, volume or biomass. Published studies show that good results can be achieved with 65% to 90% of correct tree counts within conifer stands while for broadleaf trees the results are less accurate and an underestimation of volume is often reported.

Using canopy delineation from airborne lidar data, parameters such as biomass can be determined and spatially located in relation to their surroundings. This also allows the inhibiting effects on biomass accumulation introduced by larger trees over more disadvantaged neighbours to be observed.

3.3 Summary

In operational terms, current usage of airborne lidar data is generally confined to statistical analysis at stand/forest scales (Næsset, 1997a, b). Although the method has undergone minor adaptations by several authors, this approach has become a standard practice for large-scale inventories in some European countries (Finnish_Forest_Association, 2007; Hollaus et al., 2009; Næsset, 2004).

Airborne lidar point clouds can be understood and interpreted intuitively and vegetation can be placed in the context of terrain, access routes and neighbouring competitors. The ability to remove the vegetated surface to reveal the terrain beneath provides valuable information for management purposes and topographic assessment.

Regression equations for biomass estimation may be site specific, however good relationships have been demonstrated using broad class distinctions such as broadleaves, conifers and mixed stands which may be more readily available using optical data or land cover type maps. This enables the spatial distribution of biophysical parameters to be represented over a forest scale which would be impossible using traditional field measurements.

The cost of airborne coverage may prohibit repeat lidar campaigns and so at best, such detailed information is likely to be available infrequently to monitor growth. However, looking to the future, several European countries have undertaken campaigns for partial or complete nation-wide coverage of airborne lidar data including The Netherlands (Duong et al., 2009), Norway, Austria, Switzerland (Swisstopo, 2011) and Finland (Finnish_Forest_Association, 2007). A similar commitment of lidar acquisition is in progress in Spain on a region-by-region basis where the value-added benefit of such a resource has been recognised for economic and social invigoration (Dielmo, 2011).

Whilst generally of mid-low point density by necessity of cost and not necessarily intended specifically for vegetation applications, nevertheless, such previously-unobtainable national vertical profile datasets offer potential applications for a multitude of applications including mapping and reducing uncertainty of biomass distribution. Furthermore, as outlined above, airborne lidar data are already playing a significant role in large-scale forest inventory efforts.

4. Terrestrial laser scanning

4.1 Characteristics

The small and portable terrestrial lidar systems can be mounted on a static tripod or transported on a moving vehicle and therefore can be easily taken into the field. GPS measurements also allow these scans to be geolocated. In contrast to the viewing perspective from above provided by satellite and airborne sensors, terrestrial lidar provides a clear view of the tree stem, understorey and ground surface (Figure 7).

This measurement of a relatively small area within viewing distance of the scanner can be considered to replicate field plot measurements, however additionally provides an understanding of context which would not be possible from field data. The upward looking approach often leads to difficulty in detecting tree tops, however representation of tree stems, ground surface roughness and understorey vegetation offer a level of detail which cannot be retrieved using airborne instruments.

This approach causes only the side of stems facing the scanner to be detected in any one scan and also obscures the view of trees which are behind those closer to the instrument. Therefore the combination of several scans is often required to accumulate adequate information for analysis of the scene within a plot. Low vegetation and heavy branching can affect the quality of the data and in some studies were removed before the site was scanned. This difficulty is especially important with the trees closest to the scanner as they cause the most occlusion.

Research into the detection and reconstruction of stems (e.g. Huang *et al.*, 2011) and branches (e.g. Bucksch & Fleck, 2011) is a current area of development which is of great interest. For forestry purposes this would permit the quality of timber to be determined more easily, such as stem straightness, branch number and branch angle. The ability to reconstruct the tree geometry from terrestrial laser scanning is unprecedented. Whilst for the purposes of forest inventory, the ability to detect the ground surface, height above the ground along the tree stem and to determine the size of the stem allows diameter at breast height to be directly measured, which is one of the most fundamental operational parameters collected by foresters in the field.

Fig. 7. Below-canopy plot sampling using terrestrial laser scanning. Images courtesy of Forest Research, UK.

4.2 Applications for biomass estimation

Terrestrial laser scanning has been applied within the commercial forestry sector. An example of which is TreeMetrics Ltd., Cork, Ireland, who have adapted this technology and developed purpose built software called AutoStem Forest™ to process laser scans into forest inventory-specific data. The principle behind this methodology is to extract greater value from the resource.

The application is intended principally to complement existing inventory methods by creating and measuring individual tree volumes, their straightness and calculating their potential end products. The technology may be considered as a virtual timber harvesting machine that gathers precise data as it scans trees. By positioning laser scan plots throughout a forest, it is possible to measure the variability within the forest and to describe the forest as a Timber Warehouse™ for commercial purposes or to map biomass distribution for inventory applications.

Aside from forestry management objectives, the use of terrestrial lidar can potentially complement traditional field data collection by improving the efficiency and accuracy of survey approaches. Whereas the time required to measure the trees within an inventory plot by a team of surveyors may be quite considerable, the combination of lidar scans of a few minutes each could substantially reduce this whilst providing additional contextual information which could not be achieved with field measurements.

Terrestrial laser scanning measurements are restricted to small area sampling, similar to typical field data collection. However, this permits the plot to be 'revisited' visually and analytically for multiple purposes without returning to the field, allowing the scene to be reconstructed to enable trees to be placed in the context of their immediate surroundings. For management purposes, this could be invaluable to determine optimum thinning or harvesting times or to assess growth trends against model predictions through the measurement of diameter increments.

Once diameter distributions within measured plots are calculated, in the same way as with field approaches, allometric relationships allow wider stand-level forest attributes to be inferred. If applicable species groupings are known, general DBH-based regression equations can be applied to estimate forest stand-level biomass. For sites within the USA, Jenkins *et al.*, 2003, 2004 present diameter-driven allometric equations for biomass of North American species whilst for Europe, a similar resource is provided by Zianis *et al.*, 2005.

5. Further applications of lidar remote sensing

5.1 Forest growth models

Forest growth models such as the Ecosystem Dynamics model, ED, (Hurtt *et al.*, 2004) and the Tree and Stand Simulator, TASS, (Goudie & Stearns-Smith, 2007) enable forest growth scenarios to be predicted. Remote sensing analysis can be used as a valuable tool to provide observational inputs to models and in order to produce detailed inventories for long-term scenario modelling.

Airborne lidar stand level analysis can be used to produce statistically-derived model inputs. This approach is being undertaken as part of the NASA Carbon Monitoring System (NASA, 2010) using an interpolated surface 80th percentile canopy height model as an input to the ED model (e.g. Hurtt *et al.*, 2004). This method can potentially be applied across large areas and could be achieved with relatively low density lidar data such as might be acquired for regional or national campaigns.

Alternatively, Suárez, 2010, used a tree list generated from individual tree delineation as baseline inventory data from which to predict future scenarios and demonstrate processes at work within stands using the distance-dependent model TASS. These processes include competition, establishment of a dominance hierarchy and recovery from catastrophic events such as wind damage or thinning. This means that the biological principles behind such models adapt them to local conditions, unlike empirical models and suggesting wider application may be possible.

The temporal dimension provided by the TASS simulations provides a valuable insight into the long-term effects of each stand intervention or natural disturbance. Not only growth increments, but timber products can also be predicted with this method. In addition, management practices can be balanced by the constraints introduced by the future risk of wind damage. The scale of analysis and the possibility of creating future scenarios contribute to a substantial reduction in the level of uncertainty associated with forest management.

Lidar data therefore provide a useful contribution as a baseline input position from which future scenarios can be determined. Subsequent lidar campaigns or observations of landcover disturbance from optical data (Huang *et al.*, 2010) could furthermore allow model predictions to be validated or calibrated to closer match observed growth trends.

5.2 Prospects for global modelling

Vegetation plays a significant role in global climate, water, energy and biogeochemical cycles, particularly concerning carbon, with approximately one quarter of atmospheric carbon dioxide fixed annually as gross primary production. To accurately model this and other land surface processes in General Circulation Models (GCM), properties such as radiation absorption, plant physiology, surface characteristics and climatology are required. These models require multi-temporal global datasets that can only be obtained from remotely sensed sources.

Computer-generated models of the biosphere provide a valuable means to improve understanding of the immensely complex interactions between interdependent systems affecting the Earth. By their very nature, models function as generalisations of reality and a series of component models replicating the interplay of systems often provide input to complex broader-themed Biosphere models.

Dynamic Vegetation Models are particularly valuable in enabling prediction of the carbon balance under changing ecosystem structure and composition brought about by climatic changes. Vegetation is often represented within each grid cell as generalised Plant Functional Types and climate-driven habitat changes are used to model vegetation succession and plant lifecycle. Where vegetation height is currently considered as static over time, models could benefit from future global lidar observations of vegetation height (e.g. Lefsky, 2010, Los *et al.*, 2011) or biomass, particularly if signal sensitivity permits growth over the sensor lifetime to be observed. Furthermore, the use of lidar could inform validation of LAI, fractional canopy cover or NDVI products (Los *et al.*, 2008, 2011) which are produced using indirect relationships with optical reflectance properties.

6. Emerging technologies

6.1 Waveform simulation and multispectral lidar

Previous work has demonstrated the value of lidar modelling of vegetation for discrete return lidar (Disney *et al.*, 2010) and full waveform systems (Ni-Meister *et al.*, 2001; North *et al.*, 2010; Sun & Ranson, 2000). Using a simulation approach, the sensitivity of lidar data to surface structural and optical properties can be explored to improve our understanding and interpretation of the estimation of lidar-derived biophysical parameters.

Recent discussions have turned to the prospects of multispectral lidar sensors for vegetation analysis (cArbomap, 2011; Woodhouse *et al.*, in press 2011). This project is led by Dr Iain Woodhouse at the University of Edinburgh and the concept has been considered in a simulation study by Morsdorf *et al.*, 2009. The authors demonstrated the opportunity of detecting seasonal and vertical change in normalised difference vegetation index (NDVI) which would allow canopy and ground signals to be distinguished. The variability of chlorophyll content during the growing season was also detected thereby indicating the amount of photosynthetically active biomass.

Additionally, Hancock, 2010 has demonstrated the potential offered by dual wavelength lidar using wavelengths selected either side of the electromagnetic spectrum red edge. A reflectance ratio is calculated and this profile allows the signals within the waveform from ground and vegetation to be differentiated. This would offer a valuable response to the challenging situation of combined vegetation and ground signals within large footprint waveforms on sloped surfaces.

Once issues of eye sensitivity at optical wavelengths and energy requirements are fully addressed, multispectral lidar concepts could offer the opportunity for enhanced vegetation analysis using lidar systems.

6.2 Photon counting lidar systems

The emerging technology of photon counting lidar offers the potential for low energy expenditure and potential high altitude operation allowing extended laser lifetime and large area coverage. This newest type of lidar technology is currently generally operated at green wavelengths (532 nm), in some airborne systems due to a greater efficiency of the detector and, in the case of NASA's ICESat II, as a result of technical readiness. Low laser energy output ensures eye safety of these instruments despite operating at a visible wavelength. A high pulse repetition rate and photon detection probability produces a high point density even whilst flying at greater altitudes whilst a narrow pulse duration (~1ns) allows photons to be located with greater vertical precision.

One significant factor is that photons returned from the emitted pulse cannot be distinguished from ambient noise. Acquiring data at night or dusk would minimise the difficulties of noise posed by solar background illumination, and sensor specifications such as the use of a small detector instantaneous field of view would also assist this.

Initial analysis within NASA's Carbon Monitoring System initiative (NASA, 2010) using the 3D Mapper single photon scanning lidar developed by Sigma Space Corporation, USA, suggests that promising results may be obtained from small footprint photon counting sensors for the generation of vegetation products. The greater point density of the point cloud which is produced, in excess of that which is typically collected by discrete return airborne lidar data, aims to improve the characterisation of vegetation canopies and offers the opportunity for established analysis techniques to be applied to this new technology.

The Slope Imagining Multi-polarisation Photon-counting Lidar (SIMPL) is an example of an airborne small footprint photon-counting profiling lidar which operates at both 1064nm and 532 nm wavelengths (Dabney et al., 2010). A single pulse is emitted which is split into four beams, each with four channels for green and NIR wavelengths, each of which at parallel and perpendicular polarisations. The two polarisations respectively identify photons which have been reflected from a single surface or which have undergone multiple scattering. The four beams are distanced approximately 5 metres apart, producing four profile 'slices' through the canopy. The laser repetition rate of 11.4kHz and an aircraft speed of 100m/second may be expected to produce 5-15 detected pulses per square metre.

Using SIMPL, Harding et al., in press 2011, have explored the influence of lidar wavelength on the ability to determine standard waveform metrics which may be employed to predict biomass. By aggregating detected photons over a distance along the transect, the authors calculated a cumulative height distribution (such as that used for waveform or discrete return analysis). Height of median energy (HOME) and canopy cover metrics were compared and little difference was found between the two wavelengths, suggesting that lidars using 532nm could produce comparable biomass estimates to those obtained by current 1064nm systems.

NASA's forthcoming ICESat II mission is due for launch in early 2016 (GSFC, 2011). In contrast to ICESat I, its successor will carry a medium footprint, photon-counting profiling lidar operating at 532nm wavelength. This instrument is named ATLAS, the Advanced Topographic Laser Altimeter System. The current planned configuration is for a single emitted pulse which is split into six beams, arranged as three adjacent pairs. Each pair will have a stronger and a weaker beam (100μJ and 25μJ respectively) which aims to address issues of detector sensitivity when alternating between bright and dark surfaces such as ice and water. A distance of 3.3km is anticipated between each pair and members of the pair will be separated by 90m. The high repetition rate of 10kHz from an altitude of ~496km will produce overlapping footprints of 10m diameter which will be distanced at 0.7m intervals. 1-3 photons are anticipated to be detected per footprint and, although the spatial location of photons within the footprint will be unknown, the aggregation of returns along the ground tracks will allow a vertical profile to be created. Although, like with its predecessor, the primary objective of ICESat II is not the retrieval of vegetation, one of its science objectives is measuring vegetation height as a basis for estimating large-scale biomass and biomass change (GSFC, 2011). This new technology will offer a new perspective of the world and open opportunities for different approaches to global vegetation analysis.

7. Discussion and conclusion

Laser altimetry is currently the only technique capable of measuring tree heights in closed canopies and therefore offers a remote and non-destructive means of estimating vegetation volume, biomass or carbon content to account for vegetation distribution. This avoids difficulties posed by inaccessibility, time or cost-intensive field campaigns.

The replacement of current field-based methods is not contemplated as a realistic option, however, data collection in the field can be made more effective and targeted as a result of lidar-based inventories. This is already happening in Norway for example, where 90% of stand inventories are being made in relation to lidar surveys (E. Næsset, personal communication).

At present, the retrieval of stand and individual tree parameters is highly dependent on field data collection for the calibration and validation of a sensor's estimates. However, the most efficient use of lidar will require a deeper understanding of the phenomenology of tree interception of the laser hits and how this relates to the physical characteristics of the vegetation being monitored. This understanding can be improved using lidar simulation models. This may offer the possibility to construct more widely applicable height and diameter recovery models using current allometric relationships derived from models or by observations from a network of nationwide permanent sample plots.

The recognition of the importance of biomass mapping and the significant contribution of lidar data for this purpose are demonstrated by the investment and commitment by the US Congress to research in this field at both county and national scales through the NASA-led Carbon Monitoring System initiative (NASA, 2010). This project integrates the use of multiple datasets to generate national and county level biomass products. Elsewhere, the investment in airborne lidar by several governments for national scale campaigns further demonstrates the important role that this technology can play in forest inventory and monitoring.

Such means of identifying areas of forest biomass change can offer important contributions to efforts to inform and encourage practices of Reducing Emissions from Deforestation and forest Degradation in developing countries – REDD (Asner et al., 2010; FAO et al., 2008) and to report on Land Use, Land Use Change and Forestry – LULUCF (IPCC, 2003).

As emerging technologies such as photon counting or multispectral lidar sensors come into operation, the capacity for wider coverage and increasingly accurate lidar-derived applications for biomass assessment will further expand.

8. References

Abshire, J., Sun, X., Riris, H., Sirota, J., McGarry, J., Palm, S., Yi, D. and Liiva, P. (2005). Geoscience Laser Altimeter System (GLAS) on the ICESat Mission: On-orbit measurement performance., *Geophysical Research Letters*, 32: L21S02.

Asner, G., Powell, G., Mascaro, J., Knapp, D., Clark, J., Jacobson, J., Kennedy-Bowdoin, T., Balaji, A., Paez-Acosta, G., Victoria, E., Secada, L. and Valqui, M. (2010). High-resolution forest carbon stocks and emissions in the Amazon, *PNAS*, 107(38): 16738-16742.

Baccini, A., Laporte, N.T., Goetz, S.J., Sun, M. and Dong, H. (2008). A first map of tropical Africa's above-ground biomass derived from satellite imagery, *Environmental Research Letters*, DOI: 10.1088/1748-9326/3/4/045011.

Baltsavias, E.P. (1999). Airborne Laser Scanning - Basic Relations and Formulas, *ISPRS Journal of Photogrammetry and Remote Sensing*, 54: 199-214.

Blair, J.B., Rabine, D.L. and Hofton, M.A. (1999). The Laser Vegetation Imaging Sensor: a medium-altitude, digitisation-only, airborne laser altimeter for mapping vegetation and topography, *ISPRS Journal of Photogrammetry and Remote Sensing*, 54: 115-122.

Boudreau, J., Nelson, R., Margolis, H., Beaudoin, A., Guindon, L. and Kimes, D. (2008). Regional aboveground biomass using airborne and spaceborne LiDAR in Québec, *Remote Sensing of Environment*, 112: 3876-3890.

Brenner, A., Zwally, H., Bentley, C., Csatho', B., Harding, D., Hofton, M., Minster, J.-B., Roberts, L., Saba, J., Thomas, R. and Yi, D. (2003). Derivation of Range and Range Distributions from Laser Pulse Waveform Analysis for Surface Elevations, Roughness, Slope and Vegetation Heights; Algorithm Theoretical Basis Document Version 4.1. NASA Goddard Space Flight Center.

Bucksch, A. and Fleck, S. (2011). Automated Detection of Branch Dimensions in Woody Skeletons of Fruit Tree Canopies, *Photogrammetric Engineering and Remote Sensing*, 77(3): 229-240.

cArbomap (2011). cArbomap; Multispectral lidar. Available online at: carbomap.com.

Dabney, P., Harding, D., Abshire, J., Huss, T., Jodor, G., Machan, R., Marzouk, J., Rush, K., Seas, A., Shuman, C., Sun, X., Valett, S., Vasilyev, A., Yu, A. and Zheng, Y. (2010). The Slope Imaging Multi-polarization Photon-counting Lidar: development and performance results, *IEEE International Geoscience and Remote Sensing Symposium*, 11686732, DOI 10.1109/IGARSS.2010.5650862, pp. 253-256.

Dielmo (2011). Online data server for airborne lidar. Available online at: www.dielmo.com.

Disney, M.I., Kalogirou, V., Lewis, P., Prieto-Blanco, A., Hancock, S. and Pfeifer, M. (2010). Simulating the impact of discrete-return lidar system and survey characteristics over young conifer and broadleaf forests, *Remote Sensing of Environment*, 114: 1546-1560.

Dubayah, R.O., Sheldon, S.L., Clark, D.B., Hofton, M.A., Blair, J.B., Hurtt, G.C. and Chazdon, R.L. (2010). Estimation of tropical forest height and biomass dynamics using lidar remote sensing at La Selva, Costa Rica, *Journal of Geophysical Research*, 115.

Duong, H., Lindenbergh, R. and Vosselman, G. (2009). ICESat Full-Waveform Altimetry Compared to Airborne Laser Scanning Altimetry Over The Netherlands, *IEEE Transactions on Geoscience and Remote Sensing*, 47(10): 3365-3378.

Edwards, P.N. and Christie, J.M. (1981). Yield Models for Forest Management; Booklet 48. The Forestry Commission, Edinburgh.

FAO, UNDP and UNEP (2008). United Nations Collaborative Programme on Reducing Emissions from Deforestation in Developing Countries. [online] available from http://www.undp.org/mdtf/UN-REDD/docs/Annex-A-Framework-Document.pdf accessed May 2009.

Fatoyinbo, T.E. and Simard, M. (in press 2011). Mapping of Africa's mangrove forest extent, height and biomass with ICESat/GLAS and SRTM data fusion, *International Journal of Remote Sensing*.

Finnish_Forest_Association (2007). More exact information on forest resources by remote sensing. Available online at:
http://www.forest.fi/smyforest/foresteng.nsf/10256a7420866fd0c2256b0300248b0 d/2406c3e05b1ee4f7c22572b4002fc660?OpenDocument&Highlight=0,laser.

Goudie, J.W. and Stearns-Smith, S. (2007). TASS-TIPSY advance growth and yield modelling in British Columbia, *LINK*, 8: 7-8.

GSFC (2010). Laser Vegetation Index Sensor; System design.
https://lvis.gsfc.nasa.gov/index.php?option=com_content&task=view&id=29 (accessed May 2010).

GSFC (2011). ICESat-2. Available online at: http://icesat.gsfc.nasa.gov/icesat2/index.php.

Hancock, S. (2010). Understanding the measurement of forests with waveform lidar, PhD thesis, University College London, Available online at:
http://eprints.ucl.ac.uk/20221/.

Harding, D., Dabney, P. and Valett, S. (in press 2011). Polarmetric, two-color, photon-counting laser altimeter measurements of forest canopy structure, SPIE Proceedings LIDAR and RADAR 2011, Nanjing, China.

Harding, D.J., Blair, J.B., Rabine, D.R. and Still, K. (1998). SLICER: Scanning Lidar Imager of Canopies by Echo Recovery Instrument and Product Description. NASA; June 1998.

Hollaus, M., Dorigo, W., Wagner, W., Shadauer, K., Höfle, B. and Maier, B. (2009). Operational area-wide stem volume estimation based on airborne laser scanning and national forest inventory data, *International Journal of Remote Sensing*, 30(19).

Huang, C., Goward, S.N., Masek, J.G., Thomas, N., Zhu, Z. and Vogelmann, J.E. (2010). An Automated Approach for Reconstructing Recent Forest Disturbance History

Using Dense Landsat Time Series Stacks, *Remote Sensing of Environment*, 114: 183-198.

Huang, H., Zhang, L., Gong, P., Cheng, X., Clinton, N., Cao, C., Ni, W. and Wang, L. (2011). Automated Methods for Measuring DBH and Tree Heights with a Commercial Scanning Lidar, *Photogrammetric Engineering and Remote Sensing*, 77(3): 219-227.

Hurtt, G.C., Dubayah, R., Drake, J.B., Moorcroft, P.R., Pacala, S. and Fearon, M. (2004). Beyond potential vegetation: Combining lidar remote sensing and a height-structured ecosystem model for improved estimates of carbon stocks and fluxes, *Ecological Applications*, 14(3): 873-883.

IPCC (2003). Good Practice Guidance for Land Use, Land Use Change and Forestry. Institute for Global Environmental Strategies (IGES) for the Intergovernmental Panel on Climate Change. [online] available from: http://www.ipcc-nggip.iges.or.jp/public/gpglulucf_contents.html accessed May 2009.

Jenkins, J., Chojnacky, D., Heath, L. and Birdsey, R. (2003). National-Scale Biomass Estimators for United States Tree Species, *Forest Science*, 49(1): 12-35.

Jenkins, J., Chojnacky, D., Heath, L. and Birdsey, R. (2004). Comprehensive database of diameter-based regressions for North American tree species, United States Department of Agriculture, Forest Service. General Technical Report NE-319, pp. 1-45.

Kraus, K. and Pfeiffer, N. (1998). Determination of terrain models in wooded areas with airborne laser scanner data, *ISPRS Journal of Photogrammetry and Remote Sensing*, 54: 193-203.

Lefsky, M. (2010). A global forest canopy height map from the Moderate Resolution Imaging Spectroradiometer and the Geoscience Laser Altimeter System, *Geophysical Research Letters*, 37.

Lefsky, M., Harding, D., Keller, M., Cohen, W., Carabajal, C., Del Bom Espirito-Santo, F., Hunter, M., de Oliveira Jr., R. and de Camargo, P. (2005). Estimates of forest canopy height and aboveground biomass using ICESat, *Geophysical Research Letters*, 32 L22S02.

Lefsky, M.A., Keller, M., Pang, Y., de Camargo, P. and Hunter, M.O. (2007). Revised method for forest canopy height estimation from the Geoscience Laser Altimeter System waveforms, *Journal of Applied Remote Sensing*, 1: 1-18.

Los, S.O., Rosette, J.A. and North, P.R.J. (2008). Observational evidence for links between increased drought severity and land-cover change. In: R.A. Hill, J. Rosette and J. Suárez (Editors), Proceedings of *SilviLaser 2008: 8th international conference on LiDAR applications in forest assessment and inventory*, Edinburgh, UK. ISBN: 978-0-85538-774-7.

Los, S. O., Rosette, J. A. B., Kljun, N., North, P. R. J., Suárez, J. C., Hopkinson, C., Hill, R. A. , Chasmer, L. , van Gorsel, E., Mahoney, C. and Berni, J. A. J. (2011). Vegetation height products between 60° S and 60° N from ICESat GLAS data, Geosci. Model Dev. Discuss., 4, 2327-2363, doi: 10.5194/gmdd-4-2327-2011.

Maltamo, M., Eerikäinen, K., Pitkänen, J., Hyyppä, H. and Vehmas, M. (2004). Estimation of timber volume and stem density based on scanning laser altimetry and expected tree size distribution functions, *Remote Sensing of Environment*, 90(3): 319-330.

Matthews, R.W. and Mackie, E.D. (2006). Forest Mensuration; A handbook for practioners. Forestry Commission Publications, Edinburgh.

Morsdorf, F., Meier, E., Kötz, B., Itten, K.I., Dobbertin, M. and Allgöwer, B. (2004). Lidar-based geometric reconstruction of boreal type forest stands at single tree level for forest and wildfire management, *Remote Sensing of Environment*, 3(92): 353-362.

Morsdorf, F., Nichol, C., Malthus, T. and Woodhouse, I.H. (2009). Assessing forest structural and physiological information content of multi-spectral LiDAR waveforms by radiative transfer modelling, *Remote Sensing of Environment*, 113: 2152-2163.

Næsset, E. (1997a). Determination of mean tree height of forest stands using airborne laser scanner data, *ISPRS Journal of Photogrammetry and Remote Sensing*, 52.

Næsset, E. (1997b). Estimating timber volume of forest stands using airborne laser scanner data, *Remote Sensing of Environment*, 61.

Næsset, E. (2002). Predicting forest stand characteristics with airborne scanning laser using a practical two-stage procedure and field data, *Remote Sensing of Environment*, 80: 88-99.

Næsset, E. (2004). Accuracy of forest inventory using airborne laser scanning: evaluating the first nordic full-scale operational project, *Scandinavian Journal of Forest Research*, 19(6).

NASA (2010). NASA Carbon Monitoring System Initiative. Available online at: http://cce.nasa.gov/cce/cms/index.html.

Nelson, R. (2010). Model Effects on GLAS-based Regional Estimates of Forest Biomass and Carbon, *International Journal of Remote Sensing*, 31(5): 1359-1372.

Nelson, R., Næsset, E., Gobakken, T., Ståhl, G. and Gregoire, T. (2008). Regional Forest Inventory using an Airborne Profiling LiDAR, *Journal of Forest Planning*, 13(Special Issue 'Silvilaser'): 287-294.

Nelson, R., Parker, G. and Hom, M. (2003). A Portable Airborne Laser System for Forest Inventory, *Photogrammetric Engineering and Remote Sensing*, 69: 167-273.

Nelson, R., Ranson, K.J., Sun, G., Kimes, D.S., Kharuk, V. and Montesano, P. (2009). Estimating Siberian timber volume using MODIS and ICESat/GLAS, *Remote Sensing of Environment*, 113(3): 691-701.

Nelson, R., Short, A. and Valenti, M. (2004). Measuring Biomass and Carbon in Delaware Using an Airborne Profiling Lidar, *Scandinavian Journal of Forest Research*, 19: 500-511.

Ni-Meister, W., Jupp, D.L.B. and Dubayah, R. (2001). Modeling Lidar Waveforms in Heterogeneous and Discrete Canopies, *IEEE Transactions on Geoscience and Remote Sensing*, 39(9): 1943-1958.

North, P.R.J., Rosette, J.A.B., Suárez, J.C. and Los, S.O. (2010). A Monte Carlo radiative transfer model of satellite waveform lidar, *International Journal of Remote Sensing* 31(5): 1343-1358.

NSIDC (2010a). ICESat Laser Operation Period Attributes. Available online at: http://nsidc.org/data/icesat/docs/glas_laser_ops_attrib.xls National Snow and Ice Data Center, Colorado, USA.

NSIDC (2010b). ICESat/ GLAS Data at NSIDC, Available online at: http://nsidc.org/daac/icesat/ National Snow and Ice Data Center, Colorado, USA

Popescu, S., Wynne, R. and Nelson, R. (2003). Measuring individual tree crown diameter with lidar and assessing its influence on estimating forest volume and biomass, *Canadian Journal of Forest Research*, 29(5): 564-577.

Rosette, J.A.B., North, P.R.J. and Suárez, J.C. (2008). Vegetation Height Estimates for a Mixed Temperate Forest using Satellite Laser Altimetry, *International Journal of Remote Sensing*, 29(5): 1475-1493.

Schutz, B., Zwally, H., Shuman, C., Hancock, D. and DiMarzio, J. (2005). Overview of the ICESat Mission, *Geophysical Research Letters*, 32: L21S01.

Schutz, B.E. (2002). Laser Footprint Location (Geolocation) and Surface Profiles, Algorithm Theoretical Basis Document Version 3.0. NASA Goddard Space Flight Center.

Simard, M. (2011). Combining Lidar and Radar for Remote Sensing of Land Surfaces Available online at: http://lidarradar.jpl.nasa.gov/index.html.

Sithole, G. and Vosselman, G. (2004). Experimental comparison of filter algorithms for bare earth extraction from airborne laser scanning point clouds, *International Archives of Photogrammetry and Remote Sensing*, 59: 85-101.

Suárez, J., Ontiveros, C., Smith, S. and Snape, S. (2005). Use of airborne LiDAR and aerial photography in the estimation of individual tree heights in forestry, *Computers and Geosciences*, 31(2): 253-262.

Suárez, J.C. (2010). An Analysis of the Consequences of Stand Variability in Sitka Spruce Plantations in Britain using a combination of airborne LiDAR analysis and models, PhD, University of Sheffield, Sheffield.

Sun, G. and Ranson, K.J. (2000). Modeling lidar returns from forest canopies, *IEEE Transactions on Geoscience and Remote Sensing*, 38(6): 2617-2626.

Sun, G., Ranson, K.J., Kimes, D.S., Blair, J.B. and Kovacs, K. (2008a). Forest vertical structure from GLAS: An evaluation using LVIS and SRTM data, *Remote Sensing of Environment*, 112(2008): 107-117.

Sun, G., Ranson, K.J., Masek, J., Guo, Z., Pang, Y., Fu, A. and Wang, D. (2008b). Estimation of Tree Height and Forest Biomass from GLAS Data, *Journal of Forest Planning*, 13(Special Issue 'Silvilaser'): 157-164.

Swisstopo (2011). The Federal Office of Topography (swisstopo). Available online at: www.swisstopo.ch.

Wehr, A., Lohr, U. and Baltsavias, E. (1999). Editorial: Theme Issue on Airborne Laser Scanning, *ISPRS Journal of Photogrammetry and Remote Sensing*, 54: 61-63.

WHRC (2011). Pan-tropical Forest Carbon Mapped with Satellite and Field Observations. Available online at: http://www.whrc.org/mapping/pantropical/modis.html.

Woodhouse, I.H., Nichol, C., Sinclair, P., Jack, J., Morsdorf, F., Malthus, T. and Patenaude, G. (in press 2011). A Multispectral Canopy LiDAR Demonstrator Project, *IEEE Geoscience and Remote Sensing Letters*.

Wulder, M., Bater, C., Coops, N.C., Hilker, T. and White, J. (2008). The role of LiDAR in sustainable forest management, *The Forestry Chronicle*, 84(6).

Zhang, K., Chen, S., Whitman, D., Shyu, M., Yan, J. and Zhang, C. (2003). A Progressive Morphological Filter for Removing Nonground Measurements from Airborne Lidar Data, *IEEE Transactions on Geoscience and Remote Sensing*, 41(4): 872-882.

Zianis, D., Muukkonen, P., Mäkipää, R. and Mencuccini, M. (2005). Biomass and stem volume equations for tree species in Europe, *Silva Fennica*, Monographs 4: 1-63.

Biomass Prediction in Tropical Forests: The Canopy Grain Approach

Christophe Proisy[1], Nicolas Barbier[1], Michael Guéroult[2], Raphael Pélissier[1],
Jean-Philippe Gastellu-Etchegorry[3], Eloi Grau[3] and Pierre Couteron[1]

[1]*Institut de Recherche pour le Développement (IRD), UMR AMAP*
[2]*Institut National de la Recherche Agronomique (INRA), UMR AMAP*
[3]*Université Paul Sabatier, UMR CESBIO*
France

1. Introduction

The challenging task of biomass prediction in dense and heterogeneous tropical forest requires a multi-parameter and multi-scale characterization of forest canopies. Completely different forest structures may indeed present similar above ground biomass (AGB) values. This is probably one of the reasons explaining why tropical AGB still resists accurate mapping through remote sensing techniques. There is a clear need to combine optical and radar remote sensing to benefit from their complementary responses to forest characteristics. Radar and Lidar signals are rightly considered to provide adequate measurements of forest structure because of their capability of penetrating and interacting with all the vegetation strata.

However, signal saturation at the lowest radar frequencies is observed at the midlevel of biomass range in tropical forests (Mougin et al. 1999; Imhoff, 1995). Polarimetric Interferometric (PolInsar) data could improve the inversion algorithm by injecting forest interferometric height into the inversion of P-band HV polarization signal. Within this framework, the TROPISAR mission, supported by the Centre National d'Etudes Spatiales (CNES) for the preparation of the European Space Agency (ESA) BIOMASS program is illustrative of both the importance of interdisciplinary research associating forest ecologists and physicists and the importance of combined measurements of forest properties.

Lidar data is a useful technique to characterize the vertical profile of the vegetation cover, (e.g. Zhao et al. 2009) which in combination with radar (Englhart et al. 2011) or optical (e.g. Baccini et al. 2008; Asner et al. 2011) and field plot data may allow vegetation carbon stocks to be mapped over large areas of tropical forest at different resolution scales ranging from 1 hectare to 1 km². However, small-footprint Lidar data are not yet accessible over sufficient extents and with sufficient revisiting time because its operational use for tropical studies remains expensive.

At the opposite, very-high (VHR) resolution imagery, i.e. approximately 1-m resolution, provided by recent satellite like Geoeye, Ikonos, Orbview or Quickbird as well as the forthcoming Pleiades becomes widely available at affordable costs, or even for free in certain regions of the world through Google Earth®. Compared to coarser resolution imagery with pixel size greater than 4 meters, VHR imagery greatly improves thematic information on forest canopies. Indeed, the contrast between sunlit and shadowed trees crowns as visible on such images (Fig. 1) is potentially informative on the structure of the forest canopy. Furthermore, new promising methods now exist for analyzing these fine scale satellite observations (e.g. Bruniquel-Pinel & Gastellu-Etchegorry, 1998; Malhi & Roman-Cuesta, 2008; Rich et al. 2010). In addition, we believe that there is also a great potential in similarly using historical series of digitized aerial photographs that proved to be useful in the past for mapping large extents of unexplored forest (Le Touzey, 1968; Richards, 1996) for quantifying AGB changes through time. This book chapter presents the advancement of a research program undertaken by our team for estimating above ground biomass of mangrove and *terra firme* forests of Amazonia using canopy grain from VHR images (Couteron et al. 2005; Proisy et al. 2007; Barbier et al., 2010; 2011). We present in a first section, the canopy grain notion and the fundamentals of the Fourier-based Textural Ordination (FOTO) method we developed. We then introduce a dual experimental-theoretical approach implemented to understand how canopy structure modifies the reflectance signal and produces a given texture. We discuss, for example, the influence of varying sun-view acquisition conditions on canopy grain characteristics. A second section assesses the potential and limits of the canopy grain approach to predict forest stand structure and more specifically above ground biomass. Perspectives for a better understanding of canopy grain-AGB relationships conclude this work.

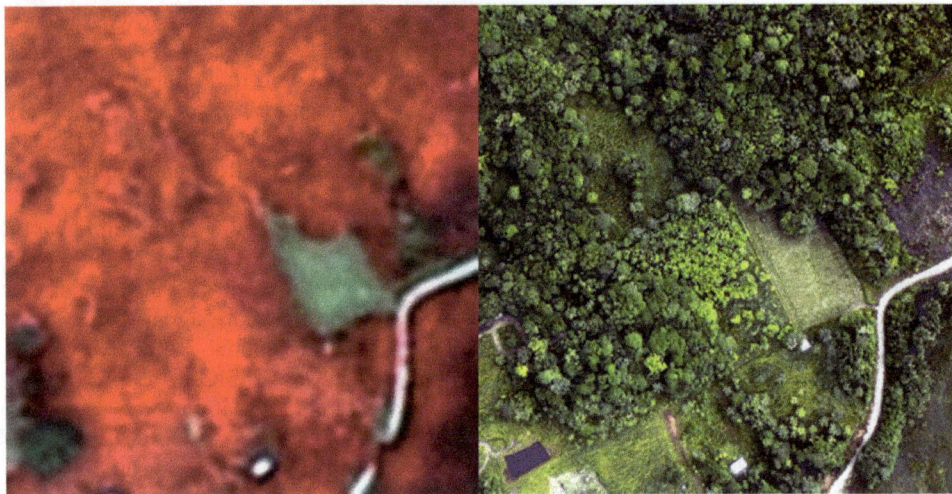

Fig. 1. Differences of canopy grain perception between two 300 m square subset images of different spatial resolution over a mixed savanna forest-inhabited area, French Guiana. Left: a 2.5-m SPOT5 Fusion image acquired in October 2010. Right: a 20-cm aerial photograph acquired in July 2010 (© L'Avion Jaune).

2. The canopy grain approach

2.1 Notion of canopy grain

The notion of canopy grain needs to be clarified. In the context of this study, it refers to the aspect of the uppermost layer of the forest, i.e. the top canopy. It emerges from the images as soon as the contrast between sunlit and shadowed tree crowns becomes perceptible. This property increases with the fineness of image spatial resolution (Fig. 1) that explains why, in VHR images, the tropical forest no longer appears as a continuous homogenous layer, or 'red carpet', as it is the case on medium resolution images with pixel size greater than 5 meters (Fig. 1). Intuitively, the canopy grain depends on both the spatial distribution of trees within a scene and the shapes and dimensions of their crowns. The question is then how to derive quantitative measurements of such canopy grain texture. Following Rao and Lohse (1993), who explained that repetitiveness is the most important dimension of human perception for structural textures, our idea is to measure the degree of repetitiveness expressed in canopy grain within a forest scene. Two dimensional (2D) Fourier or wavelet transforms proved to be well adapted for this purpose (e.g. Couteron, 2002; Ouma et al., 2006) because they allow shifting canopy grain properties from the spatial domain to the frequency domain. Though of larger potential application, we focus in this paper on the 2D Fourier-based frequency spectra as a mean for relating tropical forest canopy grain to above-ground biomass (AGB).

2.2 The FOTO method
2.2.1 Workflow up to forest AGB prediction
The well-known Fourier transform is highly suitable for analyzing repetitiveness of canopy grain as it breaks down an intensity signal into sinusoidal components with different frequencies. We built on this principle the development of the Fourier-based Textural Ordination (FOTO) method to primarily explore the potential of digitized aerial photographs and VHR satellite images for predicting tropical forest stand structure parameters including AGB (Couteron et al., 2005; Proisy et al. 2007). We summarize, hereafter, the flow of operations that yield AGB predictions from FOTO outputs.

A prerequisite of the method is to mask non-forest areas, such as clouds and their shadows, water bodies, savannas, crops and civil infrastructures areas (Fig. 2, step 1). The method then proceeds with the specification of a square window size in which 2D-Fourier spectra are computed (Fig. 2, step 2). To be clear, the window size WS is expressed in meters as:

$$WS = N.\Delta S \qquad (1)$$

where N is the number of pixels in the X or Y direction of the image and ΔS is the pixel size in meters. WS may influence the FOTO results as discussed in the following sub-section. Using large WS also means that spatial resolution of the FOTO outputs and subsequent biomass maps will be N times coarser than the spatial resolution of the source image(s). Although the use of a sliding window is computationally intensive, it can attenuate the effects of both spatial resolution degradation and study areas fringe erosion.

After windowing the forest images, Fourier radial spectra (or r-spectra) are computed and give for each window, the frequency vs. amplitude of a sinusoidal signal that fits the spatial arrangement of pixels grey levels (Fig. 2, step 3) as described in the next paragraph. The r-spectra may be then stacked into a common matrix in which each row corresponds to the r-spectrum of a given window, whereas each column contains amplitude values. This table is then submitted to multivariate analysis techniques (ordinations/classifications). With this approach, the study can concern as many images as necessary, providing they have the same spatial resolution. The resulting table can, for instance, be submitted to a standardized principal component analysis (PCA; Fig. 2, step 4). Window scores on the 3 most prominent

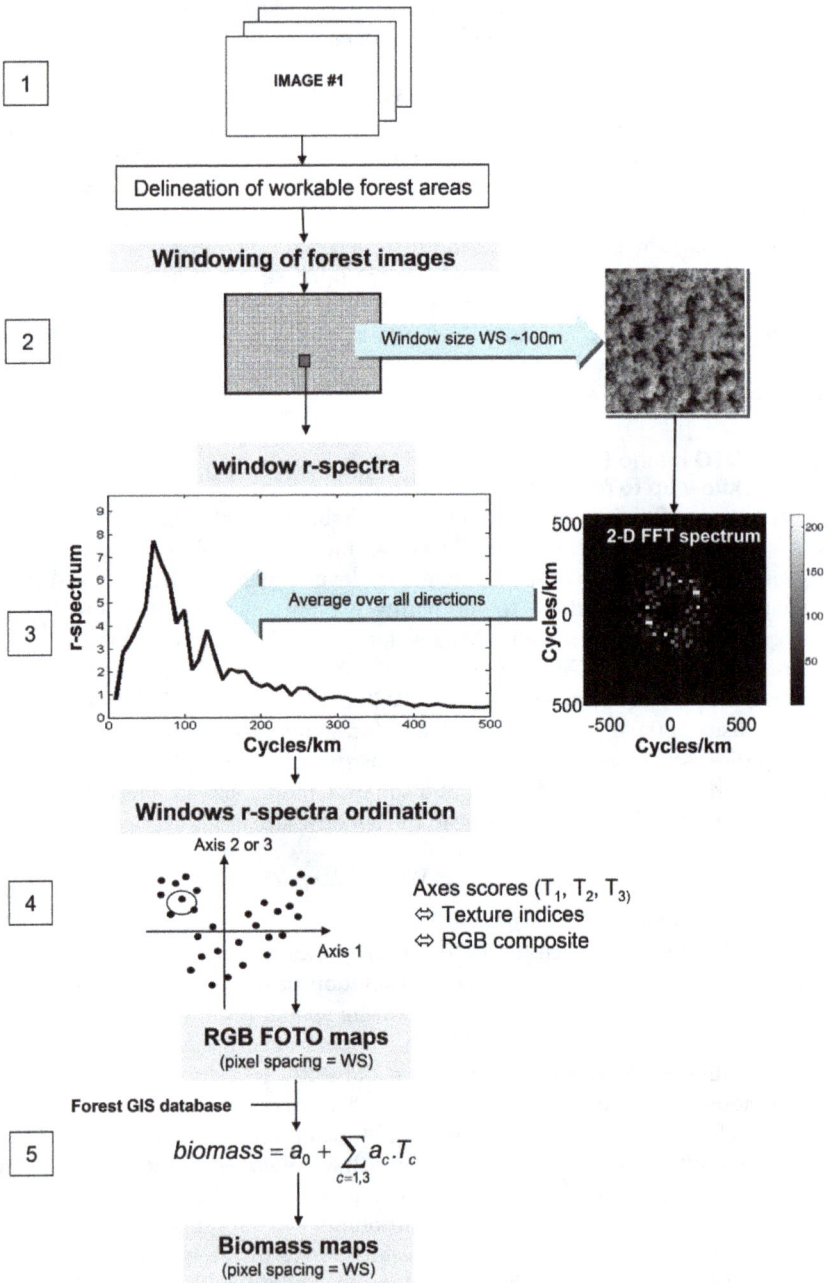

Fig. 2. Flow of operations involved in the FOTO analysis up to biomass prediction

axes are used as texture indices (the so-called FOTO indices) that are mapped by composing red–green–blue (RGB) images expressing window scores values against first, second and third axes, respectively. Such FOTO maps have a spatial resolution equal to the window size *WS*. The final step (Fig. 2, step 5) is to relate ground truth forest plot biomass to FOTO indices using a linear model of the form:

$$AGB = a_0 + \sum_{c=1}^{3} a_c T_c \qquad (2)$$

where a_0 and a_c are the coefficients of the multiple regression of *AGB* onto the texture indices *T* obtained from the first three PCA axes.

2.2.2 Computing radial spectra of forest plots

The computation of radial spectra has to be detailed because such frequency signatures are essential components of the canopy grain analysis. It is to note that the calculation of r-spectra is also possible for any single image extract centered on one forest plot as illustrated in the numerous examples provided hereafter.

Each image extract is subjected to the two dimensional discrete fast Fourier transform algorithm implemented in most of the technical computing software. Image intensity expressed in spatial *XY* Cartesian referential domain is transposed to the frequency domain. Power spectrum decomposing the image variance into frequency bins along the two Cartesian axes is then obtained for each square window (Fig. 2, step 3, right). This latter was demonstrated as an efficient way to quantify pattern scale and intensity (Couteron et al. 2006) from images of various vegetation types (Couteron et al. 2002; 2006). Assuming that images of tropical forest have isotropic properties, the radial spectra are then obtained after azimuthally averaging over all travelling directions (Fig. 2, step 3, left). Frequencies are expressed in cycles per kilometer, i.e. the number of repetitions over a 1 km distance. The discrete set of spatial frequencies *f* can be also transformed into sampled wavelengths (in meters) as $\lambda = 1000/f$. For example, a frequency of 200 cycles per kilometre corresponds to a wavelength of 5 metres.

2.2.3 Principal component analysis for regional analysis

Standardized principal component analysis of the spectra table created by the stacking of all r-spectra is a mean to perform regional analysis of canopy grain variations through one or several image scenes. For illustration, a 0.5-m panchromatic Geoeye image covering (after masking non-forest areas) 11271 hectares of mangroves is analyzed (Fig. 3). The three first factorial axes of the PCA accounted for more than 81% of the total variability. The first PCA axis opposes coarse and fine canopy grain that correspond to spatial frequencies of less than 100 cycles/km ($\lambda = 10$ m) and more than 250 cycles/km ($\lambda = 4$ m), respectively. Intermediate spatial frequencies are found with high negative loadings on the second axis.

From this analysis, we coded window scores on the three main PCA axes as RGB real values (Fig. 4). Pioneer and young stages of mangroves are characterized by red–i.e. high scores on PC1 only– whereas intergrades between blue and cyan corresponded to areas with adult trees (low positive scores on PC1 and negative scores on PC2). Green color maps mature and decaying stages of mangrove with high PC2 and very low PC1 scores. Hence, coarseness/fineness gradients of thousands of unexplored hectares of mangrove can be mapped and allow to capture, at a glance, the overall spatial organization presented in the image. An equivalent result was also obtained using a 1-m panchromatic Ikonos image (Proisy et al. 2007). The FOTO analysis is confirmed of prime interest for mangrove monitoring studies and for highlighting coastal processes in French Guiana (Fromard et al. 2004) through the mapping of forest growth stages.

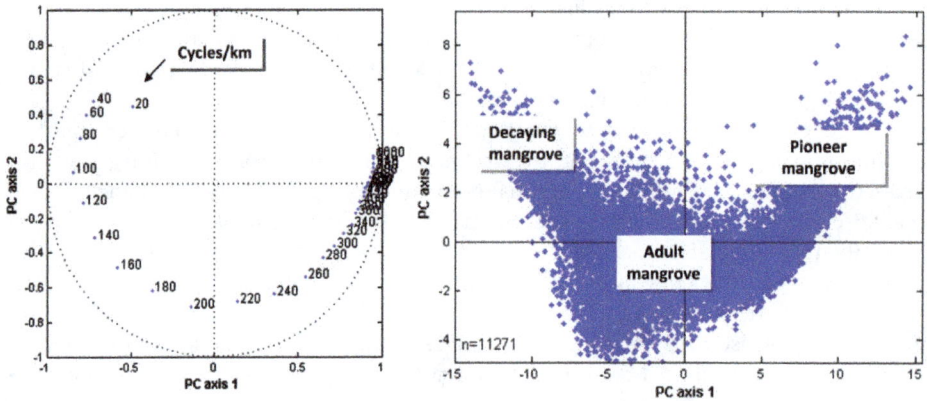

Fig. 3. Principal component analysis of Fourier spectra obtained from the FOTO analysis of a Geoeye panchromatic image covering 11271 hectares of mangroves in French Guiana. Correlation between PCA axes and spatial frequencies are shown in the left graph.

2.3 The DART modelling method

Large-scale validation of the FOTO method is highly desirable, to study both the method's sensitivity to complex variations in forest structure and to instrumental perturbations. However, it is notoriously difficult to obtain both detailed forest structure information in inaccessible tropical environments and cloudless imagery over field plots. It was therefore necessary to develop a modeling framework for testing FOTO sensitivity, in simplified but controlled conditions (Barbier et al. 2010; 2011; in press).

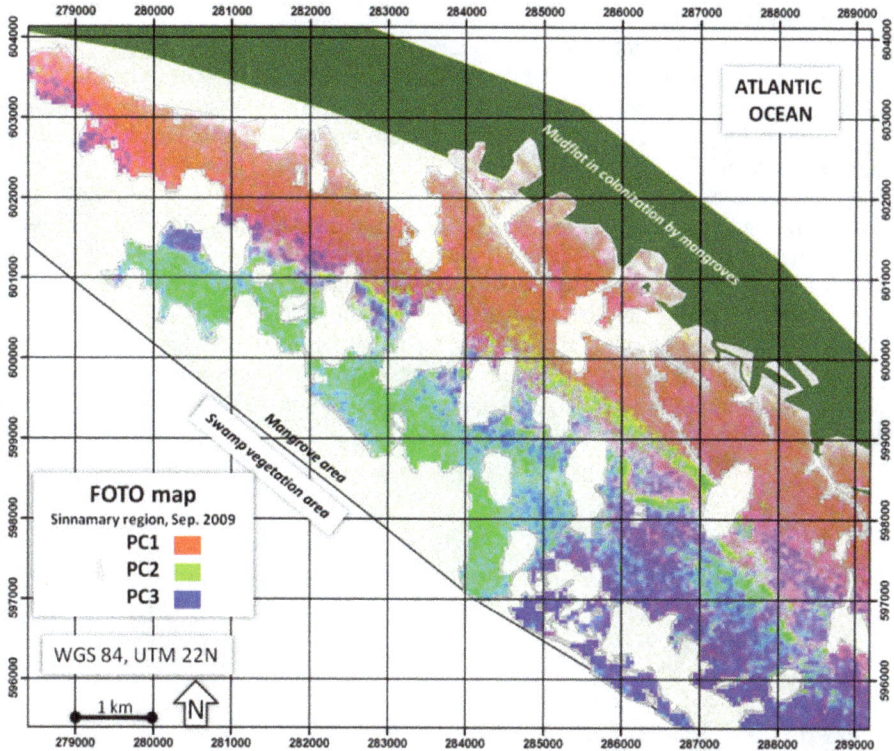

Fig. 4. Panchromatic-derived FOTO map obtained from a Geoeye panchromatic image acquired in September 2009. RGB channels code for windows scores on PCA axes. A large part of the mangrove area is masked because either under clouds or with bare mud.

2.3.1 Basic principles

The 3D Discrete Anisotropic Radiative Transfer (DART) model is a ray-tracing model that can simulate, simultaneously in several wavelengths of the optical domain, remotely sensed images of heterogeneous natural and urban landscapes with or without relief, using 3D generic representations of these landscapes for any sun direction, any view direction or any atmosphere (Gastellu-Etchegorry et al., 2004). The model is freely downloadable from *http://www.cesbio.ups-tlse.fr/fr/dart.html* for scientific studies, after signing a charter of use. In the case of forests, a DART scene, namely a 'maket', is a three-dimensional representation of a forest stand within a voxel space. Transmittance and phase functions (the optical properties) associated to each voxel depend on the voxel type (leaves, trunk, soil, etc.). Leaves cells are modelled as turbid media with volume interaction properties whereas others voxel types are taken as solid media with surface properties. Others structural characteristics within the cell (e.g. LAI, leaf and branches angle distribution) can be taken into account. The scattering of rays from each cell is simulated iteratively in a discrete number of directions. We keep the maket size 10% larger than the FOTO window or the forest plot sizes in order to avoid border effects. The final DART image is a sub-scene of equal dimensions as the reference window or plot.

2.3.2 3D forest templates

A first step within this modeling framework is to reproduce biologically realistic 3D templates of forests. Depending on the level of detail and biological realism one is to obtain, different approaches can be considered to build 3D forest mock-ups. For instance, the Stretch model (Vincent & Arja, 2008) allows accounting for dynamic crown deformations through various mechanisms and levels of plant plasticity. However, for our present purpose, we focus on variations in size-frequency distributions of trees, without entering too much into architectural (i.e. structural and dynamic) details. For this reason, we developed the Allostand model (Barbier et al. in press), a simple Matlab® algorithm using a DBH distribution, established DBH-Crown-height allometric relationships, and an iterative hard-core point process generator, to reproduce 'lollipop stands', that is a 3D arrangement of trunk cylinders bearing ellipsoid crowns. This forest template matches the DART maket requirements, e.g. a list of trees with parameters of their 3D geometry. Such simulation framework is particularly well adapted to the study of mangroves forest in which few species grow rapidly over areas with no relief (Fig. 5).

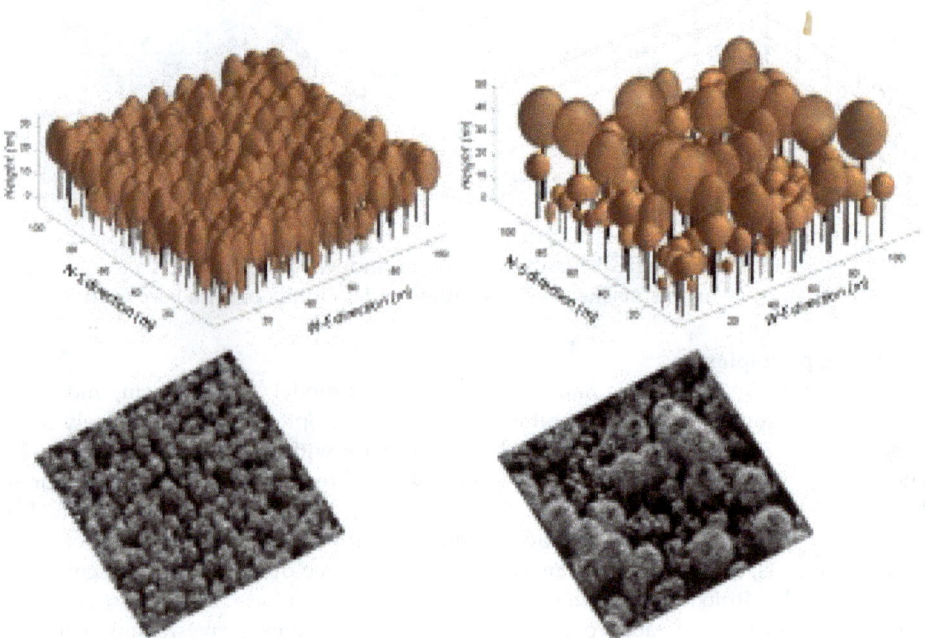

Fig. 5. Examples of 110 x 110m mockups obtained for a young Avicennia mangrove of 159 tDM.ha^{-1} (top left) and a mixed adult mangrove of 360 tDM.ha^{-1} (top right). Associated 1-m pixel DART images simulated at 0.75 µm are shown below.

2.3.3 Virtual canopy images

In this work, we only simulated mono-spectral images in the visible domain on flat topography without taking into account atmospheric effects (Fig. 5). Standard optical profiles of reflectance for soil, trunks and leaves are selected from the DART database using, for instance, '2D soil-vegetation', '2D bark_spruce' and '3D leaf_decidous' files. Such oversimplified images of virtual forest stands composed of trees with 'lollipop-shaped' crowns produce homogeneous texture dominated by few frequencies. The FOTO analysis of 330 DART images however demonstrated their potential for benchmarking textural gradient of real forest canopies throughout the Amazon basin (cf. Fig. 3 in Barbier et al. 2010).

2.4 Influence of instrumental characteristics
2.4.1 Window size and spatial resolution

Large windows may include features characterizing landforms such as relief variations rather than canopy grain (Couteron et al., 2006) whereas small windows may be unable to adequately capture large canopy features observable in mature growth stages. However, whatever the window size taken within a reasonable range of variations, i.e. 75 to 150 m for tropical forest, spatial frequencies should display more or less the same patterns of contribution to PCA axes (Couteron et al. 2006). The influence of spatial resolution on the sensitivity of r-spectra to capture canopy grain of different forest types was highlighted using 1-m panchromatic and 4-m near infrared (NIR) Ikonos images in Proisy et al. (2007).

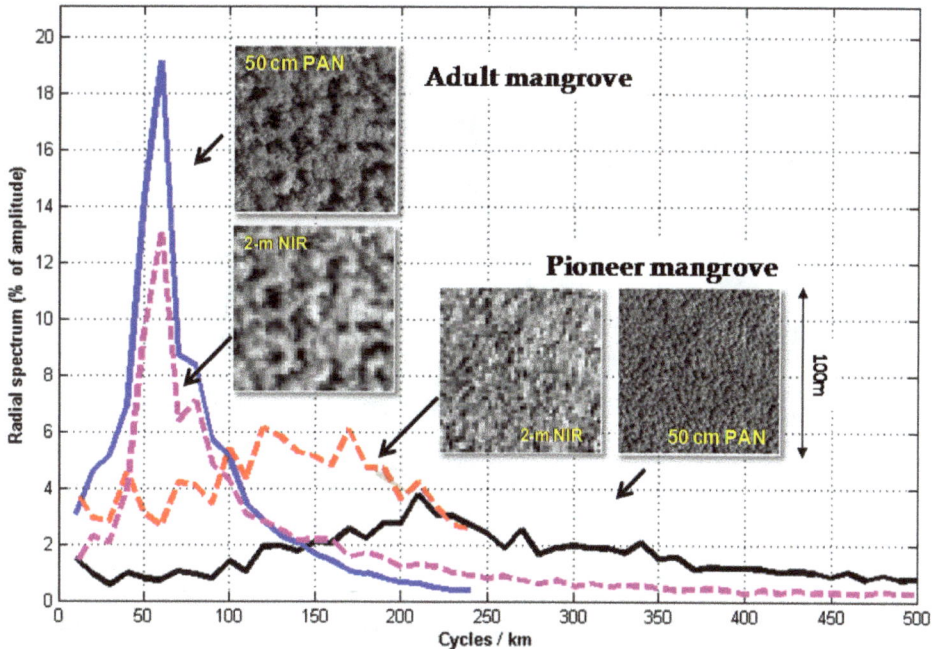

Fig. 6. Radial spectra of 2 different mangrove growth stages using 0.5-m and 2-m panchromatic and near infrared Geoeye channels.

The loss of sensitivity to the finest textures was also observed using 2-m NIR channel of Geoeye image (Fig. 6). Whereas r-spectra of 0.5-m and 2-m image extracts displayed the same behaviour with an identical dominant frequency, they did not exhibit the same profiles for the pioneer stage consisting of a very high density of trees with 2-3 m crown diameters. This limitation was also observed for the same forest growth stages after comparison of 1-m and 4-m Ikonos channels (see Fig. 4 in Proisy et al. 2007). As the limitation with regard to the youngest stages appeared using 2-m channels, it was recommended to privilege the use of panchromatic satellite images with metric and sub-metric pixels.

2.4.2 Sun and viewing angles: The BTF

Parameters of VHR image acquisitions such as sun elevation angle θ_s, viewing angle from nadir θ_v and azimuth angle Φ_{s-v} between sun and camera can vary significantly as illustrated in Fig. 7. We introduced the bidirectional texture function (BTF; Barbier et al. 2011) diagrams to map the influence of different acquisitions conditions in terms of texture perception (Fig. 8). The finest textures are perceived in the sun-backward configuration whereas the coarsest are observed when sun is facing the camera (the forward configuration) due to the loss of perception in shadowed areas. These findings show that to ensure a coherent comparison between scenes, one must either use images with similar acquisition conditions, or use a BTF trained on similar forest areas or derived from a sufficiently realistic physical simulations to allow minimizing these effects (Barbier et al. 2011).

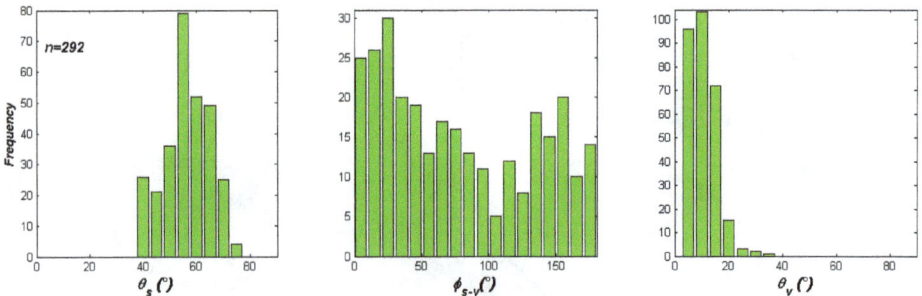

Fig. 7. Variation of acquisition parameters through a dataset of 292 images. The dataset includes 270 Quickbird, 8 Geoeye, 9 Ikonos and 5 Orbview images acquired over tropical forest of Bangladesh, Brazil, Cameroun, Central African Republic, French Guiana, India, Indonesia, Democratic Republic of Congo.

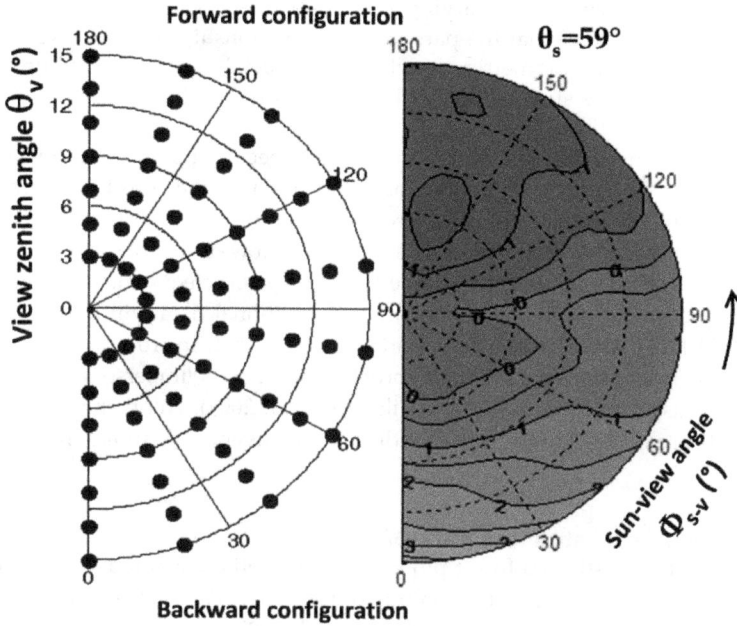

Fig. 8. Example of discrete sampling of θ_v and $\Phi_{s\text{-}v}$ acquisition angles with θ_s =59° (left) to generate the Bidirectional Texture Function (BTF). The BTF diagram (right) is computed from the mean PC1 scores resulting from the FOTO analysis of numerous DART images and 3D forest templates. Brighter intensities values imply finer perceived canopy textures.

3. From canopy grain to AGB

3.1 Requirements for forest data

The canopy grain approach must be calibrated at the forest plot scale i.e. by conducting forest inventories from which above ground biomass will be estimated. Areas of about one hectare are necessary to take into account structural diversity within the forest plot. This area of inventory can possibly be reduced for simpler forest stands and plantations, but this is basically dependent on the size of the canopy trees since the computation of FOTO indices should be meaningful at plot scale (Couteron et al. 2005). AGB estimation for each plot will be taken as the AGB of reference to correlate with FOTO indices. Since very labor-intensive destructive measures are necessary to acquire biomass values, reference field AGB values are generally computed indirectly using pre-established allometric functions predicting tree AGB from the measure of the tree diameter at breast height (DBH) as explained, for example, by Chave et al. (2005). On this basis prediction of stand AGB in reference field plots can be computed by measuring DBH>5cm in young forest and DBH>10cm in adult forest. Allometric equation between DBH and tree biomass are established from few cut trees that are weighed on site (e.g. Fromard et al. 1998 for mangroves and Brown et al, 1989 for tropical moist forest). Due to the extreme difficulty of achieving this kind of field work, relationships are often limited to trees with DBH<40cm whereas DBH histograms in tropical forest show values above 150 cm.

Additionally, for a given species varying tree heights and crowns dimensions may yield important mass differences that the parsimonious relationships cannot take into account. Selecting an appropriate allometric model is then crucial and the sampling uncertainty relative to the size of the study plot should also be addressed carefully (e.g. Chave et al. 2004).

Tree location, crown shape, tree height and wood specific gravity also constitute useful information that will contribute to the characterization of the forest structure typology. Although it remains unrealistic in heterogeneous forests without the help of skilled botanists, identification of tree species is advisable in low-diversified situations, since the inclusion of a specific wood gravity parameter into allometric equations proved to improve significantly the model (Chave et al. 2005). Such additional data will also be valuable for initializing 3D forest templates. It is important to note that, in tropical forest, tree height measurements from the ground are problematic and cumbersome explaining the enthusiasm aroused by Lidar data (e.g. Gillespie et al. 2004). Another important point to improve AGB prediction would be to conduct forest inventories simultaneously to image acquisitions.

3.2 Sensitivity to forest structure and AGB

Assuming that the constituted forest plots dataset is well distributed within the acquired scene(s), Fourier r-spectra can be computed for windows centred on each plot. For example, when applied to 1-m Ikonos (Proisy et al. 2007) or 0.5-m Geoeye panchromatic images (Fig. 9) r-spectra permit good discrimination of a wide array of canopy structures of mangroves (Fig. 9). Furthermore pre-adult, mature and decaying mangrove forests show contrasted signatures with dominant frequencies around 180, 80, 50 and 30 cycles per kilometre.

Inverting FOTO indices (the three first PCA axes) into AGB of forest plots distributed over two different sites (i.e. two different images) yielded good correlations and low errors, as presented in Fig. 10. Compared to estimations provided by the P-band HV polarisation channel, FOTO-derived AGB did not show saturations over the whole range of mangrove biomass (Fig. 9), i.e. up to 500 tDM.ha^{-1} and rmse error remains acceptable (33 tDM.ha^{-1}). This result suggests that, in the case of closed canopies with sub-strata of low biomass (e.g. the mangrove ecosystem in French Guiana), the canopy grain approach is suitable to map AGB because crown size and spatial distribution are directly correlated to standing biomass of the dominant trees. However, one do not forget that the remote sensing-based model of AGB is assessed with respect to allometric predictions of "true" AGB, i.e. the aboveground dry mass of trees, from dendrometric data, so that the quality of the allometric model is potentially an additional source of bias (Chave et al. 2004; 2005).

Good correlations were also obtained between the first axis and tree density (r^2= 0.8) or mean quadratic DBH (r^2=0.71) in tropical evergreen *terra firme* forest Couteron et al (2005). However, forest heterogeneity and presence of relief makes the canopy approach to be used carefully, that is one must analyze visually whether the relief influences or not some of the PCA axes (e.g. Ploton, 2010). Only axes immune to relief influence should be used for biomass prediction otherwise the result may be biased or highly context-dependent. Moreover, due to the diversity of forest stand structures in tropical *terra firme* forests, a sufficient number of studies in diversified locations and contexts are still needed before general conclusions can be reached about the robustness of such correlations. Independent ongoing studies suggest that the correlation with density is highly context-specific while the correlation with the mean quadratic diameter may be a more robust feature.

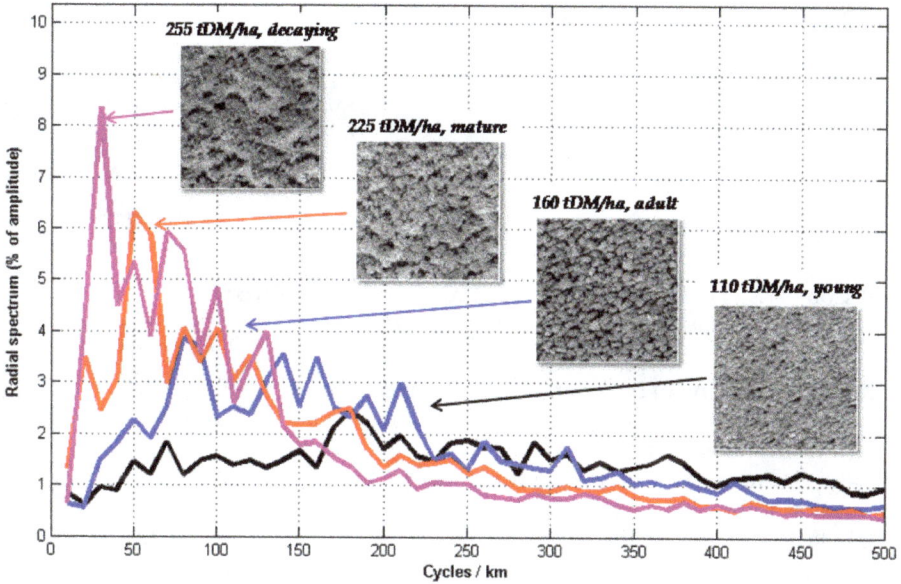

Fig. 9. Radial spectra and associated 100 x 100 m images of different mangrove growth stages using a 0.5 m panchromatic Geoeye image acquired in 2009. Forest inventories dated of 2010 and 2011. Note the r-spectra of the open canopy decaying stage. A photograph of this plot is available in Fig. 11.

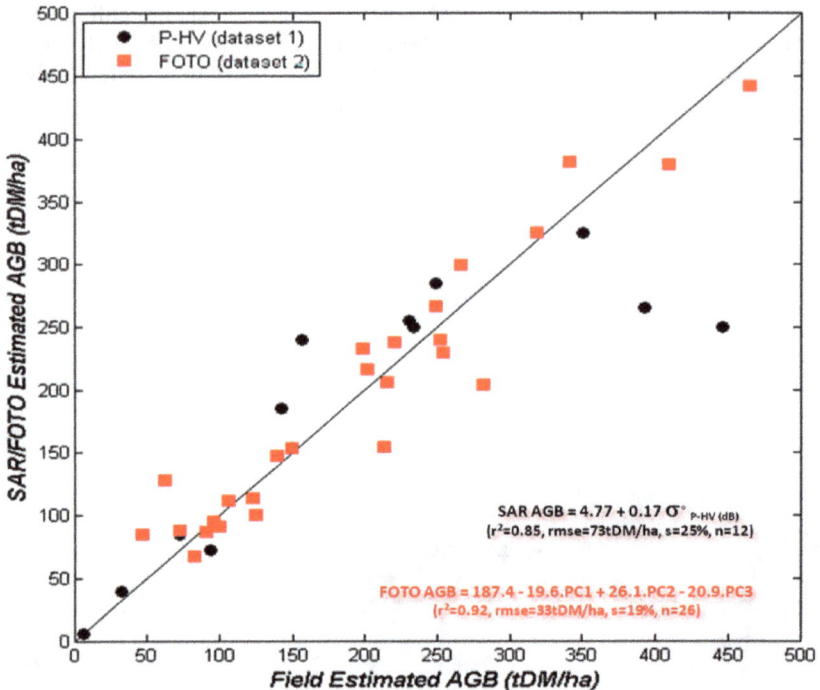

Fig. 10. Comparison of FOTO- (Proisy et al. 2007) and P-HV-derived (from Mougin et al. 1999) biomass estimates in mangroves of French Guiana

3.3 Present limitations of the methods and prerequisite

In tropical forest, both gaps and multi-strata organization are often observed. Gaps are due to accidental tree falls or natural decaying of some canopy trees (Fig. 11, left). In presence of gaps, r-spectra tend to be skewed towards low frequencies and this may be erroneously interpreted as if the canopy contained large tree crowns (Fig. 9, r-spectrum of the decaying stage). In fact, gap-influenced r-spectra cannot be automatically related to the same biomass levels and must be removed from the PCA analysis to avoid biases in the AGB-FOTO relationship. Identically, the method was so far tested principally on evergreen forests. Further studies are needed regarding deciduous forests, not only because of the seasonal changes of the canopy aspect, but also because biomass of understorey vegetation often found in such forest type is not necessarily negligible. As spectral properties of the understorey may influence the overall reflectance of the corresponding pixels, this may be all more confusing if there is no intermediate stratum beneath the highest deciduous trees. An example of this is provided by the so-called *Maranthaceae* forest in Africa (Fig. 11, right), which presents a fairly closed albeit deciduous canopy and a very scarce intermediate tree storey. Such a structure allows the development of a dense herbaceous cover. Without relevant field information, results of the FOTO approach may be confusing in those forests. Their standing biomass is probably less than for evergreen closed forests since woody intermediate storey is missing, whereas both canopies are dominated by trees with large crowns. At least, statistical relationships between FOTO indices and AGB should be analyzed after separating deciduous and evergreen forests than may be simultaneously present in a given region. Appropriate regional pre-stratification using multispectral satellite data and/or L- or P-band polarized signatures (Proisy et al. 2002) may help towards this purpose.

Fig. 11. Two examples of specific forest structures for which canopy grain and total AGB relationships cannot be safely derived without prior-stratification of the main forest types. Left: Decaying mangrove, with both large surviving trees and large canopy gaps, French Guiana © C. Proisy. Right: *Maranthaceae* understorey, overtopped by a fairly continuous albeit deciduous forest canopy referred to as *"Maranthaceae forests"* in Cameroun, Africa, note the absence of any intermediate tree strata © N. Barbier.

4. Conclusion

The canopy grain approach is largely original. It combines common techniques, i.e. Fourier transform and principal component analysis to characterize tropical canopy aspect and beyond forest structure from images of metric resolution. It can be implemented without prior radiometric correction, such as reflectance calibration or histogram range concordance. Regarding the increasing availability of metric to sub-metric optical images, the FOTO canopy grain analysis demonstrated its potential to capture gradients of forest structural characteristics in tropical regions. Within this context, the possible contribution of the canopy grain approach to the challenging task of estimating tropical above-ground biomass is worth being assessed at very broad scale. Such aim requires conducting simultaneously observational and simulation studies aiming at better understanding how canopy grain is sensitive to forest structure or biomass in various types of forests under various conditions of image acquisitions. There is particularly an important field of research in simulating multi-spectral and metric reflectance images from realistic forest 3D templates to identify, for instance, the range of conditions for which inversing above ground biomass of tropical forests appears possible. Considering the extreme complexity of most the tropical forests, it would be illusory to believe that only one remote sensing technique can provide all the information required to the AGB inversion. We thus believe that combining canopy grain analysis with low frequencies radar-based studies can provide new insights on this problem.

5. Acknowledgment

This work is supported by the Centre National d'Etudes Spatiales (CNES) for the preparation of the 'Pleiades' mission and joins the *Infolittoral-1* project funded by the French "Unique Inter-ministerial Fund" and certified by the "Aerospace Valley competitiveness cluster" (http://infolittoral.spotimage.com/). Nicolas Barbier has a Marie Curie (UE/IEF/FP7) grant. Research in central Africa is supported by the *Programme Pilote Régional* (PPR FTH-AC) of IRD. We thank J-L. Smock and Michel Tarcy for their strong motivation in mangrove field measurements. We also thank Bruno Roux and Michel Assenbaum for their kind support in providing us for free several Avion Jaune© images (http://www.lavionjaune.fr).

6. References

Asner, G. P., G. V. N. Powell, J. Mascaro, D. E. Knapp, J. K. Clark, J. Jacobson, T. Kennedy-Bowdoin, A. Balaji, G. Paez-Acosta, E. Victoria, L. Secada, M. Valqui and R. F. Hughes High-resolution forest carbon stocks and emissions in the Amazon. Proceedings of the National Academy of Sciences Vol.107, No.38 (September 2010), pp.16738-16742, ISSN: 1091-6490

Baccini, A., Laporte, N., Goetz, S. J., Sun, M. & Dong, H. (2008). A first map of tropical Africa's above-ground biomass derived from satellite imagery *Environmental Research Letters,* Vol.3, No.4, (October-December 2008), pp.1:9, ISSN: 1748-9326

Barbier, N., Couteron, P., Proisy, C., Malhi, Y. & Gastellu-Etchegorry, J.-P. (2010). The variation of apparent crown size and canopy heterogeneity across lowland Amazonian forests. *Global Ecology and Biogeography,* Vol.19, No.1, (January 2010), pp.72-84, ISSN 1466-8238

Barbier, N., Proisy, C., Véga, C., Sabatier, D. & Couteron, P. (2011). Bidirectional texture function of high resolution optical images of tropical forest: An approach using LiDAR hillshade simulations. *Remote Sensing of Environment,* Vol.115, No.1, (January 2011), pp.167-179, ISSN 0034-4257

Barbier, N., Couteron, P., Gastellu-Etchegorry, J. P. & Proisy, C. (xxxx). Linking canopy images to forest structural parameters: potential of a modeling framework. Annals of Forest Science, In press, ISSN: 1297-966X. DOI: 10.1007/s13595-011-0116-9

Brown, S., Gillespie, A. J. R. & Lugo, A. E. (1989). Biomass estimation methods for tropical forests with applications to forest inventory data. *Forest Science,* Vol.35, No.4, (December 1989), pp.881-902, ISSN : 0015-749X

Bruniquel-Pinel, V. & Gastellu-Etchegorry, J. P. (1998). Sensitivity of Texture of High Resolution Images of Forest to Biophysical and Acquisition Parameters. *Remote Sensing of Environment,* Vol.65, No.1, (July 1998), pp.61-85, ISSN: 0034-4257

Chave, J., Condit, R., Aguilar, S., Hernandez, A., Lao, S. & Perez, R. (2004). Error propagation and scaling for tropical forest biomass estimates. *Philosophical Transactions of the Royal Society of London, Series B,* Vol.359, (March 2004), pp.409-420, ISSN: 0962-8436

Chave, J., Andalo, C., Brown, S., Cairns, M. A., Chambers, J. Q., Eamus, D., Folster, H., Fromard, F., Higuchi, N., Kira, T., Lescure, J.-P., Nelson, B. W., Ogawa, H., Puig, H., Riéra, B. & Yamakura, T. (2005). Tree allometry and improved estimation of carbon stocks and balance in tropical forests. *Oecologia*, Vol.145, No.1, (August 2005), pp.87-99, ISSN: 1432-1939

Couteron, P. (2002). Quantifying change in patterned semi-arid vegetation by Fourier analysis of digitized aerial photographs. *International Journal of Remote Sensing*, Vol.23, No.17, (October 2002), pp.3407-3425, ISSN: 1366-5901

Couteron, P., Pélissier, R., Nicolini, E. & Paget, D. (2005). Predicting tropical forest stand structure parameters from Fourier transform of very high-resolution remotely sensed canopy figures. *Journal of Applied Ecology*, Vol.42, No.6, (December 2005), pp.1121-1128, ISSN: 1365-2664

Couteron, P., Barbier, N. & Gautier, D. (2006). Textural ordination based on Fourier spectral decomposition: a method to analyze and compare landscape patterns. *Landscape Ecology*, Vol.21, No.4, (May 2006), pp.555-567, ISSN: 1572-9761

Fromard, F., Vega, C. & Proisy, C. (2004). Half a century of dynamic coastal change affecting mangrove shorelines of French Guiana. A case study based on remote sensing data analyses and field surveys. *Marine Geology*, Vol.208, No.2-4, (15 August 2004), pp.265-280, ISSN: 0025-3227

Englhart, S., Keuck, V. & Siegert, F. (2011). Aboveground biomass retrieval in tropical forests -- The potential of combined X- and L-band SAR data use. *Remote Sensing of Environment*, Vol.115, No.5, (May 2011), pp.1260-1271, ISSN: 0034-4257

Fromard, F., Puig, H., Mougin, E., Marty, G., Betoulle, J. L. & Cadamuro, L. (1998). Structure, above-ground biomass and dynamics of mangrove ecosystems: new data from French Guiana. *Oecologia*, Vol.115, No.1, (June 1998), pp.39-53, ISSN: 0029-8549

Gastellu-Etchegorry, J. P., Martin, E. & Gascon, F. (2004). DART: a 3D model for simulating satellite images and studying surface radiation budget. *International Journal of Remote Sensing*, Vol.25, No.1, (January 2004), pp.73-96, ISSN: 0143-1161

Gillespie, T. W., Brock, J. & Wright, C. W. (2004). Prospects for quantifying structure, floristic composition and species richness of tropical forests. *International Journal of Remote Sensing*, Vol.25, No.4, (February 2004), pp.707-715, ISSN: 1366-5901

Imhoff, M. L. (1995). Radar backscatter and biomass saturation: ramifications for global biomass inventory. *IEEE Transactions on Geoscience and Remote Sensing*, Vol.33, No.2, (March 1995), pp.511-518, ISSN: 0196-2892

Letouzey, R. (1968), *Etude phytogéographique du Cameroun*. Lechevalier Eds., Paris.

Malhi, Y. & Román-Cuesta, R. M. (2008). Analysis of lacunarity and scales of spatial homogeneity in IKONOS images of Amazonian tropical forest canopies. *Remote Sensing of Environment*, Vol.112, No.5, (May 2008), pp.2074-2087, ISSN: 0034-4257

Mougin, E., Proisy, C., Marty, G., Fromard, F., Puig, H., Betoulle, J. L. & Rudant, J. P. (1999). Multifrequency and multipolarization radar backscattering from mangrove forests. *IEEE Transactions on Geoscience and Remote Sensing*, Vol.37, No.1, (January 1999), pp.94-102, ISSN: 0196-2892

Ouma, Y. O., Ngigi, T. G. & Tateishi, R. (2006). On the optimization and selection of wavelet texture for feature extraction from high-resolution satellite imagery with application towards urban-tree delineation. *International Journal of Remote Sensing*, Vol.27, No.1, (January 10), pp.73-104, ISSN: 0143-1161

Ploton, P. 2010. Analyzing Canopy Heterogeneity of the Tropical Forests by Texture Analysis of Very-High Resolution Images - A Case Study in the Western Ghats of

India. *Pondy Papers in Ecology*, 10: 1-71, Available from <http://hal.archives-ouvertes.fr/hal-00509952/fr/>

Proisy, C., Mougin, E., Fromard, F., Trichon, V. & Karam, M. A. (2002). On the influence of canopy structure on the polarimetric radar response from mangrove forest. *International Journal of Remote Sensing*, Vol.23, No.20, pp.4197-4210, ISSN: 0143-1161

Proisy, C., Couteron, P. & Fromard, F. (2007). Predicting and mapping mangrove biomass from canopy grain analysis using Fourier-based textural ordination of IKONOS images. *Remote Sensing of Environment*, Vol.109, No.3, (August 2007), pp.379-392, ISSN: 0034-4257

Rao, A. R. & Lohse, G. L. (1996). Towards a texture naming system: Identifying relevant dimensions of texture. *Vision Research*, Vol.36, No.11, (June 1996), pp.1649-1669, ISSN: 0042-6989

Rich, R. L., Frelich, L., Reich, P. B. & Bauer, M. E. (2010). Detecting wind disturbance severity and canopy heterogeneity in boreal forest by coupling high-spatial resolution satellite imagery and field data. *Remote Sensing of Environment*, Vol.114, No.2, (February 2010), pp.299-308, ISSN: 0034-4257

Richards, P. W. (August 1996). *The Tropical Rain Forest. An Ecological Study*, 2nd edition, Cambridge University Press, ISBN: 9780521421942, Cambridge

Vincent, G. & Harja, D. (2008). Exploring Ecological Significance of Tree Crown Plasticity through Three-dimensional Modelling. *Annals of Botany*, Vol.101, No.8, (May 2008), pp.1221-1231, ISSN: 1095-8290

Zhao, K., S. Popescu and R. Nelson (2009). Lidar remote sensing of forest biomass: A scale-invariant estimation approach using airborne lasers. Remote Sensing of Environment, Vol(113), No.1, (January 2009), pp. 182-196, ISSN: 0034-4257

3

Remote Sensing of Biomass in the Miombo Woodlands of Southern Africa: Opportunities and Limitations for Research

Natasha Ribeiro, Micas Cumbana,
Faruk Mamugy and Aniceto Chaúque
Faculty of Agronomy and Forestry, Eduardo Mondlane University
Mozambique

1. Introduction

Biomass and Leaf Area Index (LAI) are two important biophysical properties of vegetation as they inform about vegetation production. LAI is directly related to the exchange of energy and mass between plant canopies and the atmosphere (Fassnacht *et al.*, 1997), while biomass reflects the amount of carbon converted through photosynthesis and accumulated in the different plant components. Thus, the two variables reflect much of the potential and actual production of plant ecosystems (Kasischke *et al.*, 2004).

Fire is ubiquitous in most terrestrial ecosystems causing spatial patterning at many scales (Chapin III *et al.*, 2003). In tropical savannas in general and, in the southern African savannas in particular, much of the ecosystem functioning is largely defined by the combination of climate, fires and herbivory. Andreae (1993) estimated that fires in the African and the world savannas account, respectively, for 22% and 42% of the biomass burned globally. Moreover, the amount of CO_2 exchanged with the atmosphere in southern Africa may represent up to 20% of the regional net primary production (Scholes & Andreae, 2000).

In spite of the elevated importance of disturbances in miombo woodlands, there still is a gap in the understanding of the interaction between them and vegetation. This, results partially from the short temporal and spatial scales of observation of much of the existing studies. For example, except for the long-term experimental study carried out in Zambia for 15 years (Trapnell, 1959), the other studies are all *points* in space and time, much of them lasted less than 5-years. Moreover, they address a specific aspect of miombo woodlands functioning, which is important but not sufficient for a complete understanding of this ecosystem. Thus, measurements of large spatial- and temporal-scale variations of vegetation production, disturbances and their interaction are crucial to fulfill the existing data gaps. This is particularly important to understanding the role of this crucial ecosystem in the global carbon budget.

Remote sensing of vegetation production and disturbances is a critical measurement needed to extend the field level understanding of ecological, hydrological and biogeochemical processes to broader spatial and temporal scales in terrestrial ecosystems (Asner, 2004) and the different scales of energy, CO_2 and mass exchange between ecosystems and atmosphere

(Carlson & Ripley, 1997; Goward et al., 1985; Justice et al., 1998; Running et al., 1995; Schlesinger, 1996; Tucker et al. 1985). In areas where detailed and sufficient field data is scarce, as in much of the miombo context, the need for remote sensing data and techniques is even more important (Justice & Dowty, 1994; Malingreu & Gregoire 1996). Interpretation of the spaceborne data on land carbon stocks is needed, not only from the scientific point of view, but also within practical carbon management options mentioned in UNFCCC (Kyoto Protocol, REDD and REDD+).

To accurately measure objects on earth from the space several issues have to be considered including, the type and characteristics of remote sensing system, the spectral characteristics of the target objects, interactions between objects on earth, the statistical methods, among others. The advance of new generation of remote sensing such as IKONOS and QUICKBIRD optical sensors and, LiDAR and ALOS/PALSAR microwave sensors with high spatial resolution opens a new opportunity to improve understanding of miombo dynamics. These sensors, allow individual trees to be recognized and thus, large-scale biomass estimation in miombo woodlands. However, several constraints still exist and may limit the utilization of these data.

The aim of this chapter is to analyze the opportunities and constraints for the use of remote sensing techniques to estimate biomass (and carbon) in the miombo woodlands of southern Africa. The chapter also identifies research priorities for remote sensing of biomass in the miombo ecoregion.

2. Brief overview of miombo woodlands ecology

2.1 Geographic distribution

Miombo woodlands, herewith referred as Miombo, cover about 2.7 million km^2 within the southern sub-humid tropical zone of Africa from near the Equator to bellow the Tropic of Capricorn (Figure 1). They extend from Tanzania to the Democratic Republic of the Congo (DRC) in the north, through Angola, Zambia in the east to Malawi, Zimbabwe and Mozambique in the south (Desanker et al., 1997; Frost, 1996).

Miombo occur within a mean annual precipitation range of 650 to 1,500 mm and more than 95% of annual rainfall occurs during a single 5-7 months wet season from October/November to March/April (Cauldwell & Zieger, 2000; Chidumayo, 1997; Desanker et al., 1997; Frost, 1996). Few sites within the region receive more than 5% of their total mean annual rainfall during the dry months. Consequently, miombo is divided into dry and wet miombo according to the rainfall in the zone of occurrence (White, 1983). **Dry miombo woodlands** occur in the southern portion of the region in Malawi, Mozambique and Zimbabwe, in areas receiving less than 1,000 mm of rainfall annually. In contrast, **wet miombo woodlands** occur over much of eastern Angola, DRC, northern Zambia, south western Tanzania and central Malawi in areas receiving more than 1,000 mm rainfall per year. Variations to this pattern may occur within an area as a result of local variation in environmental factors such as altitude and precipitation.

This ecosystem occurs in geologically old and nutrient poor soils (Chidumayo, 1997; Frost, 1996). The dominant soils belong to the order Oxisol, which are highly weathered old soils with dominance of aluminum and iron oxides and low activity clays. The soils in miombo are typically acid (pH between 4.2 and 6.9), have a low Cation Exchange Capacity (CEC: 1.80-25.10 me100/g) and, are low in nitrogen (0.02-0.62%) and phosphorous (0.0-54 ppm) and Total Exchangeable Bases (TEB: 0.35-20.78 me100/g). The range of carbon content in soils is 0.3-3.8% (Chidumayo, 1989; Ribeiro & Matos, *unpubl. Data;* Sitoe *et al., unpubl. Data;* Walker & Desanker, 2004). Miombo soils have low concentration of organic matter with an average in the topsoil of 1% and 2% for dry and wet miombo, respectively (Chidumayo, 1997). This is a consequence of the abundant termite activities and frequent fire incidence (Cauldwell & Zieger, 2000; Chidumayo, 1997).

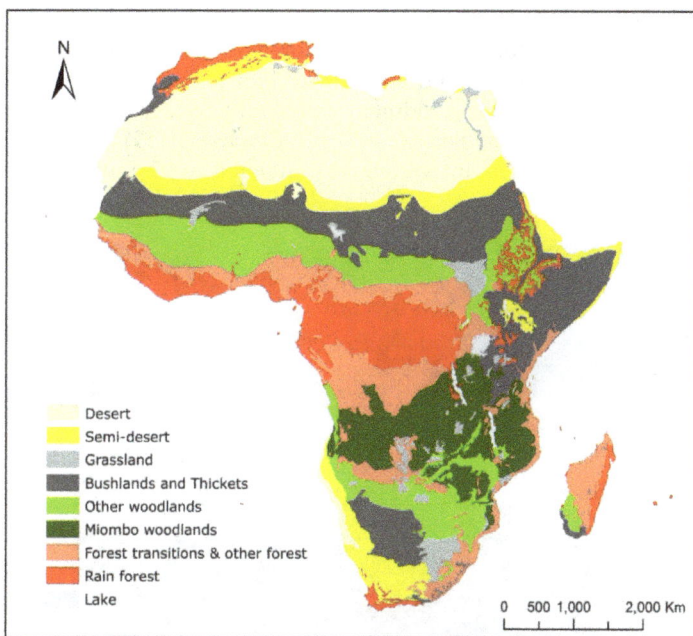

Fig. 1. Map of African vegetation, showing the miombo woodlands in dark green (Source: White, 1983).

2.2 Floristic composition and structure

Miombo have an estimated diversity of 8,500 species of higher plants, over 54% of which are endemic and 4% are tree species. Zambia is considered to be the centre of endemism for *Brachystegia* and has the highest diversity of tree species (Rodgers *et al.*, 1996). The diversity of canopy tree species is, however, low and characterized by the overwhelming dominance of trees in the genera *Brachystegia* (*miombo* in Swahili), *Julbernardia* and/or *Isoberlinia* (Campbell *et al.*, 1996). Other important tree species in miombo include *Pseudolachnostylis maprouneifolia, Burkea africana, Diplorhynchus condilocarpon* among others. In mature miombo these species comprise an upper canopy layer made of 10-20 m high trees and a scattered layer of sub-canopy trees. The understorey is discontinuous and composed of broadleaved shrubs such as *Eriosema, Sphenostylis, Kotschya, Dolichos* and *Indigofera* and suppressed saplings of canopy trees. A sparse but continuous herbaceous layer of grasses, forbs and sedges composed of *Hyparrhenia, Andropogon, Loudetia, Digitaria* and *Eragrostis* (Campbell *et al.*, 1996; Desanker *et al.*, 1997) dominate the ground-layer.

Species composition and structure of miombo vary along the rainfall gradient across the region. In the dry miombo, canopy height is less than 15 m and the vegetation is floristically poor. *Brachystegia spiciformis*, *B. boehmii* and *Julbernardia globiflora* are the dominant canopy species. The herbaceous layer varies greatly in composition and biomass and contains grasses and suppressed saplings of canopy trees. The wet miombo in turn, presents canopy heights greater than 15 m. The vegetation is floristically rich and includes nearly all of the characteristic miombo species. *B. floribunda*, *B. glaberrina*, *B. longifolia*, *B. wangermeeana* and *Marquesia macroura* are widely distributed. The understorey comprises a mixture of grasses, bracken (*Pteridium aquilinum*) and shrubs.

Biomass distribution is uniform across the ecoregion, with the woody component comprising 95-98% of the aboveground biomass in undisturbed stands (Figure 2); grasses and herbs make up the remainder (Chidumayo, 1997). However, biomass production is a function of rainfall and nutrients. For example, Chidumayo (1997) showed a variation of aboveground biomass of 53 t/ha in western dry miombo to 93 t/ha in wet miombo, in Zambia.

Fig. 2. Beginning of the wet season in mature miombo woodlands in Niassa National Reserve, northern Mozambique. A conserved stand where the homogeneous canopy layer is evident (Photo by: Ribeiro, N.).

The key indicator of the linkage between rainfall and miombo production is the observed structural and compositional variations following the rainfall gradient, from the drier fringes of the miombo to the wetter core area (Desanker *et al.*, 1997). The nutrient cycling seems to follow also the rainfall gradient across the region revealing that nutrients limitation is a function of moisture regime. Local variations are expected to be much higher and strongly affected by disturbances, especially fires and herbivory. The main structural characteristics of miombo woodlands are summarized in Table 1.

Dambos are distinctive features of the miombo and occupy seasonally waterlogged shallow valley depressions across the prominent catenas in the region (a regular alternation of two or more types of vegetation) (Campbel *et al.*, 1996; Scholes, 1997). *Dambos* are small *islands* of hygrophilous treeless grasslands emerged in the miombo landscape, which can make up to 40% of the landscape. They have a particular importance for the ecology of miombo, especially as habitat for animal species including some herbivores.

Structural characteristic	Range of values	Source
Woody density	1,500-4,100 stems/ha	Campbell *et al.*, 1995; Chidumayo 1997; Grundy 1995; Guy 1981; Ribeiro *et al.*, 2008a; Strang, 1974; Trapnell, 1959.
Tree density	380-1,400 trees/ha	Banda *et al.*, 2006; Campbell *et al.*, 1995; Chidumayo, 1985; Grundy, 1995; Guy, 1981; Strang, 1974; Trapnell, 1959.
Stand basal area	7-24 m²/ha	Banda *et al.*, 2006 ; Backéus *et al.*, 2006; Freson *et al.*,1974 ; Lawore *et al.*, 1994;
Aboveground biomass	1.5 Mg/ha (3-6 years old coppice) – 144 Mg/ha (Mature wet miombo); 53-55 Mg/ha in dry and sub-humid miombo	Chidumayo, 1991; Chidumayo, 1997; Guy, 1981; Malaisse & Strand, 1973; Ribeiro *et al.*, 2008b; Sitoe *et al.* (*unpubl. Data*)
Aboveground herbaceous biomass	0.1-4.0 Mg/ha (2-5% of the total aboveground biomass)	Frost, 1996; Ribeiro and Matos (*unpubl. Data*); Sitoe *et al.* (*unpubl. Data*)

Table 1. Summary of structural characteristics of miombo woodlands.

2.3 Ecological role of disturbances in miombo woodlands

Changes in the landscape of many types of woodland in Africa have been attributed directly to the interactive effect of elephants and fire (Buechner & Dawkins, 1961; Guy, 1981, 1989; Laws, 1970; Mapaure & Campbell, 2002; Ribeiro *et al.*, 2008a; Sukumar, 2003; Walpole *et al.*, 2004). In general, the pattern of change is the same: as elephants over-browse the woodlands, laying waste to mature trees, there is an increase in the low woody vegetation and grass cover as well as a dramatic increase in fuel load. This allows fire to become progressively more intense. Fiercer and frequent fires affect both large trees and saplings lowering species diversity. Debarking of large trees by elephants may further expose inner tissues to fire damage and death (Laws, 1970; Sukumar, 2003).

Nearly 90% of fires in miombo are anthropogenic and associated with several human activities in the woodland including: hunting, honey collection and shifting agriculture (Figure 3). They occur every 1 to 3 years in the dry season from May to October/November with a peak in the late dry season (August-October). They are largely fuelled by grasses and take place in the understorey with flame heights generally low (Gambiza *et al.*, 2005; Trollope *et al.*, 2002). Thus fire intensity and frequency is linked through grass production to the previous season rainfall, the intensity of grazing and the extent of woody plant cover (Frost, 1996).

Fire frequency in miombo is expected to be locally highly variable according to fuel accumulation rates, the proximity to sources of ignition and interannual climatic variations (Chidumayo, 1997; Frost, 1996; Kikula, 1986; Ribeiro 2007; Trapnell, 1959). The impact of fire on plants depends on its intensity, frequency, seasonality and interaction with herbivory (Bond & Van Wilgen, 1996; Frost, 1984; Ribeiro, 2007; Trollope, 1978). The effect of seasonality was studied by Chidumayo (1989), indicating that stem mortality measured over a two-year period in wet miombo was only 4-3% when both woodland and coppice plots were burned in early dry season but 18% and 40% respectively when burned in late dry season.

Fig. 3. Agriculture and associated fires in the Gorongosa Mountain, Central Mozambique (Photos by Ribeiro, N.)

The tolerance or susceptibility of miombo plants to frequent late dry season fire is a function of their growth form, developmental stage, physiological conditions and phenological state at the time of burning. For instance, grasses and many non-woody herbs tolerate intense, late dry season fires better than most woody plants. During this time of the year most tree species produce new leaves, which make them less tolerant to intense fires. Frequent late dry season fires also destroy young trees and shrubs, or their aboveground parts, preventing the development of taller, more fire resistant size classes (Brookman-Amissah *et al.*, 1980; Hopkins, 1965).

Scientific accounts of elephant damage to vegetation in African savannas and woodlands emerged during the 1960s. Buechner & Dawkin (1961) analyzed the vegetation changes in the Murchinson Falls National Park, Uganda between 1932 and 1956 and found that tree populations had halved during that period. Thomson (1975) report that nearly 67% of the 500 original mature trees of *B. boehmii* in Chizarira National Park, Zimbabwe died, and another 20% were damaged, transforming relatively dense woodlands into more open wooded grasslands.

Guy (1981) studying miombo changes in the Sengwa Wildlife Research Area, Zimbabwe found that the biomass was reduced by 54% between 1972 and 1976. The decline was associated with the decreasing number of the dominant tree species, *B. boehmii*, a species markedly selected by elephants. In 1989 the area was dominated by *J. globiflora*, *P. maprouneifolia* and *Monotes glaber* but *B. boehmii* was rare (Guy, 1989). Comparing the woodlands inside the research area with those outside (where the elephants population and fires were excluded) the author found less tree density (267 stems/ha over 334 stems/ha), lower stem area (3.56 m²/ha compared to 9.52 m²/ha) and lower biomass (8.5 t/ha compared to 26.2 t/ha). Recently Mapaure & Campbell (2002) working in the same area report an overall rate of decrease in woody cover of 0.75% per year. Elephants (significantly high negative correlation coefficient, $r=-0.90$, $p<0.05$), and fire ($r=-0.35$, $p=0.61$) were pointed out as the main cause of decline. The authors noticed that 82% reduction in the elephant population by culling between 1979 and 1982, result in a noticeable vegetation recovery. However, field observations indicate that some areas were not reverting to miombo, but to *Combretaceae*-dominated thickets. This behavior was also observed by Ribeiro et al. (2008a) in the Niassa National Reserve of northern Mozambique.

Dublin *et al.* (1990) studied the effects of elephants and fires in the Serengeti-Mara woodlands in Kenya and found that the combined effect of fire and elephants caused consistently decrease in the recruitment of tree species. Recently Walpole *et al.* (2004) assessed the status of the Mara woodlands (adjacent to the Serengeti National Park). They report that species diversity is relatively low, possibly because of decline in the density and extent of woodland and thickets originated by increasing elephant population within the area.

3. Remote sensing techniques for biomass estimation in the miombo woodlands

Remote Sensing refers to the acquisition and recording of information about objects on Earth without any physical contact between the sensor and the subject of analysis. It provides spatial coverage by measurement of reflected and emitted electromagnetic radiation, across a wide range of wavebands. Remotely sensed data provide in many ways an enhanced and very feasible alternative to manual observation with a very short time delay between data collection and transmission.

Since the first launch of an earth observatory satellite, in July 1972 by the US called Earth Resources Technology Satellite (ERTS, later renamed Landsat), remote sensing has been increasingly used to acquire information about environmental processes and over the last three decades this use has had a substantial increase with the development of modern remote sensing technology as radar sensors (ERS, RADARSAT, etc), LiDAR systems and the new generation of optical sensors (IKONOS, SPOT-5, GeoEye, etc.). Thus, remote sensing techniques are a time-and-cost-efficient method of observing forest ecosystems.

The capabilities of remote sensing techniques have been lately expanded to serve monitoring efforts as well as to provide valuable information on degradation in specific areas. Remote sensing, due to its potential to estimate biophysical parameters with detention of temporal and spatial variability, becomes a powerful technique. These, in combination with field sampling methods may provide detailed estimations of forestry parameters. In forest ecosystems, estimation of biophysical parameters has been made by a succession of methods. However, the extension of these estimations in space and time has obvious limitations especially in highly variable ecosystems such as miombo. Moreover, the network of field plots across the miombo ecoregion is not sufficient enough to cover it, which imposes further limitations in the use of remote sensing for biomass estimation.

In this section we highlight the uses of remote sensing techniques to derive biomass information in miombo. Emphasis is given to optical remote sensing techniques because the majority of measurements have been concentrated on these techniques. Radar and microwave remote sensing in miombo have, in the last decade, become a useful source of information for biomass estimation, but has been barely applied in the region. Therefore, radar and microwave remote sensing are also briefly discussed in this section.

3.1 Remote sensing of biomass using optical sensors

Passive optical remote sensing, the process of imaging or sampling the interactions between electromagnetic energy and matter at selected wavelengths, has the ability to monitor terrestrial ecosystems at various temporal and spatial scales and has been widely tested for land cover mapping and various forestry applications (Patenaude *et al.*, 2004).

Remote sensing of vegetation depends on the optical properties of plant tissues (foliage, wood, litter, etc.) that determine the landscape-scale reflectance of ecosystems (Asner, 2004; Jensen, 1996). Thus, plant species composition, stand structures and associated canopy shadows and vegetation vigor (leaves, branches and bark) are important factors affecting vegetation reflectance (Lu, 2001). Figure 4 is an example of the spectral signatures of two different types of vegetation in comparison with sand and water. In this example is clear the discrepant reflectance of vegetation and other targets. The shape of the curves of the two vegetation types is similar, but they have dissimilar percent of reflectance, which is due to differences in vegetation structure and composition.

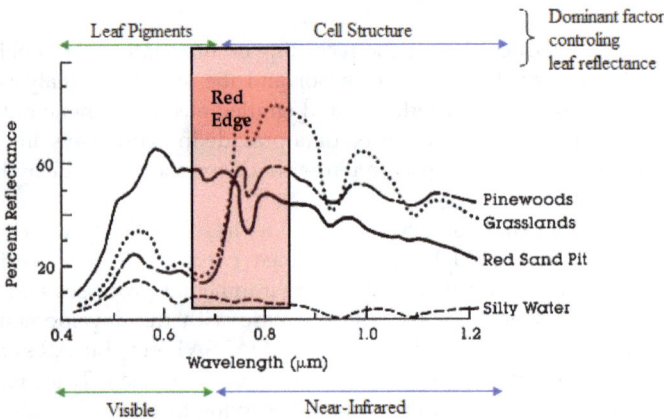

Fig. 4. Differential spectral signatures of vegetation, sand and water (adapted from http://rst.gsfc.nasa.gov/Intro/Part25.html). The Red Edge represents the basis for optical remote sensing of vegetation.

The basis for satellite measurements of vegetation from optical sensors is the differential reflection of light in the visible (0.4 to 0.7 µm) and Near-Infrared (NIR~0.8 µm) regions of the electromagnetic spectrum originated by the various leaf components (chlorophylls and β-carotene) and structure. For instance, chlorophyll strongly absorbs light in the red (~0.6 µm) while leaf's structural components highly reflect in the NIR (Jensen, 1996). In the visible part of the spectrum (Figure 4) the dominant signal detected by the sensor is the direct scattering. Multiple bounces of photons from the forest back to the sensor are small in comparison to direct scattering because the chlorophyll and other leaf pigments within the forest foliage are highly absorbing at visible wavelengths. At NIR wavelengths, however, the foliage is more transparent and multiple scattering becomes more important. Hence, this striking difference in light reflectance, known as the *red edge* (Figure 4), is sensitive to vegetation changes over a wider range of vegetation densities (Jensen, 1996; Shugart, *et al.*, 2010).

Within the same forest stand the sensor measures the aggregate reflectance as a function of the wavelength of the tree canopies and understory within a pixel. The reflectance changes seasonally with phenological changes in vegetation. For example in the dry and sub-humid miombo most species lose their leaves for 4-5 months of the year in response to the dry season. The associated spectral reflectance spectrum within a year is exploited to extract information on woodland structure, type, age, condition, biomass, leaf area, and even photosynthetic rate (Shugart *et al.*, 2010).

Differential vegetation reflectance has resulted in the development of numerous vegetation indices and biomass estimation techniques that utilize multiple measurements in the visible and NIR. Vegetation indices (VIs) are dimensionless metrics that function as indicators of relative abundance and activity of green vegetation, often including LAI [most VIs saturate at around LAI = 4, Huete *et al.* (2002)], percentage green cover, chlorophyll content, green biomass, and Absorbed Photosynthetically Active Radiation (APAR) (Jensen, 1996). The first true VI was the Simple Ratio (eq. 1.) described by Birth & McVey (1968).

$$SR = \frac{\rho NIR}{\rho Red} \tag{1}$$

Where ρNIR is the reflectance in the NIR wavelength and ρRed is the reflectance in the red wavelength.

Rouse *et al.* (1974) developed what is now known as the Normalized Vegetation Index the Normalized Differentiated Vegetation Index (NDVI), a satellite metric used to detect changes in pixel-scale vegetation greenness (Asner, 2004).

$$NDVI = \frac{(\rho NIR - \rho Red)}{(\rho NIR + \rho Red)} \tag{2}$$

(eq. 2. 17 Symbol definitions are the same as in eq. 1.) NDVI is broadly correlated with both chlorophyll and water content of canopies and consequently is linked both theoretically and experimentally to LAI, the Fraction of Photosynthetically Active Radiation (fPAR) and the biomass of plant canopies (Gamon *et al.*, 1995; Sellers, 1985). SR and NDVI have been further developed and improved in other related vegetation indices such as the Soil Adjusted Vegetation Index (SAVI, Huete, 1988; Huete & Liu, 1994) and Enhanced Vegetation Index presented in eq. 3 (EVI, Huete *et al.*, 2002).

$$EVI = G\left[\frac{\rho NIR - \rho REd}{(\rho NIR + C1\rho Red - C2\rho Blue + L)}\right] \tag{3}$$

Where C1 = 6.0, atmosphere resistance red correction coefficient; C2 = 7.5, atmosphere resistance blue correction coefficient; L = 1.0, canopy background brightness correction factor; G = 2.5, gain factor. Given that EVI considers the canopy background (L) and reflectance in the blue it is more sensible to variability of canopy structure.

These indexes might be calculated with any sensor [ETM+, MODerate Resolution Imaging Spectroradiometer (MODIS)], except for EVI, which is calibrated exclusively for MODIS data.

Most optical remote sensing techniques for biomass estimation essentially explore the relationship between field measurements of vegetation parameters with spectral VIs to estimate large-scale distribution of biomass. The results vary according to the woodland structure and diversity derived from site-specific conditions such as the disturbance regime and seasonality. Thus, we do not attempt to generalize the relationships between field measurements and VIs for miombo, but give some examples that reflect this variation.

Ribeiro *et al.* (2008)b worked in northern Mozambique in an area where fires occur annually during the dry season with higher incidence in the eastern side, where elephant density is also higher. In this area, the combination of fires and elephants imposes high variability of miombo structure (Ribeiro *et al.*, 2008a). Under these conditions, the authors found relatively low but significant relationships between woody biomass,Leaf Area Index (LAI) measured in the field and, NDVI and SR derived from Landsat ETM+ [NDVI ($r_{biomass}$ =0.30, r_{LAI} = 0.35; $p < 0.0001$) and SR ($r_{biomass}$ = 0.36, r_{LAI} = 0.40, $p < 0.0001$)].

Huete *et al.* (2002) found that the NDVI saturated in high biomass regions like the Amazon (150 Mg/ha or over 15 years of age) while the EVI was sensitive to canopy variations. NDVI ($r=0.15-0.25$) and SR ($0.12-0.24$) showed to be weakly correlated with biomass in dense stands, while the correlation coefficients were considerable ($r_{NDVI}=0.46$ and $r_{SR}=0.50$) for the less dense site. Recently a study conducted in the savannas and woodlands of the Amazon Basin detected quite good relationships between field biomass and MODIS derived NDVI with an $r^2=0.45$ (Saatchi *et al.*, 2007).

A major constraint on the use of spectral VIs in dry and sub-humid ecosystems for vegetation studies, is that VIs are likely to underestimate live biomass due to their insensitivity to nonphotosynthetic vegetation (NPV) and sensitivity to low organic matter content in soils (Okin & Roberts, 2004). The effect of NPV is particularly important in the dry miombo where tree density is low (sometimes less than 10%) and disturbances may create several open patches in which NPV (such as litter, dry woody material, dry grass in the dry season) and soils are a significant component of the surface. In these areas, the effect of the exposed soil may be pronounced, while the low organic matter content on soils makes them bright and mineralogically heterogeneous.

The combination of those factors, creates vertical and horizontal structural differences that may not be depicted using VIs techniques, especially when using low to medium resolution sensors (e.g. 30-m Landsat, 250-m and greater MODIS, etc). In this situation, high spatial resolution optical sensors such as 1-4 m IKONOS, 2.5-10 m SPOT-5, among others, can detect variations in the spatial distribution of biomass density. This new generation of sensors also offer systematic observations at scales ranging from local to global and improves the monitoring of inaccessible areas (Rosenqvist *et al.*, 2003). However, these high spatial resolution instruments are only available for small geographic areas in the miombo ecoregion and thus they are not suitable for large-scale high-resolution ecological assessments in this ecosystem.

The use of Spectral Mixture Analysis (SMA) has become an important technique to overcome the spatial and temporal heterogeneity in miombo. SMA uses a linear mixture technique to estimate the proportion of each ground's pixel area that belongs to different cover type (Okin & Roberts, 2004). SMA is based on the assumption that spectra of materials in a pixel combine linearly with proportion given by their relative abundance. For example, an NPV value of 0.50 from SMA indicates that woody tissues, surface litter and other dead vegetation occupy 50% of the surface area of the pixel. A combined spectrum can thus be decomposed in a linear mixture of its spectral endmembers, spectra of distinct material within the pixel. The weighting coefficients of each spectral endemember, which must sum up to 1, are then interpreted as the relative area occupied by each material in a pixel.

Fractional cover of photosynthetic vegetation (PV), NPV and bare substrate are the principal determinants of ecosystem composition, physiology, structure, biomass and biogeochemical stocks. These also capture the biophysical impact of disturbances caused by both natural and human drivers (Asner et al., 2005). In this case, SMA provides the fraction of per-pixel reflectance that is accounted for by the selected target. Results from SMA can be directly evaluated in field campaigns, including endmember abundance change over time. The temporal or spatial change in the NPV fraction can provide a quantitative measure of disturbance (Chambers et al., 2007).

Up to date SMA has not been applied to miombo but due to its capacity to proportionally decompose several endmembers in a pixel it is a promising remote sensing technique in this ecosystem given the spatial heterogeneity referred previously. Asner et al. (2005) applied SMA for the Brazilian Amazon and bordering Cerrado (an ecosystem structurally similar to miombo) to estimate vegetation attributes for a large-scale and multi-temporal Landsat ETM+ scene spanning the years 1999-2001. Their results showed that PV fractional covers for the cerrado range from 74% to 85% depending on the degree of woody cover, while the NPV cover varied from ~9% and 12% and bare soil between 2% and 10%. The study was able to demonstrate the capacity of this technique to separate vegetation types (forests from savannas) as well as different kinds of disturbances (logging, grazing, etc).

In a particular interesting application Asner et al. (1998) applied SMA to invert a geometrical-optical model to estimate overstory stand-density and crown dimensions. The study applied SMA to a Landsat image to provide estimates of woody cover, herbaceous, bare soil and shade fractions. The estimated cover was then used to unmix the contribution of woody cover and herbaceous canopies to AVHRR multiangle reflectance data. The angular reflectance was used with the radiative transfer (RT) model inversions to estimate LAI, PAR and carbon uptake.

There are two drawbacks of biomass estimation using optical remote sensing techniques: (i) field plots are rarely designed to be related to spaceborne data and (ii) saturation at dense leaf canopies restricts estimates to low biomass levels when passive sensor data is used (Gibbs et al., 2007). Also, biomass is a three-dimension feature of vegetation. However, the ability of optical sensors is limited to two dimensions only, i.e. the upper layers of vegetation (Anaya et al., 2009). Remote-sensing systems relying on optical data (visible and infrared light) are further limited in the tropics by cloud cover. In view of these drawbacks microwave data has become an attractive technique to estimate biomass worldwide and in miombo in particular.

3.2 Radar remote sensing of biomass

The term radar remote sensing is used to denote the active remote sensing where microwave or radio frequency radiation is transmitted to the surface. Microwaves have higher wavelengths (3 to 75 cm for vegetation studies) than solar radiation (400 to 2,500 nm) that is used by optical remote sensing techniques. Unlike optical sensors, the microwave energy penetrates clouds, rain, dust, or fog and allows collection of images, regardless of solar illumination, so that the radar images can be generated at any time under the most varied weather conditions. The microwaves penetrate into the forest canopy and scatter from large woody components (stems and branches) that constitute the bulk of biomass and carbon pool in the forested ecosystems.

The microwaves in a specific wavelength and polarization is emitted from the sensor to the Earth, where it interacts with vegetation (and other objects) and is then backscattered to the sensor. Three different bands are commonly used in radar remote sensing of vegetation: P-band (30 cm) L-band (λ=23.5 cm), C-band (λ=5.8 cm) and X-band (λ=3.1 cm). The sensitivity of backscatter measurements at different wavelengths and polarization (horizontal and/or vertical) to the size and orientation of woody components and their density makes the radar sensors suitable for direct measurements of aboveground woody biomass (carbon stock) and structural attributes such as volume and basal area.

Radar remote sensing has been widely used to map vegetation worldwide, but it is not commonly used in miombo. But, given the spatial variability of vegetation in this ecosystem, radar is a promising technique to estimate biomass. Another advantage of the radar system for miombo is its insensitivity to weather conditions, making wet season measurements possible. During that time of the year some areas are inaccessible. The constraint though is that radar imagery does not cover large areas in the miombo region nor it is being collected frequently, causing difficulties for large-scale frequent monitoring of biomass.

Pierce et al. (2003) refer that Synthetic Aperture Radar (SAR) is known to have a response directly related to the amount of living material that it interacts with. The strong relationship between field biomass and radar image intensity and the development of approaches to estimate aboveground biomass represents one of the unique applications of SAR data in ecology (Kasischke et al., 2004). SAR sensor offers the potential of rapid, accurate, high resolution and low cost mapping of the lower biomass density vegetation of Africa. Moreover, the 46-day repeat cycle of ALOS/PALSAR allow sufficient images to be captured during the year to negate any effects of seasonality and soil moisture, and allow the monitoring of landscapes for any changes in aboveground biomass (Mitchard et al., 2009).

Interferometric SAR (InSAR), which uses the phase information in signals received taken at multiple times to derive very precise measurements of movements on Earth, has been widely used. Recent innovations in orbital designs for repeat pass radar interferometry (InSAR) will allow the sensor to measure height of forest and provide a vertical dimension for accurately resolving the vegetation biomass of forests globally (Chambers et al., 2007).

The most basic approach to estimating aboveground biomass in forests is to develop a multiple-linear regression equation that estimates total biomass as a function of a combination of SAR channels or ratio of different channels (Kasischke et al., 2004). For example Pierce et al. (2003) studied the relationship between radar L-band (25 cm wavelength) and C-band (5.2 cm wavelength) combined and separately, in the regrowth forest of the Amazon Basin. The study found that L- and C-band together improve the accuracy compared to using only one frequency channel. In a recent study Saatchi et al. (2007) found high correlation between radar L-band and field biomass (r^2=0.68) in woodlands and savannas of the Amazon Basin. The authors were able to produce a multiple linear regression model that includes microwave and optical data to estimate the biomass of this vegetation type.

Ribeiro et al. (2008)b correlated contemporary 30-m C-band (5.8 cm) RADARSAT backscatter and field woody biomass and LAI data ($r_{biomass}$= 0.65 and r_{LAI}=0.57, $p<0.0001$) to, in combination with optical data (30-m Landsat ETM+), produce the aboveground biomass map for the Niassa National Reserve (NNR) in northern Mozambique. The results were satisfactory but expected to underestimate biomass in dense woodlands due to radar saturation at high biomass levels. The relationship between radar and field biomass in this area was further explored by Mitchard *et al.* (2009) by using ALOS PALSAR, an L-band sensor. The authors also studied other forested areas in Africa (Cameroon and Uganda). For all sites they found an improved relationship between field biomass and HV-backscatter ($r^2_{biomass}$=0.61-0.76, p<0.0001), but with a clear saturation between 150-200 Mg/ha of biomass. Biomass prediction was done with fairly good accuracy of ±20% for plots with less the 150 Mg/ha. According to the authors these are partly because L-band SAR does not respond directly to aboveground biomass, but to aspects of vegetation structure, partially due to spatial variability in structure and partially due to radar calibration aorthorectification and field estimation errors propagating through the analysis. Also, backscatter responds differently to differing soil and vegetation moisture conditions, and the surface topography, adding to observed prediction errors. Despite these factors the analysis was able to predict, with fairly good accuracy, aboveground biomass from radar data for very different in vegetation types. This finding suggests that utilization of L-band data should be essential for projects involving the mapping and monitoring of woodland and savanna biomass, thus having important implications for carbon-credit projects, such as those under proposed REDD schemes (Mitchard *et al.*, 2009). The comparison of Ribeiro *et al.* (2008)b and Mitchard *et al.* (2009) study for Niassa National Reserve in northern Mozambique indicate clearly that L-band is an improvement over C-band for biomass prediction in the low biomass miombo in this area.

Light Detection and Ranging (LiDAR) measurements provide the most direct estimates of canopy height and the vertical structure of canopy foliage. Together, these measurements enable ecologists to quantify the 3D distribution of vegetation at a landscape scale, to understand processes of carbon accumulation and forest succession and to improve the state of ecosystem models (Chambers *et al.*, 2007). LiDAR systems incorporate a laser altimeter to measure accurately the distance from the sensor to the canopy top and bottom elevations. The energy returned from distances between the canopy and ground provides evidence of the vertical distribution of sub-canopy strata. One application of LiDAR technology has been to map variations in canopy height with meter-level accuracy. Canopy heights can then be translated into estimates of aboveground biomass based on the allometric relationships between height, basal area and biomass (Shugart *et al.*, 2010). This airborne system has not been used in Miombo woodlands (Shugart *et al.*, 2010).

Other approaches for biomass estimation from radar sensors develop regression equations to use different channels of radar imagery for average height and basal diameter estimation and then use these parameters to estimate biomass (Kellndorfer *et al.*, 2004; Simard *et al.*, 2006). Kellndorfer *et al.* (2004) used the Shuttle Radar Topography Mission (SRTM) data in conjunction with a National Elevation Dataset (NED) to estimate pine canopy height in Southeast Georgia, USA. The study indicates that SRTM can be successfully correlated via linear regression modeling with ground-measured mean canopy height (r^2=0.79-0.86). Mean canopy height can *a posteriori* be used in allometric equation to estimate landscape biomass. In a recent study Simard *et al.* (2006) calibrated the SRTM data using Light Detection and Range (LiDAR) data and high resolution Digital Elevation Model (DEM) for the mangrove forests in Everglades National Park. The resulting mangrove tree height map (error 2.0 m) was then used in combination field data to map the spatial distribution of biomass for the entire area. This kind of approach has not been used yet for the miombo or similar woodlands but the results above indicate that the technique can be applied to miombo with some caution, due to its heterogeneity. This may imply for example that the canopy height estimation may only apply to canopy species, while the understory is not completely measured. This should be considered a topic of research for miombo, before any conclusion is raised.

Radar and LiDAR sensors provide complementary information about the forest structure. LiDAR is sensitive to eaf material and radar to structural features, which can be combined to increase accuracy of biomass and the forest structure estimates (Figure 5). However, the signatures have some level of commonality because of biophysical and structural nature of forest stands. The vertical distribution of reflective surfaces that can be inferred from LiDAR reveals the bole and branch structure supporting the leaves within a near vertical volume of the vegetation. LiDAR sensors measure this vertical profile by sampling the forest stand along its orbital tracks. Radar provides imaging capability to estimate forest height through InSAR configuration or forest volume and biomass through polarimetric backscatter power. However, the vegetation signature from radar measurements is from a slanted volume and is sensitive to both vertical and horizontal arrangement of vegetation components (leaves, branches, and stems). The combination of the two sensors has the capability of providing the vegetation three-dimensional structure at spatial resolutions suitable for ecological studies (25-100 m). However, there are limited studies to explore the fusion of the two measurements because of the lack of data over the same study areas as well as well-developed ground data (Shugart *et al.*, 2010). According to the authors this will be an ongoing research area for several years to come.

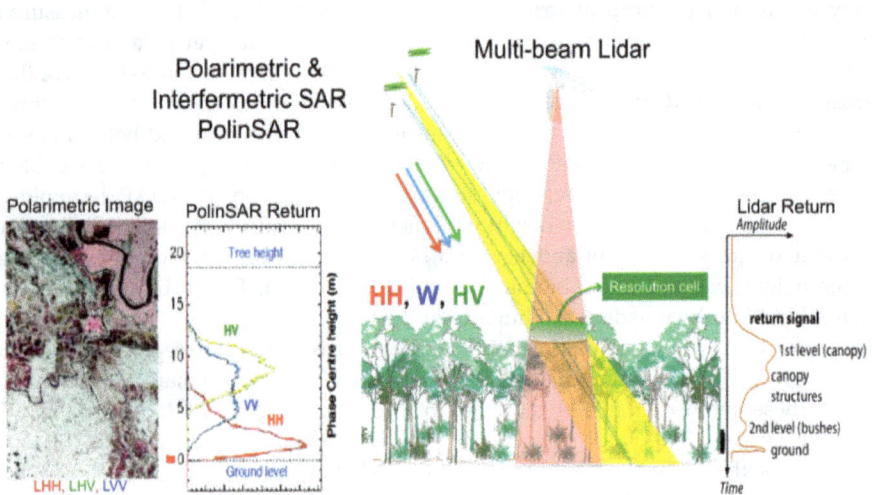

Fig. 5. Radar imager maps of observed radar energy returned at various polarization transmit-receive combinations (HH, HV, VV, where H is horizontal polarization and V is vertical polarization) which are related to the volume and biomass of forest components. The multibeam lidar sampler measures the vertical variation in the strength of the scattered laser signal, which is related to forest vertical structure profile and biomass. The two signals are combined using "fusion" algorithms to improve the accuracy of radar estimates of biomass and to extend lidar measurements of structure in both space and time (Source: Shugart *et al.*, 2010).

This technique is promising in estimating biomass and other parameters for miombo, due its high spatial variability associated to disturbances across the region. Relatively to herbivory by elephants (tree debarking and debranching) and fires (killing of juvenile trees and natural regeneration, soil degradation, killing of adult debarked trees, etc.) data fusion from these instruments may have a significant contribution in addressing the impacts of these disturbances on biomass production in the region. The immediate result will be an improved an informed management of the woodlands in particular for conservation areas.

Future satellite missions plan to make frequent measuring of standing vegetation feasible. Houghton *et al.* (2009) laid out the required specifications for these and other future satellite missions to significantly reduce the nowadays existing uncertainties. They claimed, that in order to reduce the uncertainty in the land-atmosphere fluxes to those of the next uncertain term (which is the net carbon uptake of the ocean with an uncertainty of $\pm$ 18%) a measurement error of less than 2 MgC (or an aboveground biomass uncertainty of about 4 Mg) per ha is required. It is furthermore argued, that disturbances (deforestation, fires and herbivory) are patchy on spatial scales of 100 m and less and only if remote sensing is operating on a similar scale, one can clearly identify changes in carbon storage over time and minimize sampling errors due to averaging (Kohler & Huth, 2010).

4. Remote sensing of biomass in miombo woodlands: Opportunities and limitations for research

Miombo display complex vegetation patterns in which dense vegetation alternates with sparsely populated or bare soil in response to environmental and disturbance (deforestation, fires and herbivory) factors. Low vegetation cover, in some places, and small-scale variations in others, can produce unpredictable errors in the quantification of biophysical and ecological properties of the vegetation. Ignoring this spatial variation can produce inaccurate results, even in fairly homogeneous environments (Aubry & Debouzie, 2001 cited by Hufkens *et al.*, 2008; Haining, 1990).

The use of remote sensing for estimations of vegetation biomass has proved to be of great importance to fill up data gaps and to estimate large-scale variations, especially in low accessible places. The technique lacks of precision when compared to detailed forest inventories. But, for miombo the lack of detailed field data and uncertainty of biomass stocks associated to disturbances, make remote sensing one important technique to address temporal and spatial variations in biomass.

Although repeated satellite imaging has improved in resolution over the years, it is still limited in detecting fine patterns within savanna vegetation. Some forms of remote sensing, such as the sub-meter resolution IKONOS, GeoEye and QUICKBIRD satellite sensors, allow individual trees to be recognized. Other, such as microwave remote sensing from radar (RADARSAT, ALOS-PALSAR, JERS-1, etc.) and LiDAR provide a three dimensional representation of vegetation, which is an improvement over optical remote sensing. However, both low-resolution optical and microwave scenes are currently too expensive for large-scale or regional studies and they require a substantive amount of processing capacities. These represent some of the major limitations for its use in the southern Africa region.

Discriminating between subtypes of savanna vegetation, even simply looking at structural differences, has proved a taxing undertaking, especially in places where field data is limited.

Thus, there is a pressing need to intensify studies calibrating satellite with field measurements, albeit the low accessibility of some areas in the miombo ecoregion. In addition to spatial measurement, it has long been clear that temporal surveys at varied time intervals are required, particularly in biomes that are highly dynamic such as miombo (Furley, 2010).

The problem of biomass retrieval in miombo is compounded by the fact that senescent and dead material (also known as NPV) associated with phenology and disturbances can be a major component of the total surface cover. These three factors play a major role in both abiotic and biotic dynamics of the miombo as explored in Section 2 of this chapter. NPV can represent an important carbon pool in this ecosystem. Many common methods for estimating vegetation cover and biomass use VIs that are insensitive to the presence of NPV. Thus, SMA, data fusion among others may be more appropriate techniques to address spatial and temporal variations of biomass in miombo, but they still need to be adapted and calibrated for this ecosystem.

Addressing large-scale and temporal variations of biomass in the different compartments of miombo (PV, NPV, soils and underground) has to be enhanced in the near future due to their varied contribution to global emissions of greenhouse gases - Carbon dioxide in particular. For example, NPV is the main source of fuel-load for fires, which causes large amounts of carbon to be lost in the form of carbon dioxide to the atmosphere. Because of this, the contribution of miombo to global climate changes may be significant but not completely understood yet.

The southern African region is embarking in the carbon market under the Kyoto Protocol and REDD schemes of the UNFCCC. The ability to negotiate in this market and gain a better position is dependent on the capacity of a country (or region) to estimate carbon stocks with minimal errors. Thus, the advance of remote sensing measurements of biomass from space is an important step towards that achievement (Houghton, 2005 cited by Houghton, 2010). However, it is important to acknowledge that most remote sensing techniques for measurements of biomass usually miss belowground biomass and soil carbon in tropical and sub-tropical regions, which may hold a representative fraction of carbon in the whole system. Thus, future research should focus on the use of microwave radar remote sensing (P-Band and higher) or optical techniques such as SMA and data fusion, that are able to differentiate carbon pools.

5. Conclusions

Remote sensing of biomass in miombo faces the following constraints:
1. Biomass variations at scales of less than 25 m associated with season and disturbances (fires, herbivory and slash and burn agriculture);
2. Limited network of field plots due to low accessibility of some areas;
3. Lack of contemporary field and remote sensing data;
4. Limited accessibility to improved sensors (high resolution optical, microwave and LiDAR data).

However there are some opportunities that may be explored to improve data generation and analysis, thus a better understanding of this ecosystem:
1. Improved methods to estimate biomass in different compartments of the ecosystem (SMA, data fusion, etc.);

2. There exist few sites in the region with detailed data on biomass variations that can be used to test improved remote sensing techniques. For example, permanent sample plots exist in almost all miombo countries. These plots are being frequently evaluated for several parameters including biomass and carbon in different compartments;

3. There are a growing number of remote sensing specialists in the region. In addition, several networks (miombo network, safnet, saccnet, among others) are dedicated to improve remote sensing techniques and data sharing. These involve not only regional but also international senior specialists thus representing a good way of data sharing, improve techniques, establish a link between research and decision making, etc.

In face of the limitations and opportunities for the remote sensing of biomass in miombo, there are five research areas of interest in the region that can benefit from the advance of remote sensing techniques. The core areas presented below are general and can accommodate several research topics according to site particularities. The research themes are:

1. Land use and land cover changes and its effects on miombo biodiversity and biomass;
2. Improved techniques for spatial and temporal variations of biomass and biodiversity;
3. Contribution of miombo to the global changes. This may include several topics such as: carbon stock assessment for different ecosystem compartments (vegetation, soils, NPV, belowground); fire regimes and management; vulnerability and adaptation to climate change.
4. Biomass changes and effects on the availability of resources to human population;
5. Knowledge Management: from science to policy.

6. References

Anaya, J.A.; Chuvieco, E. & Palacios-Orueta, A. (2009). Aboveground biomass assessment in Colombia: A remote sensing approach. *Forest Ecology and Management*, Vol.257, pp.1237–1246. doi:10.1016/j.foreco.2008.11.016

Andreae, M.O. (1993). The influence of tropical biomass burning on climate and the atmospheric environment, In: *Biogeochemistry of Global Change: Radiatively Active Trace Gases*, Oremland, R.S., (ed.), 113-150, Chapman & Hall, New York, NY, USA.

Asner, G.P.; Wessman, C.A. & Schimel, D.S. (1998). Heterogeneity of savanna canopy structure and function from imaging spectrometry and inverse modeling. *Ecological Applications*, vol.8, pp.926-941

Asner, G.P. (2004). Remote Sensing of terrestrial ecosystems: Biophysical remote sensing signatures in arid and semi-arid ecosystems In: *Remote Sensing for Natural Resources Management and Environmental Monitoring*, Ustin, S.L, (ed.), 53-109, John Wiley and Sons, Hoboken, New Jersey, USA

Asner, G.P.; Knapp, D.E.; Cooper, A.N.; Bustamante, M.M.C. & Olander, L.P. (2005). Ecosystem Structure throughout the Brazilian Amazon for Landsat Observationsand Auntomated Spectral Unmixing. *Earth Interactions*, Vol.9, pp.1-30

Backéus, I.; Petterson, B.; Strömquist, L. & Ruffo, C. (2006). Tree communities and structural dynamics in miombo (*Brachystegia_Julbernardia*) woodland, Tanzania. *Forest Ecology and Management*, Vol.230, pp. 171-178

Banda, T.; Schwartz, M.W. & Caro, T. (2006). Woody vegetation structure and composition along a protection gradient in a miombo ecosystem of western Tanzania. *Forest Ecology and Management*, Vol.230, pp.179-185

Birth, G.S. & McVey, G. (1968). Measuring the color of growing turf with a reflectance spectrophotometer. *Agronomy Journal*, Vol.60, pp.640-643

Bond, W. & van Wilgen, B.W. (1996). *Fire and plants: Population and community Biology*, Chapman Hall, London, UK

Brookman-Amissah, J.; Hall, J.N. & Attakorah, J.Y. (1980). A re-assessment of a fire protection experiment in North-Eastern Ghana savanna. *Journal of Applied Ecology*, Vol.17, pp.85-99

Buechner, H.K. & Dawkins, H.C. (1961). Vegetation change induced by elephants and fire in Murchison Falls National Park, Uganda. *Ecology*, vol.42, pp.752-766

Campbell, B.M.; Cunliffe, R.N. & Gambiza, J. (1995). Vegetation structure and small-scale pattern in miombo woodland, Marondera. *Bothalia*, Vol. 25, pp.121-126

Campbell, B.; Frost, P. & Byron, N. (1996). Miombo woodlands and their use: overview and key issues, In: *The Miombo in Transition: Woodlands and Welfare in Africa*, Campbell, B. (ed),1-5, CIFOR. Bogor, Indonesia.

Carlson, T.N. & Ripley, D.A. (1997). On the relation between NDVI, Fractional vegetation Cover, and Leaf Area Index. *Remote Sensing of Environment*, Vol.62, pp.241-252

Cauldwell A.E. & Zieger U. (2000). A reassessment of the fire-tolerance of some miombo woody species in the Central Province, Zambia. African. *Journal of Ecology* Vol.38, No2, pp.138-146.

Chambers, J.Q.; Asner, G.P.; Morton, D.C.; Anderson, L.O.; Saatchi, S.S.; Espírito-Santo, F.D.B.; Palace, M. & Souza Jr., C. (2007). Regional ecosystem Structure and function: ecological insights from remote sensing of tropical forests. *Trends in Ecology and Evolution*, Vol 22, No 8, pp. 414-423, ISSN

Chapin III, F.S.; Matson, P.A. & Mooney, H.A. (2002). *Principles of Terrestrial Ecosystems Ecology*, Springer Verlag, New York, USA

Chidumayo, E.N. (1985). Structural differentiation of contiguous savanna woodland types in Zambia. *Geo-Eco_Trop.*, vol.9, pp.51-66.

Chidumayo, E.N. (1989). Early post-felling response of *Marquesia* woodland to burning in the Zambian Copperbelt. *Journal of Ecology*, vol.77, pp.430-438

Chidumayo, E.N. (1991). Woody biomass structure and utilization for charcoal production in a zambian miombo woodland. *Bioresources Technology*, vol.37, pp.43-52

Chidumayo, E.N. (1997). *Miombo ecology and Management: An introduction*, Stockholm Research Institute, Stockholm, Sweden

Desanker, P.V.; Frost, P.G.H.; Frost, C.O.; Justice, C.O. & Scholes, R J. (1997). *The Miombo Network: Framework for a Terrestrial Transect Study of Land-Use and Land-Cover Change in the Miombo Ecosystems of Central Africa*, IGBP Report 41

Dublin, H.T.S.; Sinclair, A.R.E. & McGlade, J. (1990). Elephants and fires as causes of multiple stable states in the Serengeti-Mara woodlands. *Journal of Animal Ecology*, vol.59, pp.1147-1164

Fassnacht, K.S.; Gower; S.T.; MacKenzie, M.D.; Nordheim, E.V. & Lillesand, T.M. (1997). Estimating the leaf area index of North Central Wisconsin forests using the Landsat Thematic Mapper. *Remote Sensing of Environment*, vol.61, pp.229-245

Freson, R. ; Goffinet, G. & Malaisse, F. (1974). Ecological effects of the regressive succession in muhulu-miombo-savanna in upper Shaba, Zaire, *Proceedings of the first international congress of ecology, structure, functioning and management of ecosystems*, pp.365-371, PUDOC, Wageningen, Holland

Frost, P. (1984). The responses and survival of organisms in fire prone environments, In: *Ecological effect of fire in South African Ecosystems*, Booysen, PV. & Tainton, NM (Eds.), 273-309, Springer-Verlag, Berlin, Germany

Frost, P. (1996). The ecology of Miombo Woodlands, In: *The Miombo in Transition: Woodlands and Welfare in Africa*, Campbell, B. (ed.), 11-55, CIFOR, Bogor, Indonesia

Frolking, S.; Palace, M.W.; Clark, D.B.; Chambers, J.Q.; Shugart, H.H. & Hurtt, G.C. (2009). Forest disturbance and recovery: A general review in the context of spaceborne remote sensing of impacts on aboveground biomass and canopy structure. *Journal of Geophysical Research*, Vol. 114, pp.1-27, G00E02, doi:10.1029/2008JG000911

Furley, P. (2010). Tropical savannas: Biomass, plant ecology, and the role of fire and soil on vegetation. *Progress in Physical Geography*, Vol.34, No4, pp. 563-585, doi: 10.1177/0309133310364934

Gambiza, J.; Campbell, B.M.; Moe, S.R. & Frost, P.G.H. (2005). Fire behavior in a semi-arid *Baikiaea plurijuga* savanna woodland in Kalahari sands in western Zimbabwe. *South African Journal of Science*, vol.101, pp.239-244

Gibbs, H.K.; Brown, S.; O Niles, J. & Foley, J.A. (2007). Monitoring and estimating tropical forest carbon stocks: making REDD a reality. *Environmental Research Letter*, Vol.2, pp.1-13, doi:10.1088/1748-9326/2/4/045023

Gamon, J.A.; Field, C.B.; Goulden, M.; Griffin, K.; Hartley, A.; Joel, G.; Peñuelas, J. & Valentini, R. (1995). Relationships between NDVI, canopy structure, and photosynthetic activity in three Californian vegetation types. *Ecological Applications*, vol.5, No1, pp.28-41

Goward, S.N., Ticker, S.J. and Dye, D.G. 1985. North American vegetation patterns observed with the NOAA-7 Advanced High Resolution Radiometer. *Vegetatio*, vol.64, pp.3-14

Grundy, I. M. (1995). Wood biomass estimation in dry miombo woodland in Zimbabwe. *Forest Ecology and Management*, vol.72, No2-3, pp.109-117

Guy, P.R. (1981). Changes in the biomass and productivity of woodlands in the Sengwa Wildlife Research area, Zimbabwe. *Journal of Applied Ecology*, vol.18, No2, pp.507-519

Guy, P.R. (1989). The influence of elephants and fire on a *Brachystegia-Julbernardia* Woodland in Zimbabwe. *Journal of Tropical Ecology*, vol.5, pp.215-226

Hopkins, B. (1965). Observations on savanna burning in the Olokemeji Forest Reserve, Nigeria. *Journal of Applied Ecology*, vol.2, pp367-381

Houghton, R.A. (2010). How well do we know the flux of CO2 from land-usechange? *Tellus*, Vol.62B, pp.337-351, doi: 10.1111/j.1600-0889.2010.00473.x

Houghton, R. A.; Hall, F. & Goetz S. J. (2009). Importance of biomass in the global carbon cycle, *Journal of Geophysical Research*, Vol.114, G00E03, doi:10.1029/2009JG000935

Huete, A.R. (1988). A soil-adjusted vegetation index (SAVI). *Remote Sensing of Environment*, vol.25 pp.295-309

Huete, A.R. & Liu, H.Q. (1994). An Error and Sensitivity Analysis of the Atmospheric-and Soil-Correcting Variants of the NDVI for the MODIS-EOS. *IEEE Transactions on Geoscience and Remote Sensing*, vol.32, No4, pp.897-905

Huete, A.R.;, Didan, K.; Miura, T.; Rodriguez, E. P.; Gao, X. & Ferreira, L.G. (2002). Overview of the radiometric and biophysical performance of the MODIS Vegetation Indices. *Remote Sensing of Environment*, vol.83, pp.195–213

Hufkens, K.; Bogaert, J.; Dong Q.H.; Lu L.; Huang C.L.; Ma M.G.; Che T.; Li X.; Veroustraete F. & Ceulemans, R. (2008). Impacts and uncertainties of upscaling of remote-sensing data validation for a semi-arid woodland. *Journal of Arid Environments,* vol.72, pp.1490-1505, doi:10.1016/j.jaridenv.2008.02.012.

Hurtt, G.C.; Dubayah, R.; Drake, J.; Moorcroft, P.R.; Pacala, S.W.; Blair, J.B. & Fearon, M.G. (2010). Beyond Potential Vegetation: Combining LiDAR Data and a Height-Structured Model for Carbon Studies. *Ecological Applications,* Vol. 14, No3, pp.873-883

Kasischke, E.S.; Goetz, S.; Hansen, M.C.; Ozdogan, J.R.; Ustin, S.L. & Woodcock, C.E. (2004). Remote Sensing of terrestrial ecosystems: temperate and boreal forests, In: *Remote Sensing for Natural Resources management and environmental monitoring,* Ustin, S.L (ed.), 147-235, John Wiley and Sons, Hoboken, New Jersey, USA

Kellndorfer, J.M.; Walker, W.S.; Dobson, M.C.; Vona, J. & Clutter, M. (2004). Vegetation height derivation from Shuttle Radar Topography Mission data in Southeast Georgia, USA. *IEEE,* vol.7803, pp.4512-4515

Kikula, I.S. (1986). The influence of fire on the composition of miombo woodlands of Southwest Tanzania. *Oikos,* vol.46, pp.317-324

Kohler, P. & Huth, A. (2010). Towards ground-truthing of spaceborne estimates of above-ground life biomass and leaf area index in tropical rain forests. *Biogeosciences,* Vol.7, pp.2531-2543, doi:10.5194/bg-7-2531-2010

Lawore, J.D.; Abbot, P.G. & Werren, M. (1994). Stackwood volume estimations for miombo woodlands in Malawi. *Commowealth Forestry Review,* vol.73, pp.193-197

Laws, R.M. (1970). Elephants as agents of habitat and landscape change in East Africa. *Oikos,* vol.21, pp.1-15

Malaisse, F.P. & Strand, M.A. (1973). A preliminary miombo forest seasonal model, In: *Modeling forest ecosystems,* Kern, L. (ed.), 291-295. Oak Ridge, TN, USA

Jensen, J.R. (1996). *Remote Sensing of the Environment: An Earth Resource Perspective,* Prentice Hall, Upper Saddle River, USA

Justice, C. O. & Dowty, P. (1994). IGBP-DIS satellite fire detection algorithm workshop technical report. IGBP-DIS Working Paper 9. NASA/GSFC, Greenbelt, Maryland.

Justice, C.O.; Vermote, E.; Townshend, J.R.G.; Defries, R.; Roy, D.P.; Hall, D.K.; Salomonson, V.V.; Privette, J.L.; Riggs, G.; Strahler, A.; Lucht,W.; Myneni, R.B.; Knyazikhin, Y.; Running, S.W.; Nemani, R.R.; Wan, Z.; Huete, A.R.; van Leeuwen, W.; Wolfe, R.E.; Giglio, L.; Muller, J.P.; Lewis, P. & Barnsley, M.J. (1998). The Moderate Resolution Imaging Spectroradiometer (MODIS): Land Remote Sensing for Global Change Research. *IEEE Transactions on Geosciences and Remote Sensing,* vol.36, No4, pp.1228-1249

Lu, D. (2001). Estimation of forest stand parameters and application in classification and change detection of forest cover types in the Brazilian Amazon basin. PhD diss., Indiana State University, Terre Haute, Indiana, USA

Malingreau, J.P. & Gregoire, J.M. (1996). Developing a global vegetation fire monitoring system for global change studies: a framework, In: *Biomass Burning and Global Change,* Levine, J.S. (ed.), 14–24, The MIT Press, Cambridge, MA, USA

Mapaure, I.N. & Campbell, B.M. (2002). Changes in miombo woodlands cover in and around Sengwa Wildlife Research Area, Zimbabwe, in relation to elephants and fire. *African Journal of Ecology,* vol.40, pp.212-219

Mitchard, E.T.A.; Saatchi, S.S.; Woodhouse, I.H.; Nangendo, G.; Ribeiro, N.S.; Williams, M.; Ryan, C. M.; Lewis, S.L.; Feldpausch, T.R. & Meir P. (2009). Using satellite radar backscatter to predict above-ground woody biomass: A consistent relationship across four different African landscapes. *Geophysical Research Letter*, vol. 36, L23401, doi:10.1029/2009GL040692.

Okin, G.S. & Roberts, D.A. (2004). Remote Sensing in Arid Environments: Challenges and Opportunities, In: *Remote Sensing for Natural Resources management and environmental monitoring*, Ustin, S.L (ed.), 111-145 , John Wiley and Sons, Hoboken, New Jersey, USA

Patenaude, G.; Hill, R.A; Milne, R.; Gaveau, D.L.A.; Briggs, B.B.J. & Dawson T.P. (2004). Quantifying forest above ground carbon content using LiDAR remote sensing. *Remote Sensing of Environment*, Vol.93, pp. 368-380, doi:10.1016/j.rse.2004.07.016.

[a]Ribeiro, N.S.; Shugart, H.H & Washington-Allen, R. (2008). The effects of fire and elephants on species composition and structure of the Niassa Reserve, northern Mozambique. *Journal of Forest Ecology and Management*, Vol.255, pp. 1626-1636.

[b]Ribeiro, N.S.; Saatchi, S.S.; Shugart, H.H. & Washington-Allen, R.A. (2008). Aboveground biomass and leaf area index (LAI) mapping for Niassa Reserve, northern Mozambique. *Journal of Geophysical Research*, Vol.113, G02S02, doi:10.1029/2007JG000550.

Ribeiro N.S. (2007). Interaction between fires and elephants in relation to vegetation structure and composition of miombo woodlands in northern Mozambique. (PhD Thesis). University of Virginia, Charlottesville, Va, USA

Rodgers, A.; Salehe, J. & Howard, J. (1996). The biodiversity of miombo woodlands, In: *The Miombo in Transition: Woodlands and Welfare in Africa*, Campbell, B. (ed.), 12, CIFOR, Bogor, Indonesia

Rouse, J.W.; Haas, R.H.; Schell, J.A. & Deering, D.W. (1974). Monitoring vegetation systems in the Great Plains with ERTS. *Proceedings of the Third Earth Resources Technology Satellite-1 Symposium*, 3010-3017, Greenbelt, Maryland, USA

Running, S.W.; Loveland, T.R.; Pierce, L.L.; Nemani, R.R. & Hunt, Jr., E.R. (1995). A remote sensing based vegetation classification logic for global land cover analysis. *Remote Sensing of Environment*, vol.51, pp.39-48

Saatchi, S.; Houghton, R.A.; dos Santos Alvalá, R.C.; Soares, J.V. & Yu, Y. (2007). Distribution of aboveground live biomass in the Amazon Basin. *Global Change Biology*, vol.13, pp.816-837

Schlesinger,W.H. (1996). *Biogeochemistry: An Analysis of Global Change*, Academic Press, San Diego, California, USA

Scholes, M. & Andreae, M.O. (2000). Biogenic and pyrogenic emissions from Africa and their impact on the global atmosphere. *Ambio*, Vol.29, pp.23-29

Scholes, R.J. (1997). Biomes: Savannas, In: *vegetation of Southern Africa*, Cowling, RM; Richardson, DM & Pierce, S. (eds), 258-277, Cambridge University Press, Cambridge, UK

Sellers, P.J. (1985). Canopy reflectance, photosynthesis and transpiration, International *Journal of Remote Sensing*, vol.6, pp.1335-1372, doi: 10.1080/01431168508948283

Shugart, H.H.; Saatchi, S. & Hall, F.G. (2010). Importance of structure and its measurement in quantifying function of forest ecosystems. *Journal of Geophysical Research*, Vol.115, G00E13, doi:10.1029/2009JG000993

Simard, M.; Zhang, K.; Rivera-Monroy, V.H.; Ross, M.S.; Ruiz, P.L.; Castaneda-Moya, E.; Twilley, R.R. & Rodriguez, E. (2006). Mapping Height and Biomass of Mangrove Forests in Everglades National Park with SRTM Elevation Data. *Photogrammetric Engineering and Remote Sensing*, vol.72, No3, pp.299-311

Strang, R.M. (1974). Some man-made changes in successional trends on the Rhodesian highveld. *Journal of Applied Ecology*, vol.111, pp.249-263

Sukumar, R. (2003). *The Living elephants*, Oxford University Press, New York, USA Thomson, P.J. (1975). The role of elephants, fire and other agents the decline of *Brachystegia boehmii* woodland. *Journal of Southern African Wildland Management Association*, vol.5, pp11-18

Trapnell, C.G. (1959). Ecological results of woodland burning experiments in northern Rhodesia. *Journal of Ecology*, vol.47, pp.129-168

Trollope, W.S.W. (1978) Fire behavior-a preliminary study, *Proceedings of the grassland Society of Southern Africa*, vol.13, pp.123-128

Trollope, W.S.W., Trollope, L.A. & Hartnett, D.C. (2002). Fire behavior a key factor in the fire ecology of African grasslands and savannas, In: *Forest Fire Research and Wildland Fire Safety*, Viegas (Ed.), 1-15, MillPress, Rotterdam, Holland

Tucker, C.J. & Vanpraet, C. L. (1985). Satellite remote sensing of total herbaceous biomass production in the Senegalese Sahel: 1980-1984. *Remote Sensing of Environment*, vol.17, No3, pp.233-249

Walker, S.M. & Desanker, P.V. (2004). The impact of land use on soil carbon in Miombo Woodlands of Malawi. *Forest Ecology and Management*, vol.203, pp.345-360

Walpole, M.J.; Nabaala, M. & Matankory, C. (2004). Status of the Mara Woodlands in Kenya. *African Journal of Ecology*, vol.42, pp.180-188

White, F. (1983). The vegetation of Africa, a descriptive memoir to accompany the UNESCO/AETFAT/UNSO Vegetation Map of Africa (3 Plates, Northwestern Africa, Northeastern Africa, and Southern Africa, 1:5,000,000. UNESCO. Paris

4

Forest Structure Retrieval from Multi-Baseline SARs

Stefano Tebaldini
Politecnico di Milano
Italy

1. Introduction

The vertical structure of the forested areas has been recognized by the scientific community as one major element in the assessment of forest biomass. Dealing with a volumetric object, the remote sensing techniques that are best suited to infer information about the forest structure are those that guarantee under foliage penetration capabilities. One important, widely popular technology used to investigate the forest vertical structure from the above (typically on board aircrafts) is given by high resolution LIDAR (Light Detection and Ranging) sensors, whose signals penetrate down to the ground trough the gaps within the vegetation layer. In the recent years, however, the attention of the scientific community has been drawn by the use of SAR (Synthetic Aperture Radar) techniques. As opposed to LIDARs, for which high resolution is crucial, the signals emitted by SARs propagate down to the ground by virtue of the under foliage penetration capabilities of microwaves. This different way of sensing the volumetric structure of the imaged objects determines many peculiarities of SAR imaging with respect to LIDAR, some of which are advantages and other drawbacks. The most remarkable advantage is perhaps the one due to the ability of microwaves to penetrate into semi-transparent media, which makes lower wavelength (L-Band and P-Band, typically) SARs capable of acquiring data almost independently of weather conditions and vegetation density. Conversely, the most relevant drawback is that the three dimensional (3D) reconstruction of the imaged objects requires the exploitation of multiple (at least two, but preferable many) images acquired from different points of view, and hence multiple passes over the scene to be investigated.

The aim of this chapter is to discuss relevant topics associated with the employment of a multi-baseline SAR system for the reconstruction of the 3D structure of the imaged targets, with particular attention to the case of forested areas.

The first topic considered is the design of a multi-baseline SAR system for 3D reconstruction in the framework of Fourier Tomography (FT), also referred to as 3D focusing. Even though seldom used in practical applications due to poor imaging quality, FT allows to discuss quite easily the design and overall features of a SAR tomographic system, and represents the basis for all of the developments presented in the remainder of this chapter.

The next part of this chapter will focus on operative methods the generation of high-resolution tomographic imaging from sparse data-sets. This is the case of interest in practical applications, due to the costs associated with flying a high number of passes and platform trajectory accuracy. T-SAR will be cast here in terms of an estimation problem, considering both non-parametric and parametric, or model based, approaches. Non-parametric approaches provide enhanced resolution capabilities without the need for a-priori information about the targets. Model based approaches are even more powerful, providing access to a quantitative, large scale characterization of the forest structure. Nevertheless, model based approaches are affected by an intrinsic limitation, in that data interpretation can be carried out only on the basis of the model that has been adopted.

The final part of this chapter will focus on the joint exploitation of multi-polarimetric and multi-baseline data. It will be shown that under large hypotheses the second-order statistics of such data can be expressed as a Sum of Kronecker Products (SKP), which provides the basis to proceed to decomposing the SAR data into ground-only and volume-only contributions even in absence of a parametric model. Furthermore, the SKP formalism will be shown to provide a compact representation of all physically valid and data-consistent two-layered models. Such a feature will be exploited to provide a exhaustive discussion about the validity of such a class of models for forestry investigation.

Different case studies will be discussed throughout the whole chapter, basing on real data-sets from the ESA campaigns BIOSAR 2007, BIOSAR 2008, and TROPISAR.

2. SAR tomography

2.1 SAR imaging principles

RADAR (Radio Detection And Ranging) is a technology to detect and study far off targets by transmitting ElectroMagnetic (EM) pulses at radio frequency and observing the backscattered echoes. In its most simple implementation, a Radar "knows" the position of a target by measuring the delay of the backscattered echo. Delay measurements are then converted into distance, or range, measurements basing on the wave velocity in the medium (most commonly free space is assumed). Range resolution, Δr, is then directly related to the bandwidth of the transmitted pulse according to well known relation:

$$\Delta r = \frac{c}{2B} \tag{1}$$

where c is the speed of light and B the signal bandwidth.

A Synthetic Aperture Radar (SAR) system is constituted, in its essence, by a Radar mounted aboard a moving platform, such as an aircraft or a satellite. As the platform moves along its trajectory, the Radar mounted aboard transmits pulse waveforms at radio-frequency and gathers the echoes backscattered from the scene. The ensemble of the echoes recorded along the track is generally referred to as raw data, meaning that this product is not, in most cases, directly interpretable in terms of features of the illuminated scene. This happens because of the small size of the antennas usually employed by SAR sensors (few meters as order of magnitude), which results in a significantly large beam on the ground. As a consequence, each recorded echo is determined by a very large number of targets, widely dispersed along the scene. The procedure through which it is possible to separate the contributions of the targets within the Radar beam is generally referred to as SAR focusing. Such a procedure consists in combining the echoes collected along the platform track so as to *synthesize* a much longer antenna than the real one, typically ranging from tens of meters in the airborne case to tens of kilometers in the spaceborne case, and narrow the beam width on the ground to few meters. After focusing, the contribution of each target is identified by two coordinates, as shown in see Fig. (1). The first one, conventionally named azimuth, may be though of as the projection of the platform track onto the Earth's surface. The other, conventionally named slant range, denotes the distance from the platform track at any given azimuth location. The

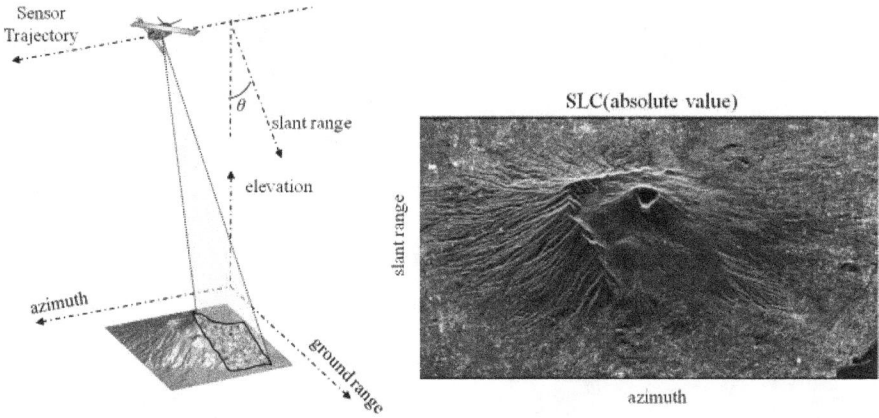

Fig. 1. SAR acquisition geometry.

result is a complex 2D signal indexed by the slant range, azimuth coordinates, usually referred to Single Look Complex (SLC) image.

Assuming a narrow bandwidth signal the azimuth resolution, $\triangle x$, can be obtained as:

$$\triangle x = \frac{\lambda}{2L_s} r \qquad (2)$$

where λ is the carrier wavelength, L_s is the length of the synthetic aperture, and r is the stand-off distance between the sensor and the target. Slant range resolution is obtained according to eq. (1). A further description of SAR image formation is beyond the scopes of this chapter. For further details, the reader is referred to (1), (2), (3), (4).

2.2 3D SAR imaging

Given the definitions above, a SAR SLC image represents the scene as projected onto the slant range, azimuth coordinates. This entails the complete loss of the information about the third dimension, i.e.: about the vertical structure of the targets within the imaged scene. The key to the recovery of the third dimension is to enhance the acquisition space so as to form a *further* synthetic aperture. The easiest way to do this is to employ a multi-baseline SAR system, where several SAR sensors, flown along parallel tracks, image the scene from different points of view, see figure (2).

Such a system offers the possibility to gather the backscattered echoes not only along the azimuth direction, but also along the cross-range direction, defined by the axis orthogonal to the Line Of Sight (LOS) and to the orbital track. Accordingly, the backscattered echoes can be focused not only in the slant range, azimuth plane, but in the whole 3D space. Therefore, the exploitation of multi-baseline acquisitions allows to create a fully 3D imaging system, where the size of the 3D resolution cell is determined by the pulse bandwidth along the slant range direction, and by the lengths of the synthetic apertures, according to eq. (2), in the azimuth and cross range directions, see Fig. (3).

Both 2D and 3D SAR imaging data may be regarded as specializations to the SAR case of the more general concept of Diffraction Tomography, widely exploited in seismic processing (5). As such, 3D focusing is analogous to the basic formulation of SAR Tomography (T-SAR), in which no particular assumption is adopted to describe the imaged scene (6).

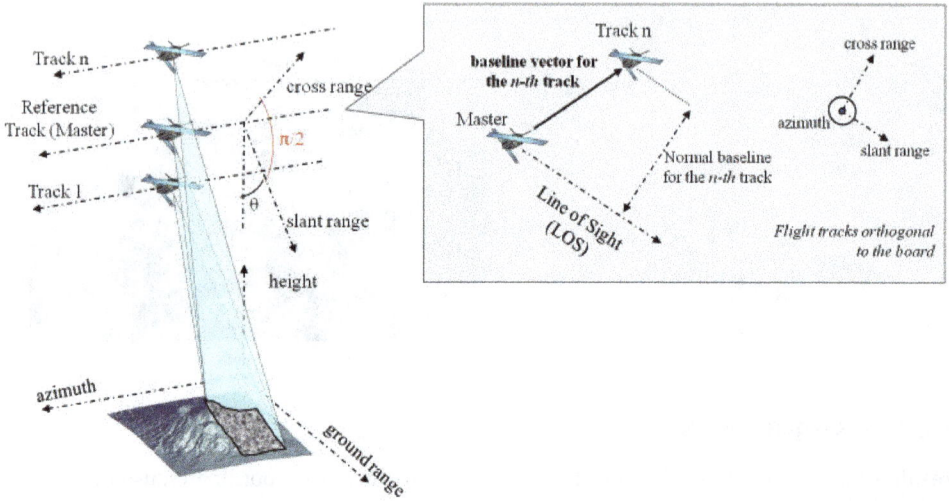

Fig. 2. A multi-baseline SAR system

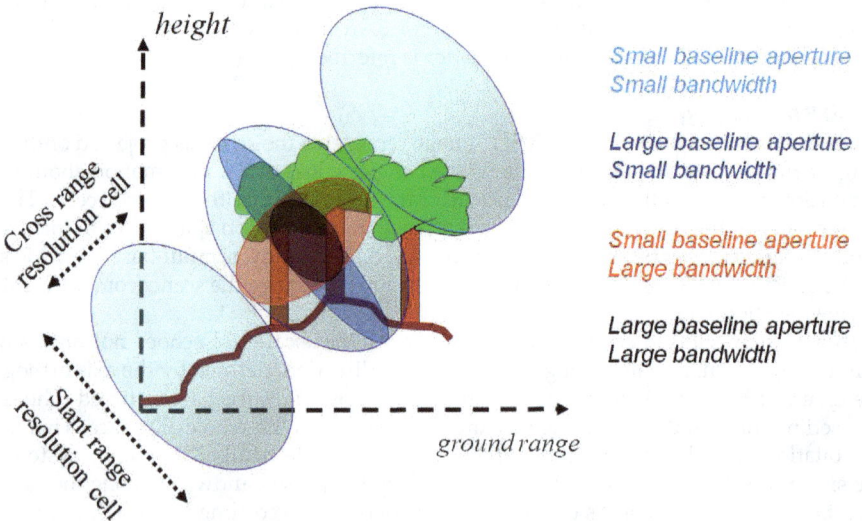

Fig. 3. Section of the 3D resolution cell in the slant range, cross range plane, represented as a function of baseline aperture and pulse bandwidth.

2.3 SAR tomography: mathematical formulation

Consider the case of a multi-baseline data-set, represented by N Single Look Complex (SLC) monostatic SAR images, properly co-registered and phase flattened (7). Let y_n denote a complex valued pixel in the $n - th$ image, the dependence on the slant range, azimuth location, (r, x), being made implicit in order to simplify the notation. After (7), each acquisition can be expressed through the following forward model:

$$y_n = \int h\left(r', x'\right) s\left(r', x', \zeta'\right) \exp\left(j\frac{4\pi}{\lambda r} b_n \zeta'\right) dr' dx' d\zeta' \tag{3}$$

where: (r', x') are the slant range, azimuth coordinates with respect to the center of the SAR resolution cell; ζ' is the cross range coordinate with respect to the center of the SAR resolution cell, given by the axis orthogonal to (r', x'); $s(r', x', \zeta')$ is a the scene complex reflectivity, namely a complex valued quantity accounting for amplitude and phase modifications due to the interactions of the transmitted pulse with each of the scatterers within the SAR resolution cell; $h(r', x')$ is the end-to-end SAR impulse response (after focusing); b_n is the normal baseline relative to the $n - th$ image with respect to a common master image.
Resolving the integral in (3) with respect to (r', x'), one gets:

$$y_n = \int P\left(\zeta'\right) \exp\left(j\frac{4\pi}{\lambda r} b_n \zeta'\right) d\zeta' \tag{4}$$

where

$$P\left(\zeta'\right) = \int h\left(r', x'\right) s\left(r', x', \zeta'\right) dr' dx' \tag{5}$$

is the target projection within the SAR resolution cell along the cross-range coordinate. Equation (4) may be thought of as the key equation of T-SAR. It states that the SAR data and the target projections form a Fourier pair. Hence, the cross-range distribution of the backscattered power, namely

$$S_\zeta\left(\zeta\right) = E\left[|P\left(\zeta\right)|^2\right], \tag{6}$$

can be retrieved by estimating the Fourier Spectrum (FS) of the data with respect to the normal baselines. The evaluation of (6) for every slant range, azimuth location results in a 3D reconstruction of the imaged scene, in the coordinate system defined by the slant range, azimuth, cross-range coordinates. Accordingly, the vertical distribution of the backscattered power can be retrieved by re-sampling the reconstructed scene from the slant range, azimuth, cross-range coordinates to the ground range, azimuth, elevation coordinates. For sake of simplicity, in this paper I will suppose that the passage from the cross-range coordinate, ζ, to elevation, z, can be simply described through the equation

$$\zeta = z/\sin\theta \tag{7}$$

where θ is the look angle. Under this assumption, the vertical distribution of the backscattered power can be obtained from the FS simply as $S(z) = S_\zeta(\zeta = z/\sin\theta)$. Notice that such a simplification does not entail any loss of generality, as the exact reconstruction of the vertical backscattered power at every ground range, azimuth location can be carried out a-posteriori through simple geometrical arguments.

2.4 Expected FS for forested areas

A brief discussion is now required about the physics of radar scattering from forested areas, in order to discuss the expected properties of the FS. Many excellent works have been done on modeling radar backscattering for forested areas, see for example (8), (9), (10). After these works, forested areas will be characterized as the ensemble of trunks, canopies, intended as the ensemble of leaves and branches, and terrain. A simple and widely exploited solution to describe the scattering behavior of a target above the ground is provided by modeling the ground as a flat half space dielectric and using image theory. Four Scattering Mechanisms arise by retaining this approach (11), (12), (9): backscatter from the target; target to ground double bounce scattering; ground to target double bounce scattering; backscatter from the target illuminated by the reflected wave, namely ground to target to ground scattering. In the following the contributions from ground to target to ground scattering will be assumed to be negligible. In the case of trunks, this assumption is supported by considering that backscatter is dominated by specular scattering, and may thus be neglected (9). For the same reason, direct backscatter from the trunks will be neglected as well. The case of the canopy is a little more delicate, since randomly oriented small objects such as branches and leaves should be treated as isotropic scatterers. Also in this case, however, it makes sense to retain that ground to target to ground contributions may be neglected, by virtue of the attenuation undergone by the wave as it bounces on the ground and propagates through the trunk layer and the canopy layer of adjacent trees.

Accordingly, higher order multiple scattering will be neglected as well. As for double bounce scattering, through simple geometric arguments it is possible to see that the distance covered by the wave as it undergoes the two consecutive specular reflections on the ground and on the target, or viceversa, is equal to the distance between the sensor and the projection of the target onto the ground. Therefore, at least on an approximately flat terrain, every double bounce mechanism is located on the ground. Assuming that trees grows perfectly vertical, the projection of the trunk onto the ground collapses into a single point. Hence, double bounce scattering due to trunk-ground interactions may be regarded as a point-like scattering mechanism, located on the ground. The projection of the canopy onto the ground, instead, gives rise to a superficial scattering mechanism located on the ground, analogous to ground backscatter[1].

Based on the model here discussed, a forest scenario is to be characterized, from the point of view of a SAR based analysis, as an ensemble of point like, superficial and volumetric scatterers. Basing on single polarimetric, multi-baseline observations, the element of diversity among the different Scattering Mechanisms (SMs) is given by their corresponding FS, see Fig. (4). Trunk-ground interactions give rise to a point-like SM whose phase center is ground locked, resulting in an isolated narrow peak in the associated FS, located at ground level. Ground backscatter and canopy-ground interactions give rise to a superficial, ground locked, SM. It follows that the associated FS is located at ground level as well, but it is spread along the cross-range axis. Canopy backscatter gives rise to a volumetric SM whose phase center is located above the ground, depending on canopy height, from which it follows that the associated FS is located above the ground, and it is spread along the cross-range axis as well.

2.5 Baseline design

After equation (4), the design of the baseline set for tomographic application may be performed basing on standard results from Signal Theory (14). In particular, as the aim is

---

[1] It has to be remarked that these conclusions do not apply in the case of bi-static acquisitions, see for example (13)

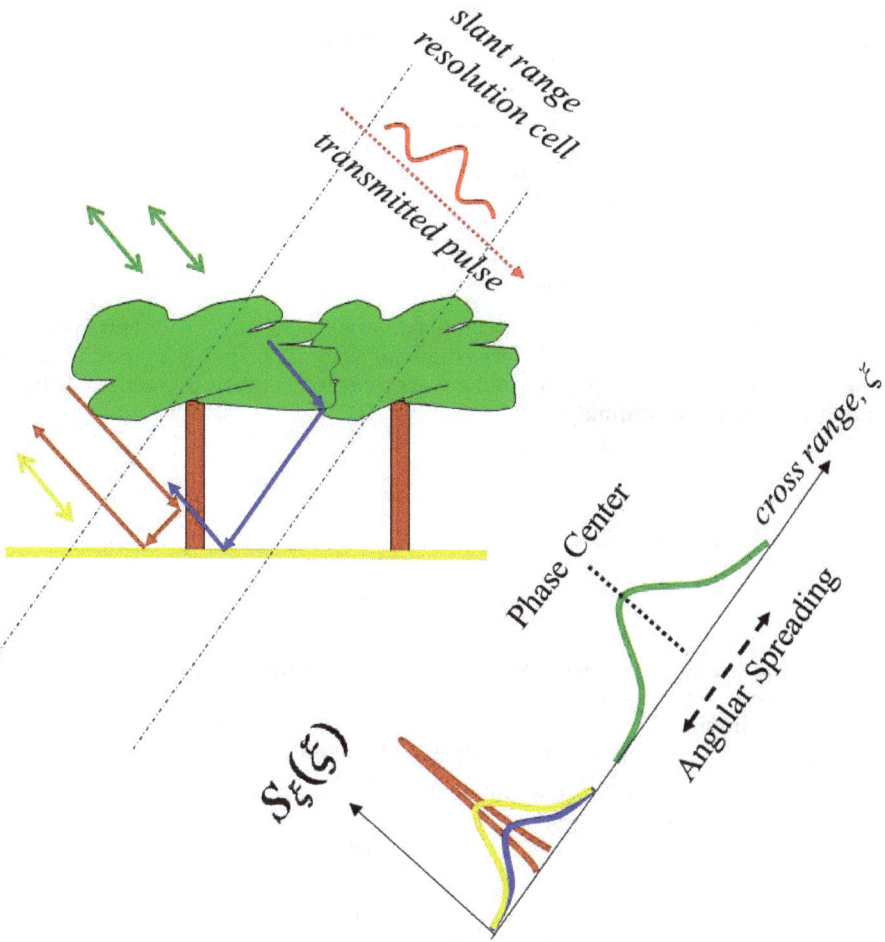

Fig. 4. Pictorial representation of the Fourier Spectra associated to the four scattering mechanisms discussed. Brown: Trunk-ground scattering; yellow: ground backscattering; green: canopy backscattering; blue: canopy-ground scattering.

to correctly represent the power spectrum of the data, the baseline set has to be designed in such a way as to avoid the arising of aliasing phenomena while allowing a sufficient resolution. As discussed above, equation (4) states that, at each slant range azimuth location, the complex valued pixel y_n associated with the $n - th$ baseline is obtained by taking the spectral component of target projection corresponding to the spatial frequency:

$$f_n = \frac{2b_n}{\lambda r} \tag{8}$$

Assume now that baseline sampling is uniform with baseline spacing $\triangle b$, i.e.: $b_n = n\triangle b$. It then follows that the reconstruction of $P(\xi)$ from y_n is unambiguous only provided that the overall extent of the target along the cross-range direction, H_ξ, is lower than the inverse of the frequency spacing, i.e.: $H_\xi \leq \frac{1}{\triangle f} = \frac{\lambda r}{2\triangle b}$. After fig. (4) it is immediate to see that:

$$H_\xi = \frac{H}{\sin\theta} + \triangle r\cos\theta \tag{9}$$

where H is forest height. Accordingly, the fundamental constraint to obtain a correct tomographic imaging of a forest is that[2]:

$$\frac{H}{\sin\theta} + \triangle r\cos\theta \leq \frac{\lambda r}{2\triangle b} \tag{10}$$

which allows to set the baseline spacing as a function of forest height and system features.

Cross range system resolution can be derived analogously to 2, simply by noting that the overall extent of the synthetic aperture in the cross direction is given by $N\triangle b$, N being the total number of tracks. Accordingly:

$$\triangle \xi = \frac{\lambda r}{2N\triangle b} \tag{11}$$

after which vertical resolution is written immediately as:

$$\triangle z = \frac{\lambda r}{2N\triangle b}\sin\theta. \tag{12}$$

Accordingly, the number of tracks is directly related to the vertical resolution of the tomographic system. Put in these terms, one would be tempted to assume that vertical resolution can be improved at will, simply by allowing a sufficient number of tracks. There are factors, though, which pose an upper limit about baseline aperture:

- if different baselines are collected at different times it has to be considered that the scene can undergo significant changes, resulting in the impossibility to coherently combine different images. This phenomenon is known in literature as temporal decorrelation (15), (16).

- It is important to keep in mind that equation (3) is valid upon the condition that the scene reflectivity is isotropic. This condition generally holds for narrow-angle (i.e.: small baseline) measurements, whereas it tends to break down in case of wide-angle measurements.

- Gathering multiple baselines is, in general, associated with high costs, due to the necessity to use multiple SAR sensors or fly the same sensor many times.

3. SAR Tomography as an estimation problem

Despite being a most powerful tool for system design, casting T-SAR in terms of a simple evaluation of the data Fourier Transform constitutes too simple an approach for practical applications. The main reason for this statement is that the capability to resolve

---

[2] This formula is given for typical geometries used in SAR imaging, where the incidence angle is usually larger than, say, 25°. The case of nadir looking Radar ($\theta = 0$) is not considered here.

targets in different positions through this simple approach is dramatically limited by the effective baseline aperture, according to equation (12). Furthermore, the vertical backscatter distribution that is yielded by Fourier based T-SAR is likely to be affected by the presence of grating lobes, that can arise in case of uneven baseline sampling (17). It follows that T-SAR can be more effectively posed as the problem of estimating the vertical distribution of the backscattered power based on a limited number of observations, that is as a Spectral Estimation problem. Many excellent works have appeared in literature about the application of spectral estimation methods to SAR tomography. A possible solution to improve the vertical resolution and ease the arising of side lobes is to exploit super-resolution techniques developed in the Direction of Arrival (DOA) framework, such as Capon adaptive filtering, MUSIC, SVD analysis, Compressive Sensing, and others. In the framework of T-SAR, such techniques have been applied in a number of works, among which (17), (18), (19), (20). The main limitation of most super-resolution techniques derives from the assumption that the scene is constituted by a finite number of point-like scatterers, which hinders their application in the analysis of scenarios characterized by the presence of distributed targets. For this reason, super-resolution techniques appear to be mainly suited for the tomographic characterization of urban areas. A different solution may be found in the works by *Fornaro et al.*, (21), (22), focused on the analysis of urban areas, and in those relative to Polarization Coherence Tomography (PCT), due to *Cloude*, (23), (24), focused on the analysis of forested areas. The common trait of those works is that super-resolution is achieved by exploiting a priori information about target location. In the case of forested areas, for example, the a priori information is represented by ground topography and canopy top height.

Alternatively, a sound characterization of the imaged scene can be given by modeling its vertical backscattering distribution, in such a way as to pose T-SAR as a Parametric Estimation problem. Model based techniques have been successfully employed in a number of works about the analysis of forested areas, see for example (25), (13), (26), (27). Nevertheless, model based approaches are affected by an intrinsic limitation, in that data interpretation can be carried out only on the basis of the model that has been adopted.

3.1 Two simple spectral estimators

Spectral estimation techniques are mostly based on the analysis of the data covariance matrix among different tracks, namely the matrix whose elements are obtained as:

$$\{ \mathbf{R} \}_{nm} = E \left[y_n y_m^* \right] \tag{13}$$

The most simple estimator of the scene backscatter distribution is given by the Periodogram, which is defined as:

$$S_p \left(\zeta \right) = \mathbf{a} \left(\zeta \right)^T \widehat{\mathbf{R}} \mathbf{a} \left(\zeta \right) \tag{14}$$

where $\widehat{\mathbf{R}}$ is the sample estimate of the data covariance matrix defined above and $\mathbf{a} \left(\zeta \right) = \left[\exp \left(j \frac{4\pi}{\lambda r} b_1 \zeta \right) \cdots \cdots \exp \left(j \frac{4\pi}{\lambda r} b_N \zeta \right) \right]^H$. The Periodogram is formally equivalent to the Fourier transform of the data with respect to the normal baseline. The vertical resolution provided by such an estimator is then given by eq. (12).

This limit may be overcome by adaptive filtering techniques. Capon filtering, for example, designs a filter $\mathbf{f}_{min}$ such that:

$$\mathbf{f}_{min} = \arg \min \left\{ \mathbf{f}^H \widehat{\mathbf{R}} \mathbf{f} \right\} \tag{15}$$

under the linear constraint

$$\mathbf{f}^H \mathbf{a} \left(\zeta \right) = 1 \tag{16}$$

This leads to the definition of the Capon Spectral Estimator as:

$$S_C\left(\xi\right) = \mathbf{f}_{min}^H \widehat{\mathbf{R}} \mathbf{f}_{min} = \frac{1}{\mathbf{a}\left(\xi\right)^H \widehat{\mathbf{R}}^{-1} \mathbf{a}\left(\xi\right)} \tag{17}$$

Such an estimator produces super-resolution capabilities and grating lobes reduction without the need for a-priori information about the targets, see for example (17), (18). The main drawback associated with the Capon Spectral Estimator is the poor radiometric accuracy, especially in presence of a small baseline aperture. Accordingly, the employment of the Capon Spectral Estimator has to be considered as a compromise. Indeed it results in non accurate backscattered power values, yet it permits to appreciate details that could not be accessible in presence of a small baseline aperture.

3.2 Representation of multiple SMs
Dealing with natural scenarios, it is sensible to assume that the target signature changes randomly from one pixel to another, either in the case of distributed or point-like targets. For this reason, the data will be assumed to be a realization of a zero-mean complex process, from which it follows that a unified mathematical treatment of all the SMs described in section 2.4 may be provided by characterizing the second order moments of the data, represented by the expected value of the interferograms. Under the hypothesis of statistical uncorrelation among the different SMs, the expected value of the $nm - th$ interferogram may be expressed as:

$$E\left[y_n y_m^*\right] = \sum_{k=1}^{K} \sigma_k^2 \gamma_{nm}^{(k)} \tag{18}$$

where: K is the total number of SMs; σ_k^2 is the backscattered power associated to the $k - th$ SM; $\gamma_{nm}^{(k)}$ is the complex interferometric coherence induced by the spatial structure (i.e.: point-like, superficial, or volumetric) associated with the $k - th$ SM in the $nm - th$ interferogram (15). After (18), the expression of the data covariance matrix is given by:

$$E\left[\mathbf{y}_{MB}\mathbf{y}_{MB}^H\right] \overset{def}{=} \mathbf{R} = \sum_{k=1}^{K} \sigma_k^2 \mathbf{R}_k \tag{19}$$

where: $\mathbf{y}_{MB} = \left[y_1\, y_2 \cdots y_N\right]^T$ is the stack of the multi-baseline data at a certain slant range, azimuth location; each matrix $\mathbf{R}_k$ is an $N \times N$ matrix whose entries are given by the interferometric coherences associated to the $k - th$ phase center: $\{\mathbf{R}_k\}_{nm} = \gamma_{nm}^{(k)}$. The role of each matrix $\mathbf{R}_k$ in (19) is to account for the spatial structure associated to each SMs. According to this interpretation, such matrices will be hereinafter referred to as *structure matrices*.

Following the arguments presented in section 2.4, the data covariance matrix will be assumed to be contributed by just two SMs, one associated with volume backscattering, and the other with the ensemble of all SMs whose phase center is ground locked, namely surface backscattering, ground-volume scattering, and ground-trunk scattering. Accordingly, eq. (19) simplifies to:

$$E\left[\mathbf{y}_{MB}\mathbf{y}_{MB}^H\right] \overset{def}{=} \mathbf{R} = \sigma_g^2 \mathbf{R}_g + \sigma_v^2 \mathbf{R}_v \tag{20}$$

the subscripts g, v standing for *ground* and *volume*.

3.3 A simple parametric model

By letting the ground and volume structure matrices depend upon some physically meaningful parameter, equation (20) provides an easy access point to an effective modeling of the problem. In this chapter I will consider a very simple model, where both the ground and the vegetation layer are represented as phase centers plus the associated decorrelation term arising from the angular spreading described in section 2.4. Hence, the phase of the structure matrices can be parametrized as:

$$\angle \left\{ \mathbf{R}_{g,v} \right\}_{nm} = \frac{4\pi}{\lambda r} \frac{b_m - b_n}{\sin\theta} z_{g,v} \tag{21}$$

where $z_{g,v}$ is the elevation of the phase center associated with ground and volume scattering. As for the amplitudes, instead, the following model will be retained:

$$\left| \left\{ \mathbf{R}_{g,v} \right\}_{nm} \right| = \rho_{g,v}^{-|b_n - b_m|/\Delta_b} \tag{22}$$

where $\rho_{g,v}$ will be referred to as spreading constant and Δ_b is the average normal baseline sampling. The role of the spreading constant is to describe in a simple fashion the angular spreading that arises from the spatial structure associated to each phase center, avoiding the dependence upon a particular physical model. Given the definition above, the spreading constant ranges from 0 to 1, corresponding to the cases of pure noise and of a perfectly coherent received signal (i.e. a point-like scatterer), respectively. For details on more sophisticated physical models, the reader is referred to (13), (27).

3.3.1 Inversion

In principle, the availability of a model for the covariance matrix suffices for solving for the unknowns through Maximum Likelihood Estimation (MLE) (28). This solution, however, would not be efficient, since it requires an exhaustive search in a significantly wide parameter space. A significant complexity reduction may be achieved by applying the Covariance Matching Estimation Method (COMET) (29), (30), (31), (32), which allows to solve the problem by minimizing a weighted Frobenius norm of the difference between the sample covariance matrix and the model . In formula:

$$\left(\hat{z}_g, \hat{\rho}_g, \hat{z}_g, \hat{\rho}_g, \widehat{\sigma_g^2}, \widehat{\sigma_v^2} \right) = \text{argmin} \left\{ trace \left(\hat{\mathbf{R}}^{-1}(\hat{\mathbf{R}} - \mathbf{M})\hat{\mathbf{R}}^{-1}(\hat{\mathbf{R}} - \mathbf{M}) \right) \right\} \tag{23}$$

where $\hat{\mathbf{R}}$ is the sample covariance matrix and $\mathbf{M}$ is the model matrix, i.e.:

$$\mathbf{M} = \sigma_g^2 \mathbf{R}_g \left(z_g, \rho_g \right) + \sigma_v^2 \mathbf{R}_v \left(z_v, \rho_v \right). \tag{24}$$

A straightforward extension to the case of multi-polarimetric data can be achieved by assuming that backscattered powers $\left(\sigma_g^2, \sigma_v^2 \right)$ only change with polarization, whereas the structural parameters $\left(z_g, \rho_g, z_v, \rho_v \right)$ stay the same. See (26) for details.

3.4 A case study: the Remningstorp forest site

This section is devoted to reporting the results of the tomographic analysis of the forest site of Remningstorp, Sweden, on the basis of a data-set of $N = 9$ P-Band, fully polarimetric SAR images acquired by the DLR airborne system E-SAR in the frame of the ESA campaign BIOSAR 2007. Prevailing tree species in the imaged scene are Norway spruce (Picea abies), Scots pine (Pinus sylvestris) and birch (Betula spp.). The dominant soil type is till with a field layer, when present, of blueberry (Vaccinium myrtillus) and narrow thinned grass

(Deschampsia flexuosa). Tree heights are in the order of 20 m, with peaks up to 30 m. The topography is fairly flat, terrain elevation above sea level ranging between 120 and 145 m. The acquisitions have been carried out from March to May 2007. The horizontal baseline spacing is approximately 10 m, resulting in a maximum horizontal baseline of approximately 80 m. The spatial resolution is approximately 3 m in the slant range direction and 1 m in the azimuth direction. The look angle varies from 25° to 55° from near to far range, resulting in the vertical resolution to vary from about 10 m in near range to 40 m in far range. See also table (2) in the remainder.

The top row of Fig. (5) shows an optical view of the Remningstorp test site captured from Google Earth, re-sampled onto the SAR slant range, azimuth coordinates, whereas the bottom row of the same figure shows the amplitude of the HH channel, averaged over the 9 images. Backscatter from open areas has turned out to be remarkably lower than that from forested areas (about 25 dB), as expected at longer wavelengths.

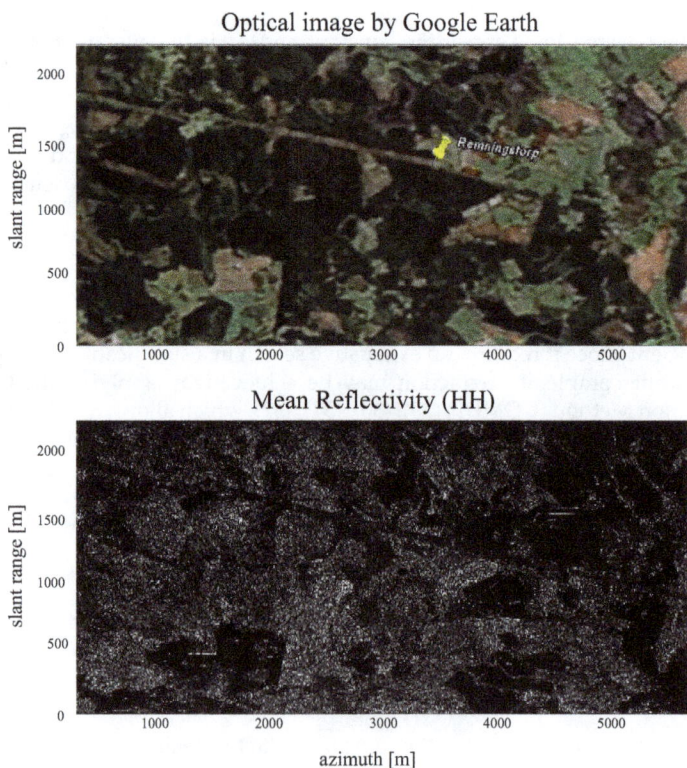

Fig. 5. Top row: Optical image of the Remningstorp forest site by Google Earth. Bottom row: Mean reflectivity of the HH channel.

3.4.1 Non tomographic analysis
An analysis of the amplitude stability of the data has been performed by computing the ratio μ/σ, where μ and σ denote the mean and the standard deviation of the amplitudes of the

SLC images. The μ/σ index is widely used in Permanent Scatterers Interferometry (PSI) as a criterion to select the most coherent scatterers in the imaged scene (33). As a result, for both the HH and VV channels the μ/σ index has resulted to be characterized by extremely high values ($\mu/\sigma > 15$), which indicates the presence of a highly stable scattering mechanism in the co-polar channels. Surprisingly, high values of the μ/σ index ($\mu/\sigma > 10$) have been observed in the HV channel as well.

Temporal decorrelation has been evaluated by exploiting the presence of additional zero baseline images. The temporal coherence at 56 days has been assessed in about $\gamma_{temp}^{HH} \simeq 0.85$ in the HH channel, $\gamma_{temp}^{VV} \simeq 0.8$ in the VV channel, and $\gamma_{temp}^{HV} \simeq 0.75$ in the HV channel, relatively to forested areas. Accordingly, the temporal stability of the scene is rather good for all the three polarimetric channels, indicating the presence of a stable scattering mechanism, especially in the co-polar channels.

The information carried by the co-polar channels has been analyzed by averaging the backscattered powers and the co-polar interferograms (i.e.: the Hermitian product between the HH and VV channels) over all the 9 tracks and inside an estimation window as large as 50×50 square meters (ground range, azimuth). The distribution of the HH and VV total backscattered power with respect to the co-polar phase, $\Delta\varphi = \varphi_{HH} - \varphi_{VV}$, has shown to be substantially bimodal, high and low energy values being concentrated around $\Delta\varphi \approx 80°$ and $\Delta\varphi \approx 0°$, respectively, see Fig.(6). The coherence between the HH and VV channels has been observed to be rather high in open areas ($\gamma_{copol} \approx 0.8$.), whereas in forested areas it has been assessed about $\gamma_{copol} \approx 0.45$.

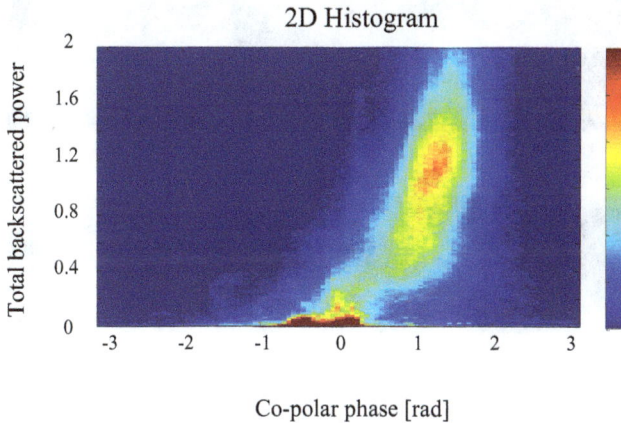

Fig. 6. Joint distribution of the total backscattered power and the co-polar phase for the co-polar channels. The total backscattered power has been obtained as the sum between the HH and VV backscattered powers. The color scale is proportional to the number of counts within each bin.

3.4.2 Tomographic analysis

A first, non parametric, tomographic analysis has been carried out by evaluating the Capon spectra of the three polarimetric channels, see for example (18), (17), reported in Fig. (7). Each spectrum has been obtained by evaluating the sample covariance matrix at each range bin on the basis of an estimation window as large as 50×50 square meters (ground range, azimuth). The analyzed area corresponds to a stripe of the data along the slant range direction, shown in the top panel of Fig. (7). Almost the whole stripe is forested except for the dark areas in near range, corresponding to bare terrain. It may be observed that the three spectra have very similar characteristics. Each spectrum is characterized by a narrow peak, above which a weak sidelobe is visible. As for the co-polar channels, this result is consistent with the hypothesis that a single scattering mechanism is dominant. It is then reasonable to relate such scattering mechanism to the double bounce contribution from trunk-ground and canopy-ground interactions, and hence the peak of the spectrum can be assumed to be located at ground level. The sidelobe above the main peak is more evident in the HV channel, but the contributions from ground level seem to be dominant as well, the main peak being located almost at the same position as in the co-polar channels. Accordingly, the presence of a relevant contribution from the ground has to be included in the HV channel too.

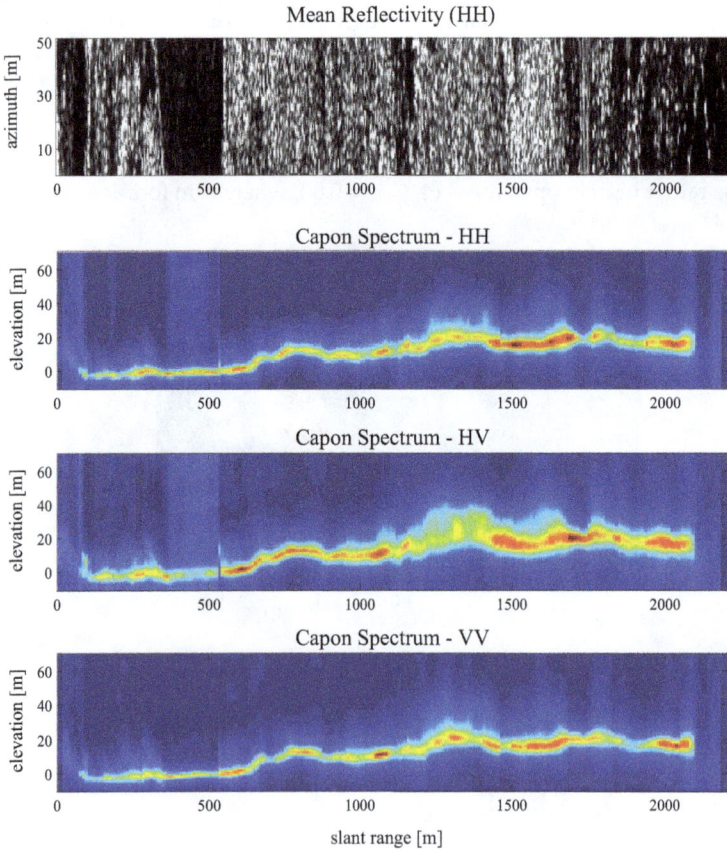

Fig. 7. Top panel: mean reflectivity of the data (HH channel) within a stripe as wide as 50 m in the azimuth direction. The underlying panels show the Capon Spectra for the three polarimetric channels. At every range bin the signal has been scaled in such a way as to have unitary energy.

3.4.3 Parametric tomographic analysis

Results shown here have been obtained by processing all polarizations at once, as suggested in section 3.3.1.

Figure (8) shows the map of the estimates relative ground elevation. The black areas correspond to absence of coherent signals, as it is the case of lakes. The estimates relative to canopy elevation are visible in Fig. (9). In this case, the black areas have been identified by the algorithm as being non-forested. It is worth noting the presence of a road, clearly visible in the optical image, see Fig. (5), crossing the scene along the direction from slant range, azimuth coordinates $(1850, 0)$ to $(1000, 5500)$. Along that road, a periodic series of small targets at an elevation of about 25 m has been found by the algorithm. Since a power line passes above the road, it seems reasonable to relate such targets to the echoes from the equipment on the top of the poles of the power line.

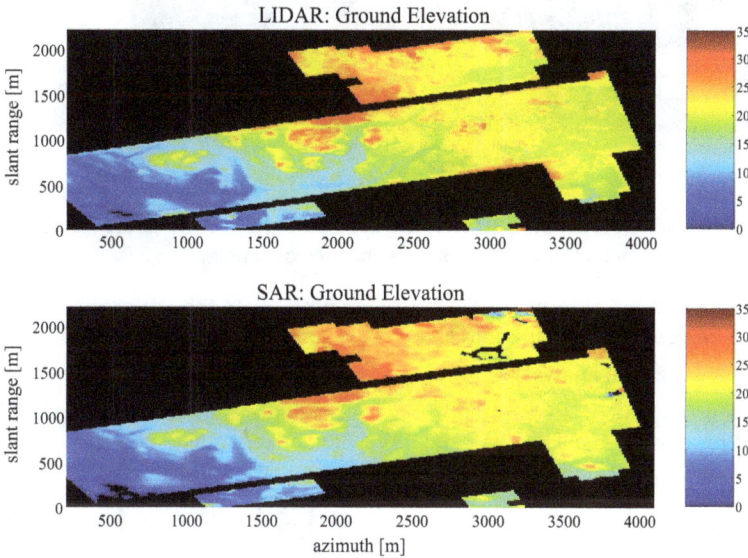

Fig. 8. Top row: ground elevation estimated by LIDAR. Bottom row: ground elevation estimated by T-SAR. Black areas correspond to an unstructured scattering mechanism.

As a validation tool, we used LIDAR measurements courtesy of the Swedish Defence Research Agency (FOI) and Hildur and Sven Wingquist's Foundation. Concerning ground elevation the dispersion of the difference $z_{SAR} - z_{LIDAR}$ has been assessed in less than 1 m. Concerning the estimated canopy elevation, the discrepancy with respect to LIDAR is clearly imputable to the fact that the estimates are relative to the average the *phase center* elevation inside the estimation window, whereas LIDAR is sensitive to the top height of the canopy. In particular, canopy elevation provided by T-SAR appears to be under-estimated with respect to LIDAR measurements, as a result of the under foliage penetration capabilities of P-band microwaves. Figure (10) reports the ratios between the estimated backscattered powers from the ground and the canopy (G/C ratio), for each polarimetric channel. As expected, in the co-polar channels the scattered power from the ground is significantly larger than canopy backscatter, the G/C ratio being assessed in about 10 dB. In the HV channel the backscattered powers

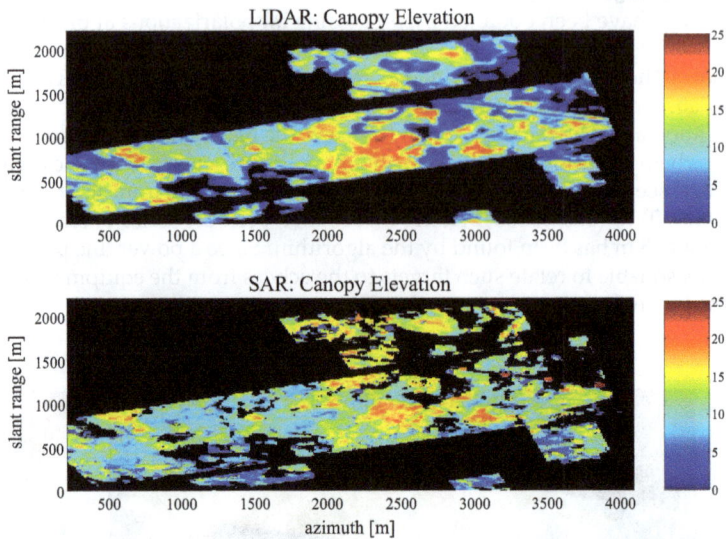

Fig. 9. Top row: canopy elevation estimated by LIDAR. Bottom row: canopy elevation estimated by T-SAR. Black areas correspond to absence of canopy.

from the ground and the canopy are closer to each other, even though ground contributions still appear to be dominant, resulting in a ground to canopy ratio of about 3 dB.

3.4.3.1 Single-polarization analysis

This last paragraph is dedicated to reporting the results relative to the elevation estimates provided by processing the HH, VV, and HV channel separately, compared to estimates yielded by the fully polarimetric (FP) tomography commented above. The joint distributions of the elevation estimates yielded by the FP tomography and by the single channel tomographies are reported in Fig. (11). It can be appreciated that, as expected, ground elevation is better estimated by processing the co-polar channel, whereas canopy elevation is better estimated by processing the HV channel. In all cases, however, the estimates are close to those provided by the FP tomography, proving that the tomographic characterization of forested areas may be carried out on the basis of a single polarimetric channel, provided that a sufficient number of acquisitions is available. It is important to note that the canopy phase center elevation yielded by the VV tomography has turned out to be slightly (between 1 and 2 meters) over estimated, with respect to both the FP tomography and to the HH and HV tomographies. This phenomenon indicates that the scatterers within the canopy volume exhibit a vertical orientation, as expected at longer wavelengths (34), (35). This result shows that effects of vegetation orientation at P-band are appreciable, and measurable by Tomography, but not so severe to hinder the applicability of the FP Tomography depicted above.

3.4.4 Discussion

Since the co-polar phase for both ground and canopy backscatter is expected to be null, (36), the value of $\Delta\varphi \approx 80°$ found in forested areas can only be interpreted as an index of the

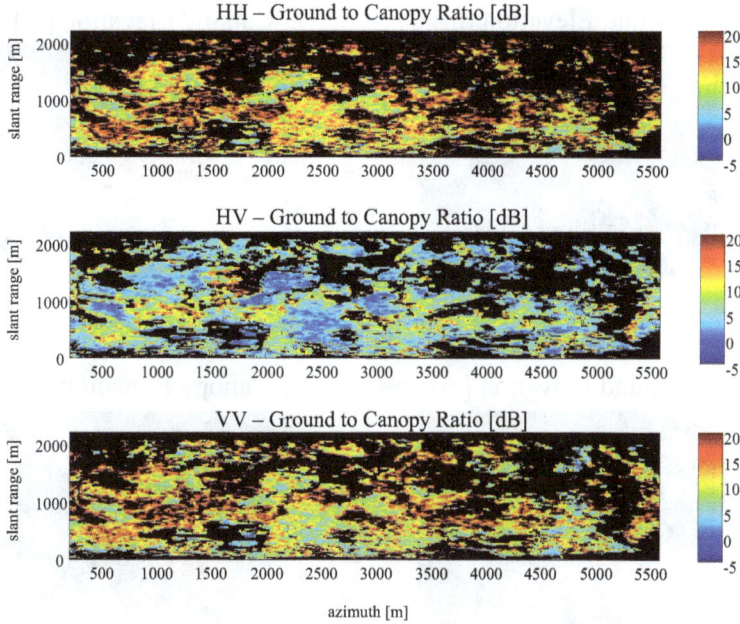

Fig. 10. Ground to Canopy Ratio for the three polarimetric channels.

presence of a dihedral contribution. Now, whereas for a perfect conducting dihedral $\Delta\varphi$ is exactly $180°$, for a lossy dielectric dihedral a lower value of $\Delta\varphi$ is expected, due to the electromagnetic properties of the trunk-ground ensemble (in (37) for example, a co-polar phase of $94°$ has been observed for trunk-ground scattering). What makes the interpretation of the co-polar signature of forested areas not straightforward is that low values of the co-polar coherence have been observed, whereas dihedral contributions to the HH and VV channels are usually assumed to by highly correlated (36). A possible explanation would be to assume that forested areas are characterized by an almost ideal dihedral scattering plus a significant contribution from canopy backscatter, which would explain both the low co-polar coherence and the reduction of $\Delta\varphi$ from the ideal value of $180°$ to $80°$. This interpretation, however, is not satisfying, since it is not consistent with the presence of highly amplitude stable targets nor with the Capon spectra in Fig. (7). At this point, a better explanation seems to be that to consider forested areas as being dominated by a *single* scattering mechanism, responsible for (most of) the coherence loss between the co-polar channels, highly coherent with respect to geometrical and temporal variations, and approximately located at ground level. From a physical point of view, it makes sense to relate such scattering mechanism to trunk-ground interactions, eventually perturbed by the presence of understory, trunk and ground roughness, small oscillations in the local topography, or eventually by canopy-ground interactions. Casting T-SAR as a parametric estimation problem has allowed to support such a conclusion by providing quantitative arguments such as the G/C ratios that indicate that not only in the co-polar channel, but also in the HV channel the contributions from the ground level dominate those from the canopy. Such result indicates that the ideal dihedral scattering model does not provide a sufficient description of the HV ground contributions under the forest, suggesting effects due to understory and volume-ground interactions.

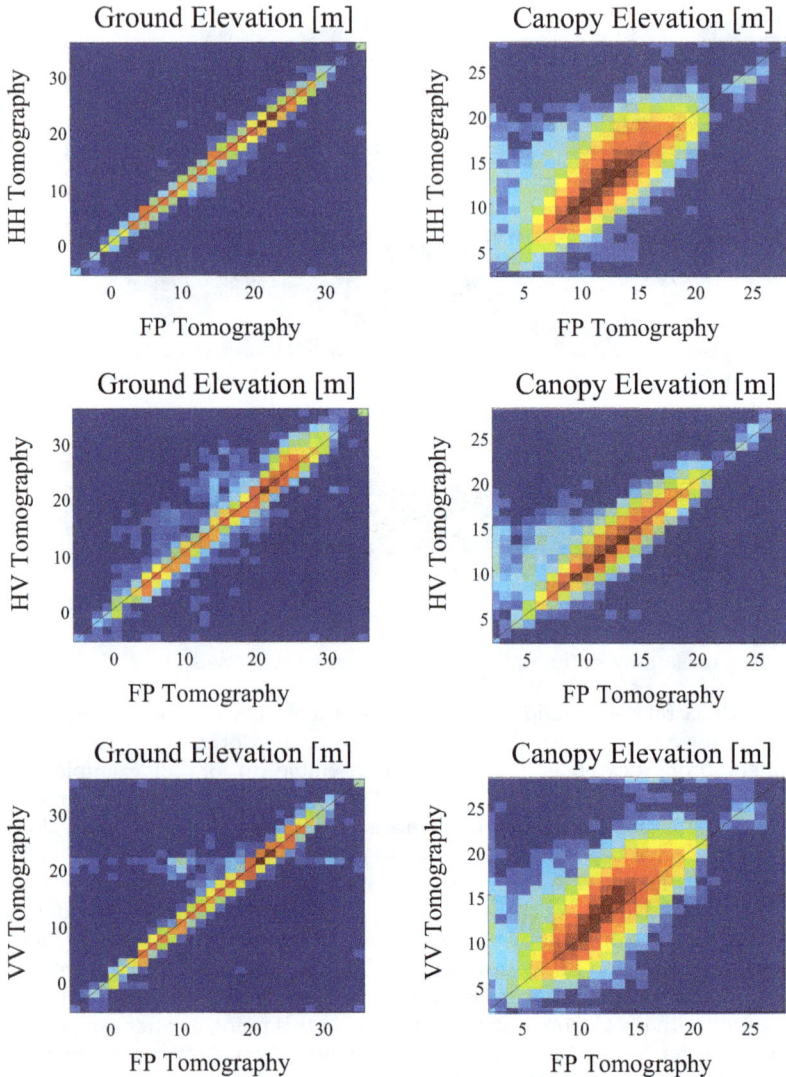

Fig. 11. Joint distribution of the phase center elevation estimates yielded by processing the single channels separately (vertical axis) and by the best tomography (horizontal axis). The black line denotes the ideal linear trend. The color scale is proportional to the natural logarithm of the number of counts within each bin.

4. Ground-volume decomposition from multi-baseline and multi-polarimetric data

4.1 Introduction

The idea that Radar scattering from forested areas can be well modeled as being constituted by two Scattering Mechanisms (SMs) has largely been retained in literature. In first place, the assumption of two SMs matches the intuitive argument that a forested area is characterized by the presence of two objects, i.e.: the ground and the vegetation layer. This idea has been formalized in literature through different physical models, considering the features of ground and volume scattering in polarimetric data (38), single-polarization tomographic data, as shown in (26) and in the previous section, or in polarimetric and interferometric (PolInSAR) data (13), (39), (27). Beside physical soundness, however, the popularity of two-layered models is also due to the fact that they provide a sufficiently simple mathematical framework to allow model inversion. This is particularly important in PolInSAR analysis, where the assumption of two layers results in the coherence loci, namely the distribution of the interferometric coherence as a function of polarization, to be given by a straight line in the complex plane (39). This simple geometrical interpretation provides the key to decompose the interferometric coherence in ground-only and volume-only contributions, after which ground and volume parameters, like terrain topographic and canopy heights are retrieved. The analysis of the shape of the coherence loci also provides a direct idea about the soundness of approximating the scene as being constituted by two SMs, which allows to assess the impact of model mismatches (35). The Sum of Kronecker Product (SKP) structure has been proposed in (40) as a general framework to discuss problem inversion in both single and multi-baseline configurations, and independently on the particular physical model adopted to represent each SM. Concerning two-layered models, the SKP formalism leads to the conclusion that the correct identification of the structural and polarimetric properties of ground and volume scattering is subject to an ambiguity, in that different solutions exist that fit the data covariance matrix up to the same error. Such an ambiguity is shown to be completely described by two degrees of freedom, which can be resolved by employing physical models. In other words, the two dimensional ambiguity following after the SKP structure represents exactly the model space, meaning that a certain physical model corresponds to a certain solution of problem ambiguity, and vice-versa. Accordingly, the SKP formalism provides a way to discuss every possible physical model, by exploring the space of ambiguous solutions. In this chapter, this methodology is applied to data from the ESA campaigns BIOSAR 2007, BIOSAR 2008 and TROPISAR. Different models are being investigated by exploring different solutions in the ambiguous space, whose features are discussed basing on polarimetric and tomographic features.

4.2 The SKP structure

We consider a scenario where Radar scattering is contributed by multiple Scattering Mechanisms (SMs), as in forested areas, and assume a data-set of $N \cdot N_p$ SAR SLC images, N_p being the number of independent polarizations (typically $N_p = 3$) and N the number of passages over the scene. Let $y_n(\mathbf{w}_i)$ denote a complex-valued pixel of the image acquired from passage n in the polarization identified by the projection vector $\mathbf{w}_i$. A simple way to model the data second order statistics is to assume that: i) different SMs are uncorrelated with one another; ii) the correlation between any two passages, say n and m, of the k-th SM alone, $r_k(n, m)$, is invariant to polarization (up to a scale factor); iii) the correlation between any two polarizations, say $\mathbf{w}_i$ and $\mathbf{w}_j$, of each SM *alone*, $c_k\left(\mathbf{w}_i, \mathbf{w}_j\right)$, is invariant to the choice of the passage (up to a scale factor), see (40) for details. Under the three hypotheses above it follows that:

$$E\left[y_n\left(\mathbf{w}_i\right) y_m^*\left(\mathbf{w}_j\right)\right] = \sum_{k=1}^{K} c_k\left(\mathbf{w}_i, \mathbf{w}_j\right) r_k\left(n, m\right) \forall \mathbf{w}_i, \mathbf{w}_j, n, m \qquad (25)$$

where K is the total number of SMs. Chosen an arbitrary polarization basis, we can recast (25) in matrix form as:

$$W = \sum_{k=1}^{K} C_k \otimes R_k \tag{26}$$

where: $\otimes$ denotes Kronecker product; W ($[N \cdot N_p \times N \cdot N_p]$) is the covariance matrix of the multi-baseline and multi-polarimetric data; the matrices C_k ($[N_p \times N_p]$) and R_k ($[N \times N]$) are such that $c_k \left(w_i, w_j \right) = w_i^H C_k w_j$ and $r_k (n, m) = \{R_k\}_{nm}$ ($\{\}_{nm}$ denoting the $nm - th$ element of a matrix).

Accordingly, model (26) allows to represent each SM through a Kronecker product between two matrices. The first, C_k, accounts for the correlation among different polarizations, and thus it represents the polarimetric properties of the $k - th$ SM, see for example (41). The second, R_k, accounts for the correlation among different baselines, therefore carrying the information about the vertical structure of the $k - th$ SM [3]. The matrices C_k, R_k will be hereinafter referred to as polarimetric signatures and structure matrices, respectively, by virtue of their physical meaning.

Following the arguments in (40), it can be shown that in most cases it can be assumed that the data covariance matrix is constituted by just two Kronecker Products, one associated with volume backscattering, and the other with the ensemble of all SMs whose phase center is ground locked, namely surface backscattering, ground-volume scattering, and ground-trunk scattering. Accordingly, eq. (26) defaults to:

$$W = C_g \otimes R_g + C_v \otimes R_v \tag{27}$$

where the subscript v refers to volume backscattering and the subscript g refers to all SMs whose phase center is ground locked. For sake of simplicity, hereinafter we will refer to the two terms in (27) simply as ground scattering and volume scattering. It is worth noting that the assumption of two Kronecker Products can be approximately retained also in the case of an Oriented Volume over Ground (OVoG), by interpreting the matrix R_v as the average volume structure matrix across all polarizations (40).

4.3 Model representation

The key to the exploitation of the SKP structure is the important result, due to Van Loan and Pitsianis (42), after which *every* matrix can be decomposed into a SKP. It is shown in (40) that the terms of the SKP Decomposition are related to the matrices C_k, R_k via a linear, invertible transformation, which is defined by exactly $K(K-1)$ real numbers. Assuming $K = 2$ SMs, as in the case of ground and volume scattering, it follows that there exist 2 real numbers (a, b)

---

[3] Neglecting temporal decorrelation and assuming target stationarity, the $nm - th$ entry of the matrix R_k is obtained as (18), (6), (26):

$$\{R_k\}_{nm} = \int S_k (z) \exp \{j (k_z (n) - k_z (m)) z\} dz$$

where z is the vertical coordinate, $S_k (z)$ is the vertical profile of the backscattered power for the $k - th$ SM, $k_z (n)$ is the height to phase conversion factor for the $n - th$ image (7). It is worth noting that eventual temporal coherence losses would be completely absorbed by the matrices R_k, as discussed in (40). Accordingly, in presence of temporal decorrelation nothing changes as for the validity of model (26), but it should be kept in mind that in this case the matrices R_k would represent not only the spatial structure, but also the temporal behavior of the $k - th$ SM.

such that:

$$\mathbf{R}_g = a\tilde{\mathbf{R}}_1 + (1 - a)\,\tilde{\mathbf{R}}_2 \tag{28}$$
$$\mathbf{R}_v = b\tilde{\mathbf{R}}_1 + (1 - b)\,\tilde{\mathbf{R}}_2$$
$$\mathbf{C}_g = \frac{1}{a - b}\left((1 - b)\,\tilde{\mathbf{C}}_1 - b\tilde{\mathbf{C}}_2\right)$$
$$\mathbf{C}_v = \frac{1}{a - b}\left(-(1 - a)\,\tilde{\mathbf{C}}_1 + a\tilde{\mathbf{C}}_2\right)$$

where $\tilde{\mathbf{R}}_k, \tilde{\mathbf{C}}_k$ are two sets of matrices yielded by the SKP Decomposition, see (40). It is very important to point out that the choice of the parameters (a, b) in (28) does affect the solution concerning the retrieval of the polarimetric signatures and structure matrices for ground and volume scattering, whereas the sum of their Kronecker products is invariant to choice of (a, b). In other words, the SKP Decomposition yields two sets of matrices that can be linearly combined so as to reconstruct exactly the structure matrices and polarimetric signatures associated with ground and volume scattering. Yet, the coefficients of such a linear combination are not known. An obvious criterion to eliminate non-physically valid solutions is to admit only values of (a, b) that yield, through (28), (semi)positive definite polarimetric signatures and structure matrices for both ground and volume scattering. We will define the set of values of (a, b) corresponding to physically valid solutions as Region of Physical Validity (RPV). Details about the RPV are provided in the remainder.

Suppose now that a certain matrix W_{mod} is the best estimate of the true data covariance matrix in a certain metric, and also suppose that $\mathbf{W}_{mod}$ can be written as the sum of 2 KPs. What equations (28) state is that there exist infinite ways of representing the matrix W_{mod}, each of which corresponding to a particular value of the parameters (a, b). The parameters (a, b) then represents the ambiguity of the ground-volume decomposition problem, meaning that by varying (a, b) within the RPV we end up with different physically valid polarimetric signatures and structure matrices, yet entailing no variations of the degree of fitness of W_{mod} with respect to the original data. In other words, each *solution* of the ambiguity associated with the choice of (a, b) within the RPV represents a particular *physically valid and data-consistent model* for both the polarimetric signatures and the structure matrices of ground and volume scattering.

4.4 Regions of Physical Validity (RPV)

In this section we discuss the shape of the RPV, under the assumption that the data covariance matrix is actually a sum of $K = 2$ KPs representing ground and volume scattering, according to model (27).

An exact procedure for the determination of the RPV has been derived in (40), by recasting equation (28) in diagonal form. Although the procedure is straightforward, its description involves quite lengthy matrix manipulations. For this reason, we will discuss here the shape of the RPV by resorting to a useful geometrical interpretation. To do this we will assume the Polarimetric Stationarity Condition to hold (43), resulting in the structure matrix of each SM to have unitary elements on the main diagonal, see(40). This property is inherited by the matrices $\tilde{\mathbf{R}}_1, \tilde{\mathbf{R}}_2$ in (28), which allows to interpret the off diagonal elements of $\mathbf{R}_g, \mathbf{R}_v$ as the complex interferometric coherence of ground and volume scattering (the elements on their diagonal being identically equal to 1). It is then immediate to see from equation (28) that, in each interferogram, ground and volume coherences are bound to lie on a straight line in the complex plane. Accordingly, the optimal choice of the parameters (a, b) is the

Branch	Boundary	Rank-deficient matrix
a	Inner	$\mathbf{C}_v$
a	Outer	$\mathbf{R}_g$
b	Inner	$\mathbf{C}_g$
b	Outer	$\mathbf{R}_v$

$$(29)$$

Table 1. Rank deficiencies at the boundaries of the region of positive definitiveness.

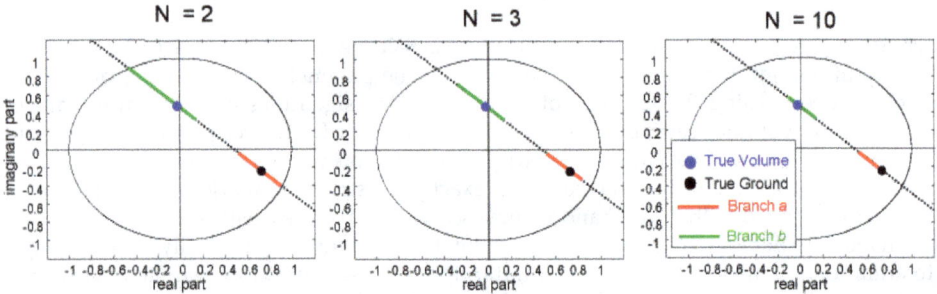

Fig. 12. Regions of physical validity for the interferometric coherences associated with ground and volume scattering in the interferometric pair between the first two tracks of the data-set. N is the total number of available tracks exploited to enforce the positive definitiveness constraint. The black and blue points denote the true interferometric coherences associated with ground and volume scattering in the considered interferometric pair. The red and green segments denote the set of all physically valid solutions obtained by varying a and b, respectively.

one that corresponds to the true[4] ground and volume coherences, whereas the region of physical validity can be simply associated with two segments along the line passing through the true ground and volume coherences, see figure 12. By definition, the points outer or inner boundaries of the two segments correspond to the case where one of the four matrices $\{\mathbf{C}_k, \mathbf{R}_k\}_{k=g,v}$ in equations (28) is singular. In particular, the outer boundaries correspond to rank-deficient structure matrices, whereas the inner boundaries correspond to rank-deficient polarimetric signature, as reported in Table 1. In the single baseline case ($N = 2$) the points at the outer boundary of both segments belong to the unit circle, indicating that physically valid ground and volume interferometric coherences are allowed be unitary in magnitude, see figure 12, left panel. This conclusion is exactly the same as the one drawn in (39), after which it follows the consistency of the SKP formalism with respect to PolInSAR. As new acquisitions are gathered, instead, the positive definitiveness constraint results in the regions of physical validity to shrink from the outer boundaries towards the true ground and volume coherences, whereas the position of the inner boundary points stay unvaried for the considered interferogram, figure 12, middle and right panels. Accordingly, the availability of multiple-baseline results not only in enhanced vertical resolution capabilities, but also in the progressive elimination of incorrect solutions.

---

[4] Assuming (27).

Campaign	BioSAR 2007
Acquisition System	E-SAR - DLR
Acquisition Period	Spring 2007
Site	Remningstorp, Central Sweden
Scene	Semi-boreal forest
Topography	Flat
Tomographic Tracks	9 - Fully Polarimetric
Band	P-Band
Slant Range resolution	2 m
Azimuth resolution	1.6 m
Vertical resolution	10 m (near range) to 40 m (far range)

Table 2. The BIOSAR 2007 data-set

4.5 Physical interpretation

A straightforward physical interpretation of the polarimetric and structural properties of models associated with different solutions can be provided by analyzing the inner and outer boundaries of the RPV:

- The inner boundary solution on branch a results in the volume polarimetric signature to be rank-deficient. This entails the existence of a polarization where volume scattering does not contribute, after which it follows that this solution is not consistent with physical model for forest scattering (44), (38). The ground structure matrix is characterized by the lowest coherence values, being contaminated by volume contributions. Accordingly, such a solution is to be discarded.

- The outer boundary solution on branch a results in the volume polarimetric signature to be full rank, consistently with physical model for forest scattering. The resulting ground structure matrix is characterized by the highest coherence values compatible with the RPV. Provided the number of tracks is sufficient, this solution yields an unbiased estimation of the ground coherences even in presence of coherence losses.

- The inner boundary solution on branch b results in the ground polarimetric signature to be rank-deficient, consistently with the hypothesis that there exists one polarization where volume only contributions are present. If this is true, the resulting volume structure corresponds to the true one. Otherwise, the result is systematically contaminated by ground contributions, resulting in apparent volume contributions close to the ground.

- The outer boundary solution on branch b results in the ground polarimetric signature to be full rank. Accordingly, this solution accounts for the presence of ground-locked contributions in all polarizations. The resulting volume structure matrix is maximally coherent. If few tracks are employed, this solution acts as an high-pass filter, resulting in the volume to appear thinner than it is. As the number of available baselines increases, this solution converges to the true structure for volume contributions in the upper vegetation layers, whereas volume contributions from the ground level are absorbed into the ground structure.

4.6 Case studies

We present here experimental results relative to three case studies based on data from the ESA campaigns BIOSAR 2007 (45), BIOSAR 2008 (46) and TROPISAR. The main features of the analyzed data are reported in tables 2, 3, 4.

Campaign	BioSAR 2008
Acquisition System	E-SAR - DLR
Acquisition Period	Fall 2008
Site	Krycklan river catchment, Northern Sweden
Scene	Boreal forest
Topography	Hilly
Tomographic Tracks	6 per flight direction (South-West and North-East) – Fully Polarimetric
Band	P-Band and L-Band
Slant Range resolution	1.5 m
Azimuth resolution at L-Band	1.2 m
Azimuth resolution at P-Band	1.6 m
Vertical resolution at L-Band	6 m (near range) to 40 m (far range)
Vertical resolution at P-Band	20 m (near range) to >80 m (far range)

Table 3. The BioSAR 2008 data-set

Campaign	TropiSAR
Acquisition System	Sethi- ONERA
Acquisition Period	August 2009
Site	Paracou, French Guyana
Scene	Tropical forest
Topography	Flat
Tomographic Tracks	6 - Fully Polarimetric
Band	P-Band
Slant Range resolution	1 m
Azimuth resolution	1 m
Vertical resolution	$\simeq$15 m

Table 4. The TropiSAR data-set

All of the result to follow have been obtained by estimating the sample covariance matrix of the multi-baseline and multi-polarimetric data by multi-looking over a 60 × 60 m (ground range, azimuth) estimation window.

4.6.1 Models for the volume structure

Figures from13 to 16 report tomographic images of the volume structures as obtained by taking three different solutions on branch b. The tomographic imaging has been performed by employing the Capon Spectral Estimator.

All the tomographic profiles behave consistently with the physical features discussed in the previous section. In particular, all profiles associated with the outer boundary solution result in very thin volumes at a high elevation, therefore accounting for the upper vegetation layer only. As the solution is moved towards the inner boundary contributions from the lower vegetation layer appear. When the inner boundary is reached the the contributions appear from the ground level up to top forest height. The following points are worth noting:

- Contributions from the ground level are observed in all cases in the inner boundary solution. As this solution corresponds by construction to the polarization where ground

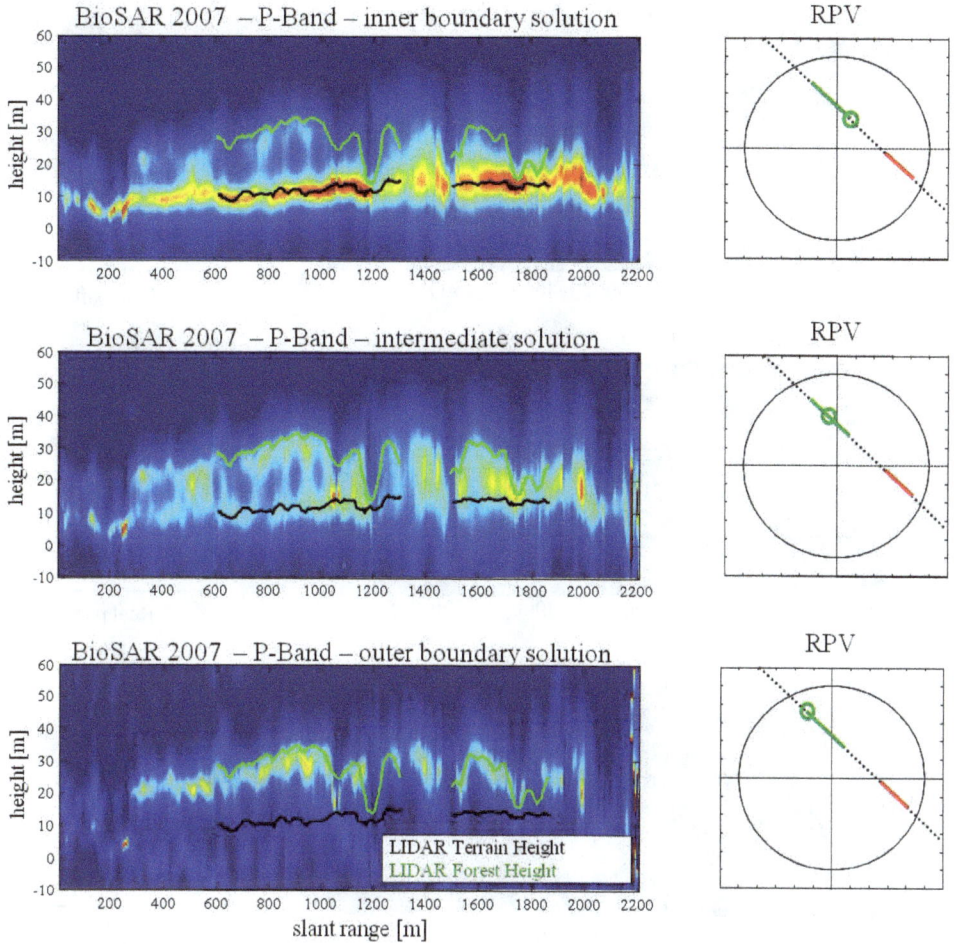

Fig. 13. Tomographic profiles along an azimuth cut corresponding to three different physically valid models for the volume structure.

scattering is not supposed to occur, we conclude that depolarized contributions from the ground level are present in all data-sets. The extent of such contributions is by far more relevant in the case of the BioSAR 2007 and BioSAR 2008 P-Band data-sets, witnessing the sensitivity of the data to both wavelength and forest structure.

- The position along the vertical axis at which the backscattered power drops down is independent of the choice of the solution, meaning that all solutions carry the same information about forest top height.

4.6.2 Fitness

The figure below reports the amount of information carried by the data that is correctly represented by assuming the data covariance matrix is a sum of 1 to 4 KPs. The measurement

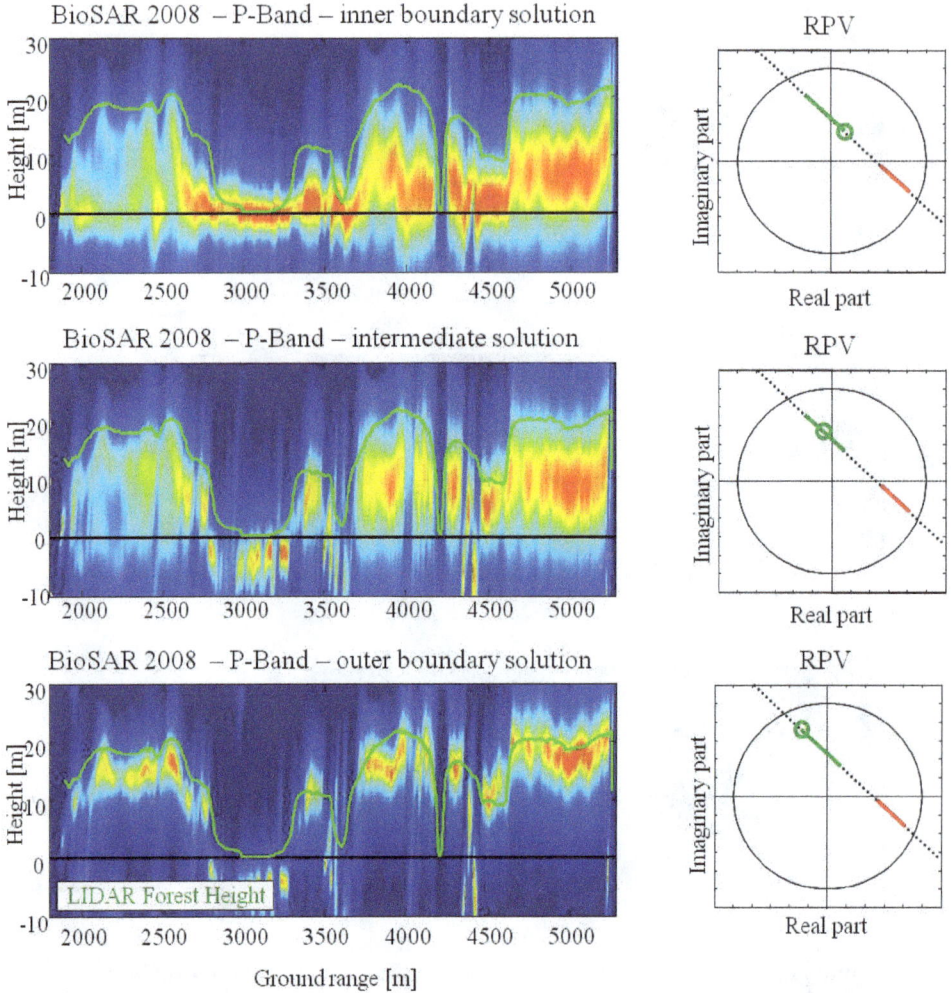

Fig. 14. Tomographic profiles along an azimuth cut corresponding to three different physically valid models for the volume structure. Note that topography has been removed such that the ground level is always at 0 m.

is carried out as $I_K = 1 - \varepsilon_K$, ε_K being defined as:

$$\varepsilon_K = \frac{\left\| \widehat{\mathbf{W}} - \widehat{\mathbf{W}}_K \right\|_F}{\left\| \widehat{\mathbf{W}} \right\|_F}$$

where $\widehat{\mathbf{W}}$ is the sample covariance matrix and $\widehat{\mathbf{W}}_K$ its best estimate in the Frobenius norm by using K KPs, see (40) for details.

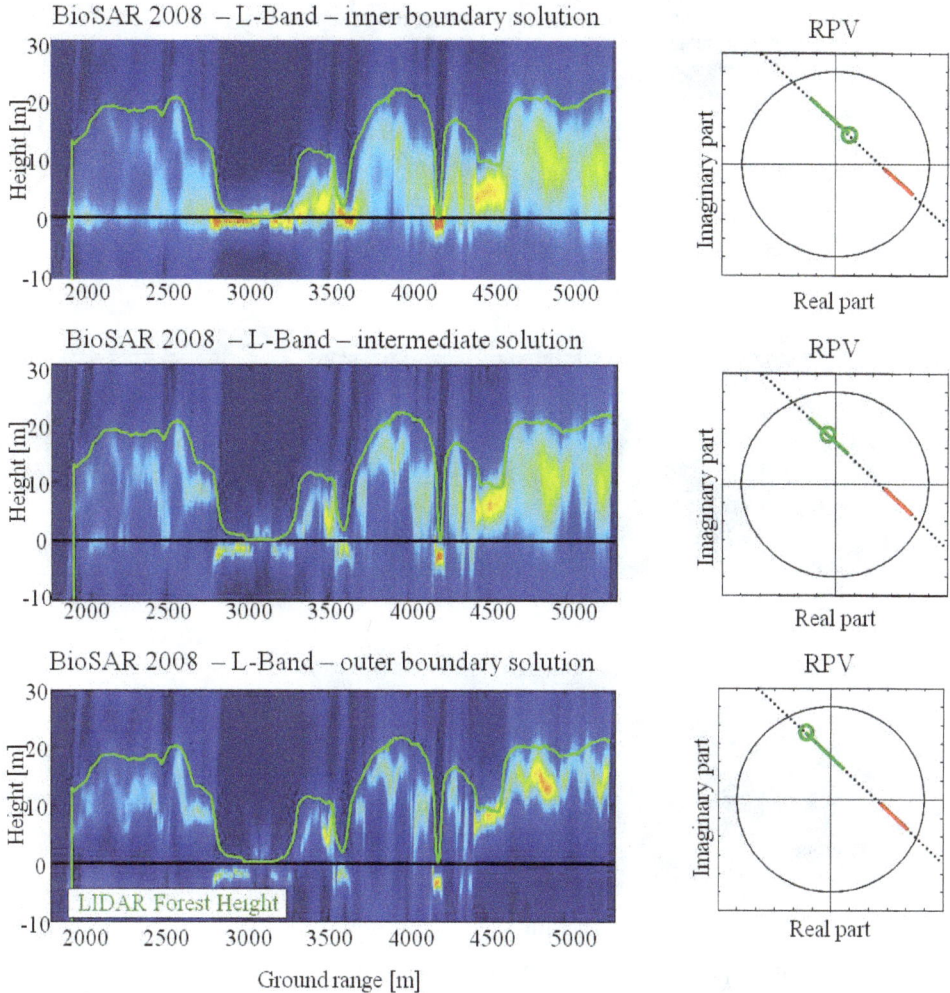

Fig. 15. Tomographic profiles along an azimuth cut corresponding to three different physically valid models for the volume structure. Note that topography has been removed such that the ground level is always at 0 m.

4.7 Discussion

On the basis of the analyzed data-sets, the following conclusions are drawn:

- The assumption of 2 KPs has turned out to account for about 90% of the information carried by the data in all investigated cases, meaning that two-layered models (ground plus volume) are well suited for forestry investigations. Accounting for other components, such as subsurface penetration or differential extinction phenomena, does not appear to be necessary as for a first-order characterization of the forest structure, their role being limited to about 10% of the total information content.

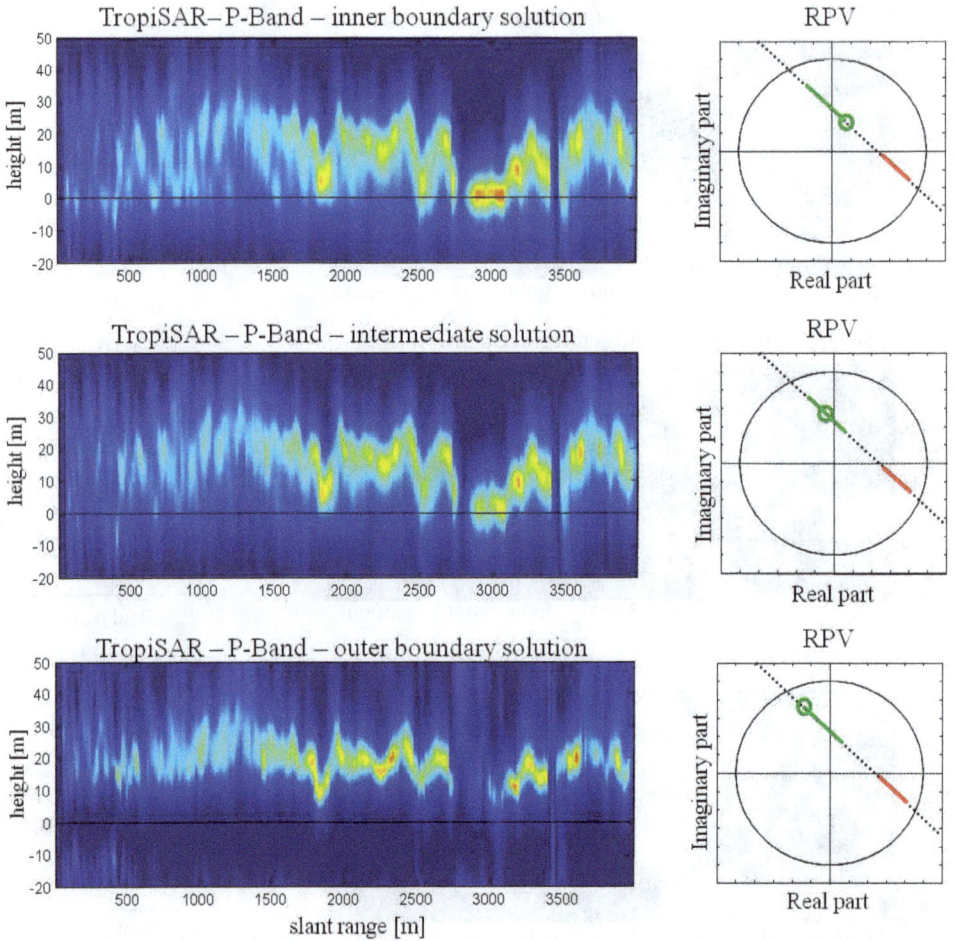

Fig. 16. Tomographic profiles along an azimuth cut corresponding to three different physically valid models for the volume structure. Note that topography has been removed such that the ground level is always at 0 m.

- The assumption of a ground-free polarization has resulted in an evenly distributed volume structure in a boreal forest at L-Band and in a tropical forest at P-Band, and in an almost ground-locked volume structure in a boreal and semi-boreal forest at P-Band, witnessing the sensitivity of Radar data to wavelength and forest structure. The extent of depolarized contributions from the ground level suggests ground-volume interaction phenomena may occur at P-Band in sparse forests. If this is the case, volume backscattering can be retrieved by allowing ground scattering to be partly entropic.

- The retrieval of forest top height is nearly invariant to the choice of the model. In this sense, forest top height appears as the most robust indicator of the forest structure as observed through microwaves measurements, providing a further and independent argument supporting the validity of PolInSAR for the remote sensing of forested scenarios.

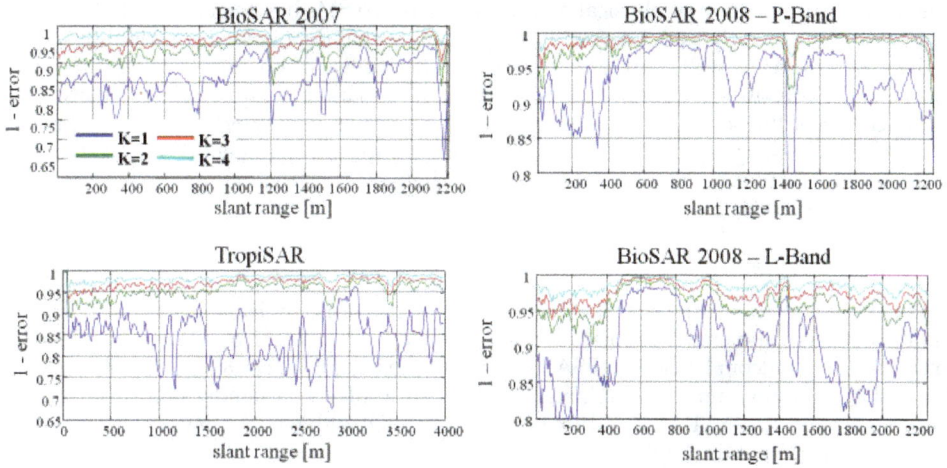

Fig. 17. Information content represented by taking K KPs for the four analyzed data-sets.

On the other hand, though, retrieving the correct forest height does not provide arguments to assess the validity of the whole model, as the same top height is consistent with different physically valid data-consistent models.

5. Conclusions

This chapter has considered the retrieval of information about the forest vertical structure from multi-baseline SARs.

Considering a purely tomographic formulation of the problem, it has been shown that the backscattered power associated with a certain depth within the vegetation layer can be retrieved with a *virtually* arbitrary resolution, simply by employing a sufficiently large baseline apertures. Besides costs, however, baseline aperture is upper bounded by physical constraints arising from the nature of the targets themselves, such as anisotropy and temporal decorrelation.

Posing T-SAR as an estimation problems greatly helps compensate for the coarse resolution arising from an insufficient baseline aperture, allowing to retrieve useful information about the forest structure even in cases where the Fourier resolution is many times the overall forest height. In particular, T-SAR has been shown to be capable of identifying bald and forested areas and estimating the elevation and backscattered power of the scattering centers representing associated with ground and volume scattering. Furthermore, it has been shown that T-SAR can be employed basing on both single and multi-polarimetric data, which makes T-SAR a valuable tool to investigate variations of the forest structure with polarization.

The availability of multi-polarimetric and multi-baseline data has been shown to provide the most information, allowing the decomposition of the data covariance matrix into ground-only and volume-only contributions even in absence of a parametric model and largely independently on baseline aperture.

The capability of the SKP formalism to represent *all* physically valid and data-consistent two-layered models has allowed an exhaustive discussion about the validity of such a class of models for the analysis of forested areas. As a result, it has been observed that two SMs account for more than 90% of the information carried by the data in all investigated cases.

The correct identification of physical models for such two SMs has been shown to be subject to an ambiguity, mostly associated with the possibility that depolarized contributions occur at the ground level as well. Furthermore, the correct retrieval of volume top height has been shown not to constitute a meaningful tool for model validation, being substantially invariant to the choice of the solution for volume scattering. In this framework it is then clear that tomographic imaging represents a most valuable tools for the assesment of physical models of the forest structure, as it allows to *see* what kind of vertical structure is actually associated with the chosen model.

6. Acknowledgment

I wish to acknowledge insights and valuable discussions with Prof. Fabio Rocca, Prof. Andrea Monti Guarnieri, and eng. Mauro Mariotti d'Alessandro at Politecnico di Milano. I also wish to thank to Dr. Malcom Davidson (ESA), Dr. Irena Hajnsek (DLR), Dr. Kostas Papathanassiou (DLR), Dr. Fabrizio Lombardini (Universitè di Pisa), Prof. Lars Ulander (FOI), Dr. Pascale Dubois-Fernandez (ONERA), Dr. Thuy Le Toan (CESBIO) with whom I have discussed many of the results within this chapter.

7. References

[1] J. C. Curlander and R. N. McDonough, *Synthetic aperture radar: systems and signal processing*. New York: John Wiley & Sons, Inc, 1991.

[2] G. Franceschetti and R. Lanari, Eds., *Synthetic Aperture Radar processing*. CRC Press, 1999.

[3] F. Rocca, "Synthetic aperture radar: A new application for wave equation techniques," *Stanford Exploration Project Report*, vol. SEP-56, pp. 167–189, 1987.

[4] R. Bamler, "A comparison of range-Doppler and wave-number domain SAR focusing algorithms," *IEEE Transactions on Geoscience and Remote Sensing*, vol. 30, no. 4, pp. 706–713, Jul. 1992.

[5] R.-S. Wu and M.-N. Toksïżœz, "Diffraction tomography and multisource holography applied to seismic imaging," *Geophysics*, vol. 52, pp. 11–+, Jan. 1987.

[6] A. Reigber and A. Moreira, "First demonstration of airborne SAR tomography using multibaseline l-band data," *IEEE Trans. on Geoscience and Remote Sensing*, pp. 2142–2152, Sep. 2000.

[7] R. Bamler and P. Hartl, "Synthetic aperture radar interferometry," *Inverse Problems*, vol. 14, pp. R1–R54, 1998.

[8] F. Ulaby, K. McDonald, K. Sarabandi, and M. Dobson, "Michigan microwave canopy scattering models (mimics)," *Geoscience and Remote Sensing Symposium, 1988. IGARSS '88. Remote Sensing: Moving Toward the 21st Century., International*, vol. 2, pp. 1009–1009, Sep 1988.

[9] G. Smith-Jonforsen, L. Ulander, and X. Luo, "Low vhf-band backscatter from coniferous forests on sloping terrain," *Geoscience and Remote Sensing, IEEE Transactions on*, vol. 43, no. 10, pp. 2246–2260, Oct. 2005.

[10] L. Thirion, E. Colin, and C. Dahon, "Capabilities of a forest coherent scattering model applied to radiometry, interferometry, and polarimetry at p- and l-band," *Geoscience and Remote Sensing, IEEE Transactions on*, vol. 44, no. 4, pp. 849–862, April 2006.

[11] Y.-C. Lin and K. Sarabandi, "Electromagnetic scattering model for a tree trunk above a tilted ground plane," *Geoscience and Remote Sensing, IEEE Transactions on*, vol. 33, no. 4, pp. 1063–1070, Jul 1995.

[12] K. Sarabandi, "Scattering from dielectric structures above impedance surfaces and resistive sheets," *Antennas and Propagation, IEEE Transactions on*, vol. 40, no. 1, pp. 67–78, Jan 1992.

[13] R.N. Treuhaft, P.R. Siqueira, "Vertical structure of vegetated land surfaces from interferometric and polarimetric radar," *Radio Science*, vol. 35, pp. 141–177, 2000.

[14] P. Stoica and R. L. Moses, *Introduction to Spectral Analysis*. New Jersey: Prentice-Hall, 1997.

[15] H. A. Zebker and J. Villasenor, "Decorrelation in interferometric radar echoes," *IEEE Transactions on Geoscience and Remote Sensing*, vol. 30, no. 5, pp. 950–959, sept 1992.

[16] I. Hajnsek, F. Kugler, S.-K. Lee, and K. Papathanassiou, "Tropical-forest-parameter estimation by means of pol-insar: The indrex-ii campaign," *Geoscience and Remote Sensing, IEEE Transactions on*, vol. 47, no. 2, pp. 481 –493, feb. 2009.

[17] F. Lombardini and A. Reigber, "Adaptive spectral estimation for multibaseline SAR tomography with airborne l-band data," *Geoscience and Remote Sensing Symposium, 2003. IGARSS '03. Proceedings. 2003 IEEE International*, vol. 3, pp. 2014–2016, July 2003.

[18] F. Gini, F. Lombardini, and M. Montanari, "Layover solution in multibaseline SAR interferometry," *Aerospace and Electronic Systems, IEEE Transactions on*, vol. 38, no. 4, pp. 1344–1356, Oct 2002.

[19] S. Sauer, L. Ferro-Famil, A. Reigber, and E. Pottier, "Multibaseline pol-insar analysis of urban scenes for 3d modeling and physical feature retrieval at l-band," *Geoscience and Remote Sensing Symposium, 2007. IGARSS 2007. IEEE International*, pp. 1098–1101, 23-28 July 2007.

[20] X. X. Zhu and R. Bamler, "Very high resolution spaceborne sar tomography in urban environment," *Geoscience and Remote Sensing, IEEE Transactions on*, no. 99, pp. 1–13, 2010.

[21] G. Fornaro and F. Serafino, "Imaging of single and double scatterers in urban areas via SAR tomography," *Geoscience and Remote Sensing, IEEE Transactions on*, vol. 44, no. 12, pp. 3497–3505, Dec. 2006.

[22] G. Fornaro, F. Lombardini, and F. Serafino, "Three-dimensional multipass SAR focusing: experiments with long-term spaceborne data," *Geoscience and Remote Sensing, IEEE Transactions on*, vol. 43, no. 4, pp. 702–714, April 2005.

[23] S. R. Cloude, "Dual-baseline coherence tomography," *Geoscience and Remote Sensing Letters, IEEE*, vol. 4, no. 1, pp. 127–131, Jan. 2007.

[24] ——, "Multifrequency 3d imaging of tropical forest using polarization coherence tomography," in *Eusar 2008*, 2008.

[25] K. Papathanassiou and S. Cloude, "Single-baseline polarimetric SAR interferometry," *Geoscience and Remote Sensing, IEEE Transactions on*, vol. 39, no. 11, pp. 2352–2363, Nov 2001.

[26] S. Tebaldini, "Single and multipolarimetric SAR tomography of forested areas: A parametric approach," *Geoscience and Remote Sensing, IEEE Transactions on*, vol. 48, no. 5, pp. 2375 –2387, may 2010.

[27] M. Neumann, L. Ferro-Famil, and A. Reigber, "Estimation of forest structure, ground, and canopy layer characteristics from multibaseline polarimetric interferometric sar data," *Geoscience and Remote Sensing, IEEE Transactions on*, vol. 48, no. 3, pp. 1086 –1104, 2010.

[28] H. V. Trees, *Detection, Estimation and Modulation Theory, Part I*. New York, NY: John Wiley and Sons, 1968.

[29] P. Stoica and T. Soderstrom, "On reparametrization of loss functions used in estimation and the invariance principle," *Signal Processing*, no. 17, pp. 383–387, 1989.

[30] A. Swindlehurst and P. Stoica, "Maximum likelihood methods in radar array signal processing," *Proceedings of the IEEE*, vol. 86, no. 2, pp. 421–441, Feb 1998.

[31] P. S. B. Ottersten and R. Roy, "Covariance matching estimation techniques for array signal processing applications," *Digital Signal Processing*, no. 8(3), pp. 185–210, 1998.

[32] B. Jansson, M.; Ottersten, "Structured covariance matrix estimation: a parametric approach," *Acoustics, Speech, and Signal Processing, 2000. ICASSP '00. Proceedings. 2000 IEEE International Conference on*, vol. 5, pp. 3172–3175 vol.5, 2000.

[33] A. Ferretti, C. Prati, and F. Rocca, "Permanent scatterers in SAR interferometry," *IEEE Transactions on Geoscience and Remote Sensing*, vol. 39, no. 1, pp. 8–20, Jan. 2001.

[34] R. Treuhaft and S. Cloude, "The structure of oriented vegetation from polarimetric interferometry," *Geoscience and Remote Sensing, IEEE Transactions on*, vol. 37, no. 5, pp. 2620–2624, Sep 1999.

[35] J. Lopez-Sanchez, J. Ballester-Berman, and Y. Marquez-Moreno, "Model limitations and parameter-estimation methods for agricultural applications of polarimetric SAR interferometry," *Geoscience and Remote Sensing, IEEE Transactions on*, vol. 45, no. 11, pp. 3481–3493, Nov. 2007.

[36] A. Freeman and S. Durden, "A three-component scattering model for polarimetric SAR data," *Geoscience and Remote Sensing, IEEE Transactions on*, vol. 36, no. 3, pp. 963–973, May 1998.

[37] D. R. Sheen, N. L. VandenBerg, S. J. Shackman, D. L. Wiseman, L. P. Elenbogen, and R. F. Rawson, "P-3 ultra-wideband SAR: Description and examples," *IEEE AES Systems Magazine*, pp. 25–29, Nov. 1996.

[38] A. Freeman, "Fitting a two-component scattering model to polarimetric SAR data from forests," *Geoscience and Remote Sensing, IEEE Transactions on*, vol. 45, no. 8, pp. 2583–2592, Aug. 2007.

[39] K. P. Papathanassiou and S. R. Cloude, "Three-stage inversion process for polarimetric SAR interferometry," *IEE Proc. Radar Sonar Navig.*, vol. 150, pp. 125–134, Jun. 2003.

[40] S. Tebaldini, "Algebraic synthesis of forest scenarios from multibaseline polinsar data," *Geoscience and Remote Sensing, IEEE Transactions on*, vol. 47, no. 12, pp. 4132 –4142, dec. 2009.

[41] S. Cloude and E. Pottier, "A review of target decomposition theorems in radar polarimetry," *Geoscience and Remote Sensing, IEEE Transactions on*, vol. 34, no. 2, pp. 498–518, Mar 1996.

[42] C. V. Loan and N. Pitsianis, "Approximation with Kronecker products," *Linear Algebra for Large Scale and Real Time Applications, M. S. Moonen, G. H. Golub, and B. L. R. De Moor, Ed. Norwell, MA: Kluwer*, pp. 293–314, 1993.

[43] L. F. Famil and M. Neumann, "Recent advances in the derivation of pol-insar statistics: Study and applications." in *Eusar 2008*, 2008.

[44] S. Cloude and E. Pottier, "An entropy based classification scheme for land applications of polarimetric SAR," *Geoscience and Remote Sensing, IEEE Transactions on*, vol. 35, no. 1, pp. 68–78, Jan 1997.

[45] S. Tebaldini and F. Rocca, "BioSAR WP43," *ESTEC Contract No. 20755/07/NL/CB*, 2008.

[46] ——, "BioSAR 2008 Data Analysis II," *ESTEC Contract No. 22052/08/NL/CT*, 2009.

Part 2

Oceans

5

Using SVD Analysis of Combined Altimetry and Ocean Color Satellite Data for Assessing Basin Scale Physical-Biological Coupling in the Mediterranean Sea

Antoni Jordi and Gotzon Basterretxea
Institut Mediterrani d'Estudis Avançats, IMEDEA (UIB-CSIC)
Spain

1. Introduction

The ocean is a highly variable system affected by a large range of processes that spans the continuous spectra of spatial and temporal scales (Wunsch, 1996). The spatial scales of variation range from basin-wide gyres (thousands of kilometers) to turbulence (less than a meter), and the time scales from those that are climate related (decades) to short-term processes (seconds). Information on this whole range of processes is required for the comprehension of the marine system dynamics. Despite the continuous advances in technology, remote sensing is the only observing platform capable of providing continuous information on biological and physical properties over vast areas of the ocean. With some limitations, the regular and repeated coverage offered by satellites is still unachievable through in situ measurements.

Because the ocean is largely opaque over much of the usable electromagnetic spectrum, the ability of satellites to capture ocean properties is generally confined to the surface. Nevertheless, satellite-borne sensors provide us with a relatively large range of measurements such as sea surface color, sea surface height, sea surface temperature, sea surface winds, sea surface salinity, waves, and to a lesser extent, current fields. The availability, for the first time, of time series expanding for several years or decades at regional and global ocean scales has changed our perception of the ocean (Barber & Hilting, 2000). A majority of these measurements is restricted to physical properties such as temperature, sea level or sea surface roughness and inferred variables (currents, winds, etc.). The only routinely acquired satellite measurement providing information on ocean biological processes is sea surface color. Since early measurements obtained by the Coastal Zone Color Scanner (CZCS), sea color sensors have provided quantitative information on the distribution of surface chlorophyll (CHL) concentration (an index of phytoplankton biomass) at regional to global scales and its variability in space and time (e.g., Abbott & Zion, 1987; Antoine et al., 2005; Behrenfeld et al., 2001). This information is relevant to estimate the ocean productivity, a key factor for understanding the dynamics of pelagic foodwebs and some aspects of climate change.

The merging of ocean color datasets with other satellite measurements providing information of ocean dynamics has potential benefits to the understanding of some aspects of the coupling between fluid driven processes and plankton dynamics. In this regard, altimetry satellite data allows the characterization of sea level anomaly (SLA), containing information on the geostrophic current fields, mesoscale eddy variability and changes in the thermocline depth (Le Traon et al., 1998; Stammer, 1997). Lehahn et al. (2007) and Leterme & Pingree (2008) investigated the effect of the geostrophic velocity derived from satellite SLA on the redistribution of satellite CHL. Also, the impact of mesoscale eddies on the spatial patterns of CHL has been analyzed by Siegel et al. (1999, 2008).

Further insight into the basin-scale dynamics affecting CHL and SLA, as an indicator of changes in the thermocline depth, in the equatorial Pacific was provided by Wilson & Adamec (2001). Correlations between CHL and SLA data using empirical orthogonal function (EOF) analysis show different responses associated with El Niño-Southern Oscillation phases. In the global ocean, direct correlations between CHL and SLA are predominately negative as lower SLA implies thermocline weakening and consequent mixing of the water column, which results in increased nutrient flux to the surface layer and phytoplankton biomass enhancement (Wilson & Adamec, 2002). However, there are areas in all ocean basins where positive correlations suggest that CHL is affected by processes other than thermocline variations. For example, Uz et al. (2001) found positive correlations between satellite CHL and SLA associated with the propagation of Rossby waves. These studies exemplify how multiple satellite observations are used to understand basin-scale dynamics and their impacts on the ocean phytoplankton biomass variability. Following these works, Jordi et al. (2009) used the singular value decomposition (SVD) analysis of the cross-covariance matrix between satellite CHL and SLA to analyze the regional scale dynamics in the northwestern Mediterranean Sea. Their results highlight the role of the water mass transported by the regional circulation on the variability of the phytoplankton biomass. The SVD analysis may be superior to EOF analysis in identifying correlated spatial patterns between pairs of spatial time series (Bretherton et al., 1992).

In the Mediterranean Sea, a semi-enclosed marginal sea with limited geographical dimensions (Fig. 1), ocean color data reveals that oligotrophic conditions prevail for most of the year (D'Ortenzio et al., 2002). Biological production is mainly regulated by physical processes enhancing nutrient supply to surface layers and by allochtonous inputs from the continents and the atmosphere (Barale et al., 2008). Satellite ocean color data in the Mediterranean Sea has demonstrated dominance of the seasonal cycle in phytoplankton biomass (Bosc et al., 2004). With some regional variations, the typical temperate-latitude cycle with maximum biomass in late winter-spring and minimum during summer stratified conditions occurs throughout most of the basin (Bricaud et al., 2002). Inter-annual variations in CHL concentrations are also noticeable both at a local scale and over the whole basin, and have been related to climatic fluctuations (D'Ortenzio et al., 2003).

Complementarily, altimetry satellite data shows that the sea level variability in the Mediterranean is a complex combination of a wide range of spatial and temporal scales (Cazenave et al., 2001; Larnicol et al., 2002). Besides the marked seasonal cycle in SLA caused by the steric effect, important intra- and inter-annual signals are observed associated with permanent or transitory oceanographic structures such as frontal currents and mesoscale eddies (Jordi & Wang, 2009; Pujol & Larnicol, 2005). The multiple driving forces including the ocean–atmosphere interaction, the phenomenology of the deep water formation and water mass hydrological properties, the frontal currents and mesoscale eddies, and the topographic and coastal influence, add complexity to the physical-biological coupling in the Mediterranean Sea.

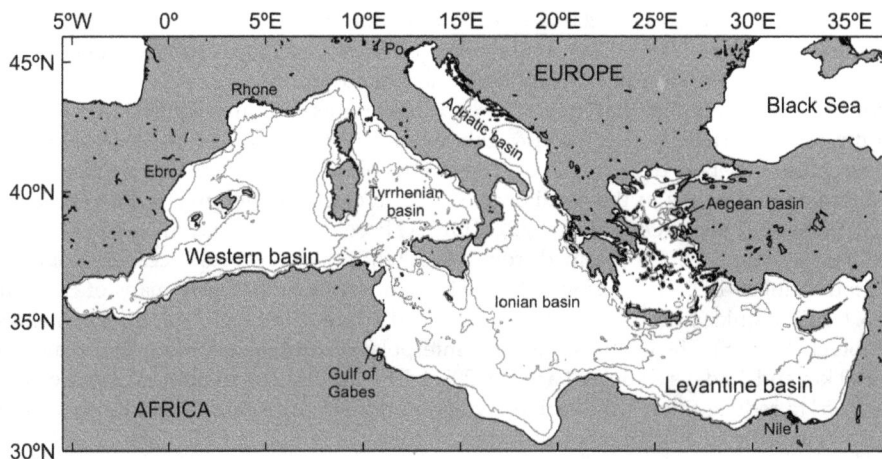

Fig. 1. Bathymetry of the Mediterranean Sea. The 200 and 2000 m isobaths are shown with gray lines.

In this work, we analyze the basin scale patterns of phytoplankton variability at inter-annual, seasonal and intra-annual scales and the associated driving forces in the Mediterranean Sea based on 12 years of concurrent ocean color and altimetry satellites data. The knowledge of the phytoplankton variability and its relation to ocean circulation is critical to understand marine ecosystem dynamics and biogeochemical cycles, with implications ranging from marine food webs to climate change. The physical mechanisms that regulate phytoplankton patterns in the Mediterranean are analogous to those in larger oceanic areas and therefore comprehension of the processes occurring therein are pertinent to the understanding of larger areas.

2. Data and methodology

2.1 Altimetry satellite data

The satellite-borne altimetry is initially designed to estimate the sea surface height (SSH) by measuring the satellite-to-surface round-trip time of a radar pulse. These measurements however include the Earth's geoid which varies by tens of meters across the ocean and is not accurately estimated (Fu et al., 1994). This unknown geoid is removed from satellite observations by subtracting a long term mean of the altimeter measurements from the observations. However, this procedure removes also the mean dynamic SSH and satellite measurements refer only to SLA, which contains information on the geostrophic current fields, mesoscale eddy variability and changes in the thermocline depth (Le Traon et al., 1998; Stammer, 1997). Uncertainties on the location of the satellite on its orbit and disturbances of the radar pulse by the atmosphere introduce additional errors in the SLA measurement. Because of these errors, the first altimetry satellites such as Seasat or Geosat did not provide very usable and useful data. In 1993, the French Centre National d'Etudes Spatiales (CNES) and the US National Aeronautics and Space Administration (NASA) launched TOPEX/Poseidon satellite, which included a very precise positioning technique. Since then, new accurate altimetry missions were launched: ERS1/2 (in 1993), Geosat Follow-On (in 2000), Jason-1 (in 2002), TOPEX/Poseidon interleaved (in 2002), ENVISAT (in 2003) and Jason-2 (in 2008). The combination of these satellites enables high-precision altimetry and improves their spatial and temporal resolution.

It is now generally accepted that at least three altimeter missions are required to resolve the ocean mesoscale variability (Le Traon & Dibarboure, 1999; Pascual et al., 2007). However, merging multi-satellite data requires consistent SLA data sets. Homogeneous and intercalibrated SLA fields in the Mediterranean Sea created by merging TOPEX/Poseidon, ERS1/2, Geosat Follow-On, Jason-1/2, TOPEX/Poseidon interleaved, and ENVISAT altimeter measurements, are obtained from AVISO (http://www.aviso.oceanobs.com/) for the period October 1997 to December 2009. The data set includes 7-day maps of SLA on a $0.125° \times 0.125°$ regular grid interpolated in time and space using a global objective analysis (Le Traon et al., 1998). The length scale of the interpolation and the e-folding time scale were set to 100 km and 10 days (Pujol & Larnicol, 2005). The SLA data is re-binned in space onto a $0.25° \times 0.25°$ to reduce small-scale variability and in time to the satellite CHL 8-day window (see below) in order to be consistent with the temporal resolution of CHL data.

2.2 Ocean color satellite data

The first instrument that demonstrated the viability of satellite ocean color measurements was the US National Oceanic and Atmospheric Administration (NOAA) and the NASA CZCS Experiment aboard the Nimbus-7 satellite (Gordon et al., 1983). Although other instruments had sensed ocean color from space, their spectral bands, spatial resolution and dynamic range were optimized for land or meteorological use, whereas every parameter in CZCS was optimized for use over water to the exclusion of any other type of sensing. The CZCS ocean color data, available from 1978 to 1986, allowed a considerable progress in the knowledge of spatial and temporal variations in surface CHL in various regions of the world ocean (Antoine et al., 1996; Behrenfeld & Falkowski, 1997; Platt & Sathyendranath, 1988).

The CZCS provided justification for future ocean color missions such as the Japanese National Space Development Agency (NASDA) Ocean Color and Temperature Scanner (OCTS) aboard the Advanced Earth Observing Satellite (ADEOS) from 1996 to 1997 (Kishino et al., 1997) or the NASA Sea Viewing Wide Field of View Sensor (SeaWiFS) aboard the Orbital Science Corporation (OSC) Orbview-II satellite from 1997 to 2010 (Hooker & McClain, 2000). Presently, the NASA Moderate Resolution Imaging Spectrometer (MODIS-A) aboard the NASA Aqua satellite (Esaias et al., 1998), and the European Space Agency (ESA) Medium Resolution Imaging Spectrometer (MERIS) aboard the ENVISAT satellite (Rast et al., 1999), both launched in 2002, provide a global monitoring of the ocean biomass. Other missions exist, with more limited coverage however, such as the Indian OCM (Chauhan et al., 2002) or the Korean OSMI (Yong et al., 1999).

To maintain the level of uncertainty of the derived products within predefined requirements, SeaWiFS and MODIS-A ocean color observations are calibrated using long-term in-situ field data (Bailey and Werdell, 2006). The calibration includes an adjustment of the overall response of the sensor, an atmospheric correction algorithm and the application of bio-optical algorithms (Gordon, 1997, 1998). The ocean color data used in this work consists of level 3 standard processed 8-day maps of CHL from SeaWiFS and MODIS-A on a 9 x 9 km regular grid in the Mediterranean Sea obtained from NASA's Ocean Color web site (http://oceancolor.gsfc.nasa.gov, see also McClain (2009)). Merging data from SeaWiFS and MODIS-A increases the coverage and reduces the uncertainties in the retrieved variables (Maritorena et al., 2010). The CHL produced by those ocean color missions are consistent over a wide range of conditions (Morel et al., 2007). We interpolate cloud-free CHL data onto the SLA grid of 0.25° x 0.25° resolution using objective analysis with a length scale of 50 km and and e-folding time scale of 10 days.

Satellite derived CHL through standard algorithms in the Mediterranean Sea is affected by a calibration problem displaying a bias when compared to in situ observations (Bosc et al., 2004; Volpe et al., 2007). This difficulty is related to the specific environmental bio-optical characteristics of the Mediterranean with respect to other oceanic regions having similar ranges of CHL. However, in the present work, we use the standard calibration algorithms because we are interested in the phytoplankton variability rather than in the absolute biomass values. The satellite derived CHL is used as a proxy for the phytoplankton biomass in the mixed layer. Although satellite derived CHL is limited to an optical depth, a reasonable correlation exists between the depth integrated and the satellite CHL (Morel and Berthon, 1989).

2.3 Seasonal cycles and inter- and intra-annual anomalies

The SLA and CHL variability in the Mediterranean are dominated by the seasonal cycle (Larnicol et al., 2002; Bosc et al., 2004). As a first step, it is necessary to remove these seasonal variations because otherwise they would dominate the resultant correlations. In our case, we calculate the seasonal cycles of SLA and CHL by averaging the value for each grid point and each 8-day window. We then subtract the seasonal cycles (8-day mean values) to the original time series for each grid point to create anomalies. Finally, we apply a low- and high-pass Lanczos filter with a cut-off period of 1 year at each grid point to compute the inter- and intra-annual anomalies, respectively, of SLA and CHL.

2.4 Correlation coefficient

The analysis of the relationships between any two satellite data sets involving large number of grid points and time series can be performed in different ways. Correlation is a simple method available when the spatial and time domains of data sets are equal. The Pearson's correlation coefficient between two time series $p(t)$ and $q(t)$ with means $\bar{p}$ and $\bar{q}$ and standard deviations σ_p and σ_q is defined as

$$r_{pq} = \frac{1}{(T-1)\sigma_p \sigma_q} \sum_{k=1}^{T} (p_k - \bar{p})(q_k - \bar{q})$$ (1)

where T is the total number of observations. We compute the Pearson's correlation coefficient for each grid point.

2.5 Singular value decomposition (SVD) analysis of the cross-covariance

A more sophisticated method to analyze the relationship between any two satellite data sets is the SVD analysis of the cross-covariance matrix between the two data sets with the same data length in time, but not necessarily the same spatial domain (Bretherton et al., 1992). The SVD analysis isolates covarying (coupled) spatial patterns of variability that tend to be linear related to one another. The SVD analysis is a generalization of empirical orthogonal function (EOF) analysis. Rather than extracting the modes that explain the greatest variance in a single data set, as in EOFs, the SVD analysis finds the covarying modes that explain as much as possible of the covariance between the two data sets.

Consider two data sets $s(x,t)$ and $c(y,t)$, consisting of SLA anomaly values at N_s grid points and CHL anomaly values at N_c grid points (possibly different), both for the same T observation times. The data time series $s(t)$ and $c(t)$ at each of the grid points can be expanded in terms of a set of $N < \min(N_s,N_c)$ vectors or patterns

$$s(t) \cong \sum_{k=1}^{N} a_k(t)\mathbf{p}_k \qquad (2)$$

$$c(t) \cong \sum_{k=1}^{N} b_k(t)\mathbf{q}_k \qquad (3)$$

The time series $a_k(t)$ and $b_k(t)$ are the expansion coefficients and the vectors $\mathbf{p}_k$ and $\mathbf{q}_k$ are the corresponding spatial patterns. The SVD spatial patterns are othonormal. Each pair of coefficients and patterns together (for a given k) make up a mode. The coefficients and patterns are chosen so that the first mode maximizes $\langle a_1(t)b_1(t) \rangle$, the cross-covariance of the expansion coefficients, where the brackets denote the time average over the T observation times. Successive pairs explain the maximum squared temporal covariance subject to orthogonality of the spatial patterns among themselves. The SVD modes are the eigensolutions of the cross-covariance matrix between the two time series.

3. Correlations between SLA and CHL

Ocean phytoplankton growth mainly depends on the availability of light and nutrients. Whereas light is rarely limiting in surface waters of the Mediterranean Sea (exceptions are some areas affected by river discharges), nutrient availability generally regulates phytoplankton growth. Since nutrient concentrations are higher in the deep ocean, physical processes that favor the supply nutrients from deeper layers into the surface euphotic zone will stimulate phytoplankton growth. Stratification and mixed layer depth changes are important factors regulating deep nutrient-rich waters supply to the upper ocean layer. SLA is indicative of changes in the thermocline depth because SLA primarily reflects the first baroclinic mode, which is related to the main thermocline (Stammer, 1997; Wunsch, 1996).

Figure 2 shows the correlation between the seasonal cycles of SLA and CHL at each grid point with shaded colors. Correlations that are not statistically significant at the 95% level are shaded white. Negative correlations are observed in most of the Mediterranean Sea, with highest (absolute) values in the northern part of the Western basin and lowest values between the Ionian and Levantine basins. Also, higher correlations are generally observed in oceanic water, off from the shelf. Inverse correlations in the seasonal cycles of SLA and CHL are typical of temperate regions where summer stratification inhibits the vertical flux of nutrients and winter mixing supplies nutrient-rich subsurface waters fueling phytoplankton growth. A few areas such as the Adriatic and Aegean basins, the entrance of the Gulf of

Gabes and the Nile River delta display positive correlations. The Adriatic basin is essentially controlled by the local winter climatic conditions, rather than the nutrient inputs from deeper layers or land sources (Santoleri et al., 2003). The Gulf of Gabes signal may be an artifact produced by direct bottom reflection in areas of shallow clear waters (Jaquet et al., 1999). The other regions with positive correlations are located over the continental shelf and receive important terrestrial inputs that may override the control of seasonal thermocline oscillation on phytoplankton production in open waters.

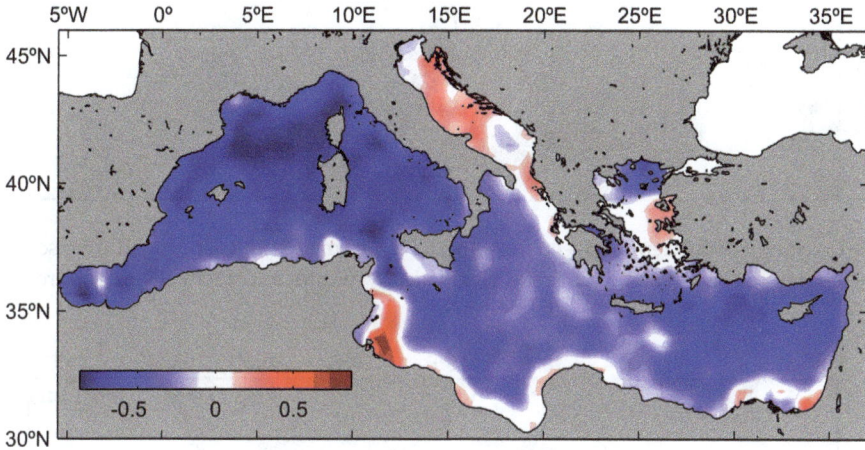

Fig. 2. Correlation coefficient between SLA and CHL seasonal cycles. The correlation in the red and blue areas is statistically significant at 95% level or more.

Although the large-scale patterns in oceanic areas observed in Figure 2 imply that SLA and related thermocline variations have key relevance on seasonal CHL, other physical processes also modulate the biological response in the Mediterranean. The correlation between SLA and CHL anomalies at inter-annual time scale is mostly negative in oceanic areas (Figure 3), suggesting the prevalence of the coupling between SLA and CHL typical of temperate areas, as observed for the seasonal cycles. However, areas showing positive correlations increase with respect to Figure 2. Water discharges from Rhone, Ebro and Nile rivers and from the Black Sea cause these positive correlations in their influence areas. Interestingly, positive correlations are also found in the southern part of the Western basin. This area is characterized by an intense mesoscale activity produced by Algerian eddies detached from the African coast and propagating to the north (Millot and Taupier-Letage, 2005). Vertical transfer of nutrients through eddy pumping is a dominant process modulating the biological activity in this region (Arnone and La Violette, 1986; Taupier-Letage et al., 2003).

The correlation between SLA and CHL anomalies at intra-annual time scales is shown in Figure 4. Correlations are not statistically significant in most of the Mediterranean Sea. Indeed, the values of the significant correlations are notably lower than the correlations between seasonal cycles and anomalies at inter-annual time scales, suggesting that direct correlation is not adequate to analyze the relationships between SLA and CHL anomalies at intra-annual time scales.

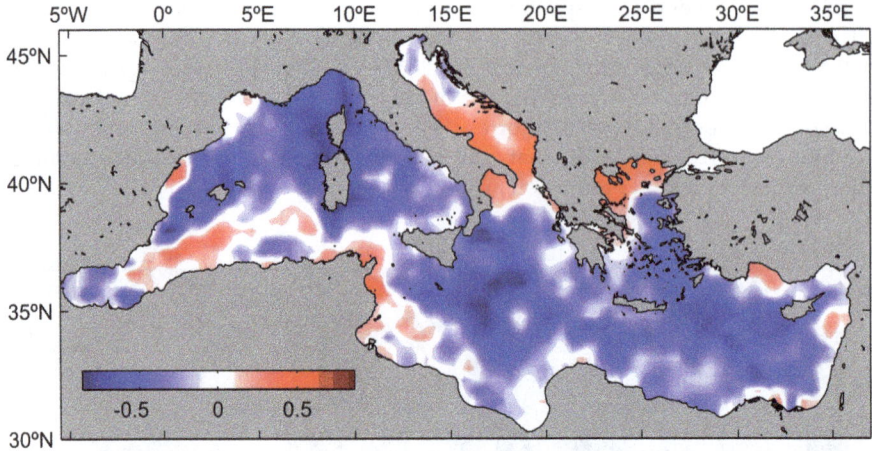

Fig. 3. Correlation coefficient between SLA and CHL anomalies at inter-annual time scales. The correlation in the red and blue areas is statistically significant at 95% level or more.

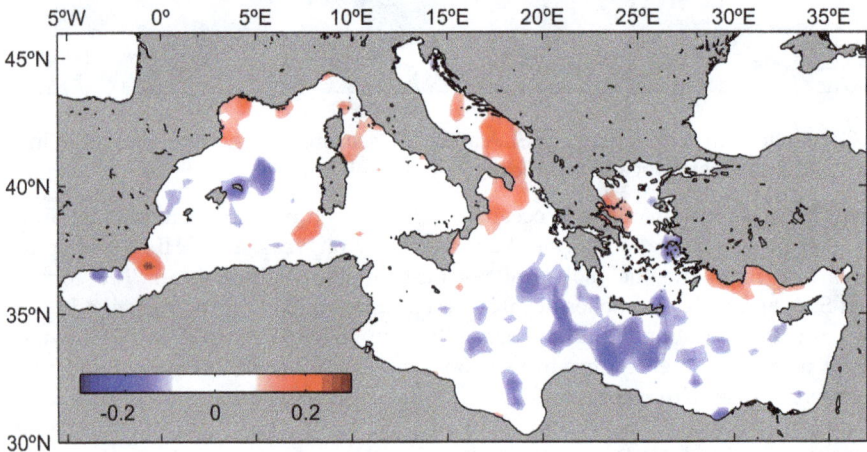

Fig. 4. Correlation coefficient between SLA and CHL anomalies at intra-annual time scales. The correlation in the red and blue areas is statistically significant at 95% level or more.

4. SVD analysis

4.1 Inter-annual anomalies

To gain further insight into the relationship between SLA and CHL anomalies at inter- and intra-annual time scales, we use SVD analysis. Figure 5 shows the spatial patterns for the first mode, which explains 41% of the covariance between the SLA and CHL inter-annual anomalies. Both patterns are scaled to represent the amplitude of SLA and CHL anomalies associated with one standard deviation of the corresponding expansion coefficients. Large areas of the Mediterranean Sea present the negative covariations (i.e. SLA increases and CHL decreases, or vice versa) typical of temperate areas, particularly in the Tyrrhenian, Ionian and Aegean basins. In the Levantine basin and in the southern part of the Western basin, areas with positive covariations are observed. These regions are characterized by high levels of mesoscale eddy variability (Pujol & Larnicol, 2005). Mesoscale eddies have important biological and biogeochemical consequences, driving vertical motions of water and lifting subsurface nutrients into the surface euphotic layer (McGillicuddy et al., 1998; Oschlies & Garçon, 1998). Positive covariations are also observed in the Adriatic basin, which is driven by the local winter climatic conditions (Santoleri et al., 2003), and in coastal areas such as Rhone, Ebro, Po and Nile river deltas, suggesting the influence of riverine inputs.

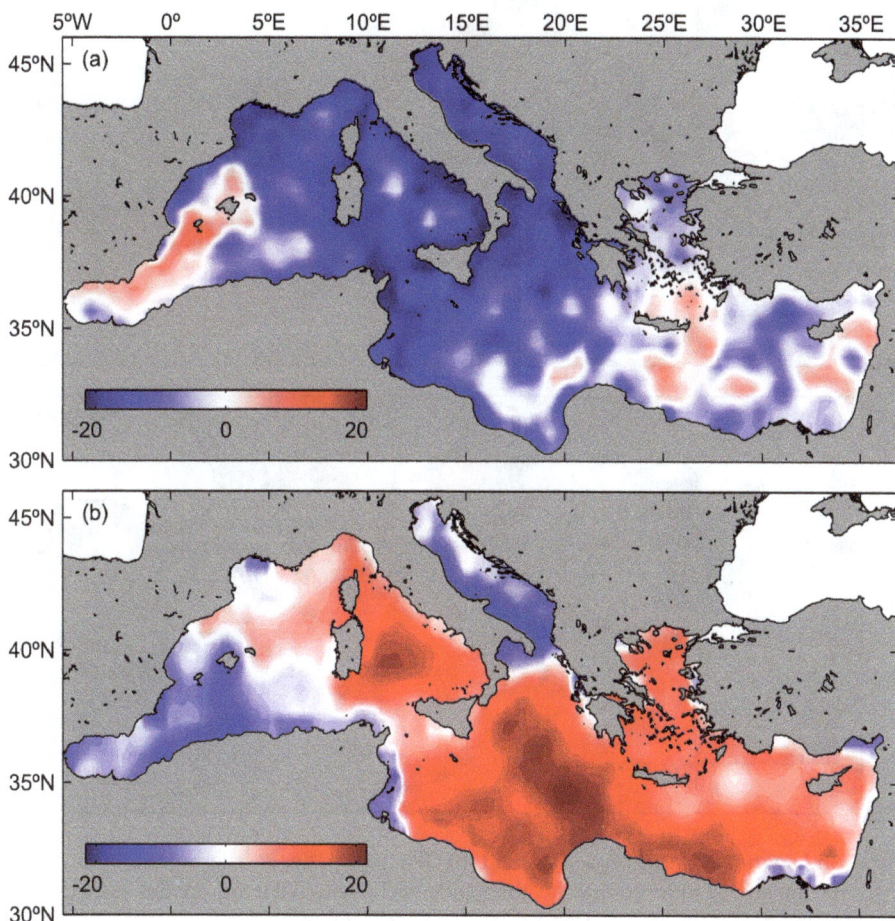

Fig. 5. First spatial patterns of (a) SLA and (b) CHL anomalies at inter-annual time scales. The patterns are scaled to represent the amplitude of SLA and CHL anomalies associated with 1 standard deviation of the first expansion coefficients.

The spatial pattern for the second mode is shown in Figure 6, accounting for 28% of the covariance between the SLA and CHL anomalies at inter-annual time scales. The areas with negative covariations dominate in the Levantine and Aegean basins. Negative covariations are also observed in the Ionian basin, with the exception of the Gulf of Gades, although the northern and southern parts behave in opposite ways (i.e. SLA increases in the north and decreases in the south, both negatively correlated with CHL). This different behavior occurs also in the Western basin, between the eastern and western parts. The Adriatic basin and the coastal areas covary positively, as observed in the first spatial pattern.

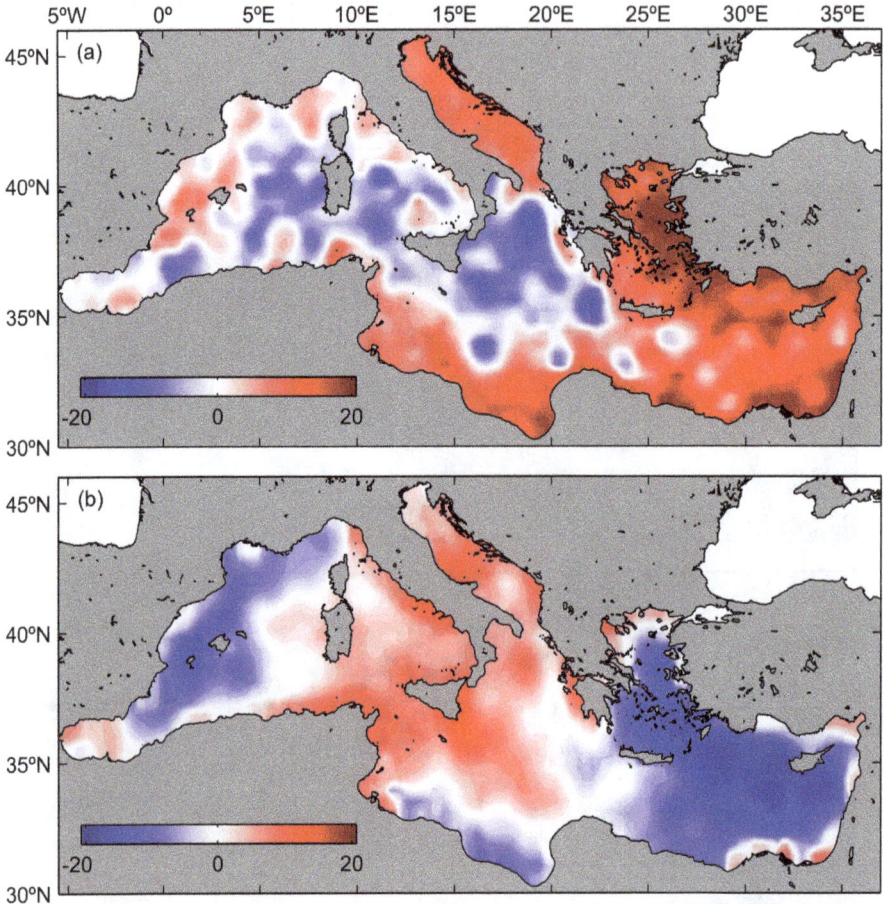

Fig. 6. Second spatial patterns of (a) SLA and (b) CHL anomalies at inter-annual time scales. The patterns are scaled to represent the amplitude of SLA and CHL anomalies associated with 1 standard deviation of the second expansion coefficients.

The time evolution of the first two modes of the SLA and CHL inter-annual anomalies is shown in Figure 7. The expansion coefficients for the first and second modes are correlated at 0.79 and 0.84 (significant at 99% level), respectively. These high values indicate that inter-

Using SVD Analysis of Combined Altimetry and Ocean Color Satellite Data for Assessing Basin Scale Physical-Biological Coupling in the Mediterranean Sea

111

annual variability is dominated by changes in vertical nutrient fluxes, that are regulated by the seasonal themocline dynamics. Mesosecale eddies plays also an important role, especially in the Levantine basin and the southern area of the Western basin. The Adriatic basin behaves completely different as its response is regulated by climatic conditions. Finally, riverine inputs influence the biological response in areas under the influence of major rivers.

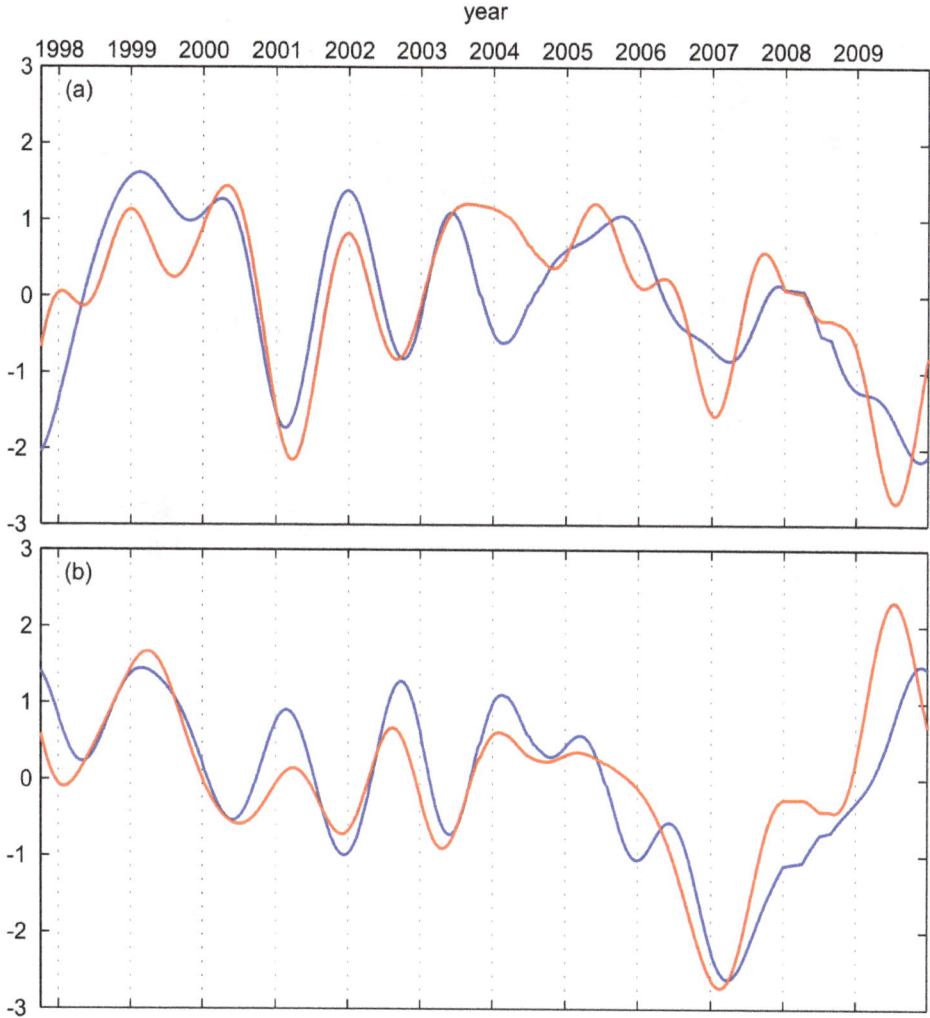

Fig. 7. Time evolution of normalized (a) first and (b) second expansions coefficients of SLA (blue line) and CHL (red line) inter-annual anomalies.

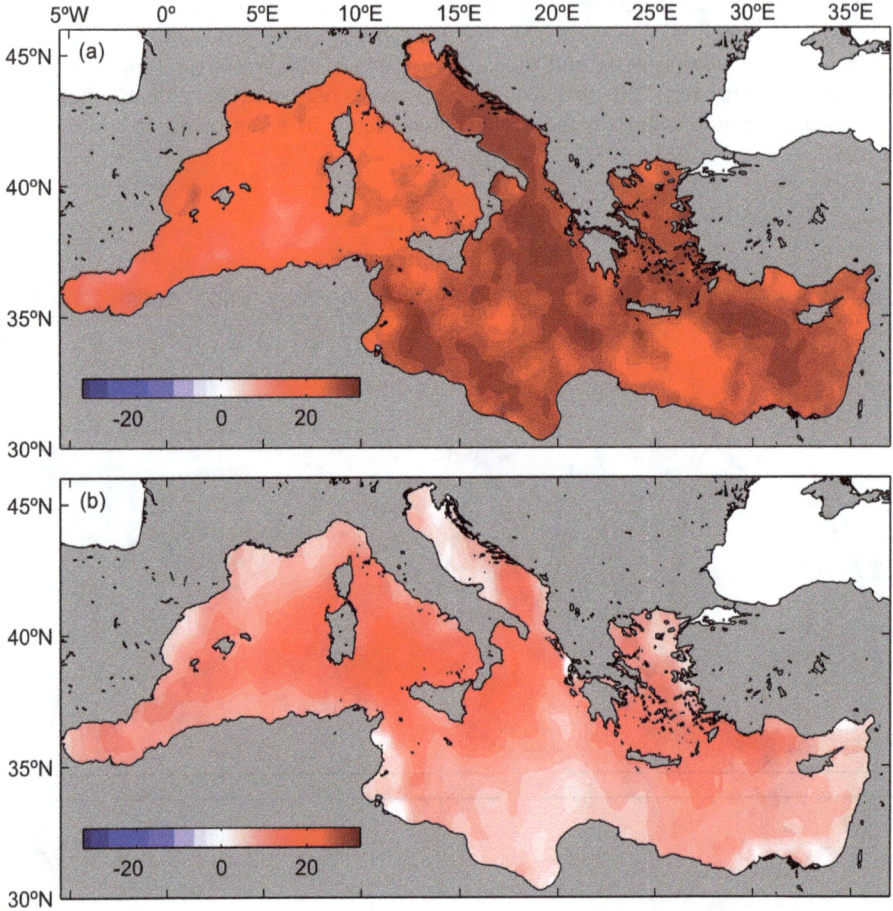

Fig. 8. First spatial patterns of (a) SLA and (b) CHL anomalies at intra-annual time scales. The patterns are scaled to represent the amplitude of SLA and CHL anomalies associated with 1 standard deviation of the first expansion coefficients.

4.2 Intra-annual anomalies

The spatial pattern for the first mode between the SLA and CHL intra-annual anomalies shows a significant positive covariation of SLA and CHL in the whole Mediterranean Sea. Note that variations in CHL at this time scale are markedly lower than in other modes. This behavior does not agree with that observed at inter-annual time scales, when the winter mixing enhances the upward transport of nutrients to the ocean surface. Therefore, other physical processes must be regarded in order to explain the biomass enhancement at this time scale. One candidate could be a mechanical effect indebted to oscillations in the thermocline. The modification of the thermocline depth due to the SLA variation could accumulate phytoplankton biomass close to the surface (or vice versa) and thus affect the CHL measured by the ocean color satellites without modifying the vertically-integrated biomass. Alternatively, the shoaling of the thermocline could extend the deeper nutrient-rich layer into the euphotic surface zone, allowing phytoplankton uptake. These compression mechanisms of the ocean surface layer would be similar dynamically to the compression caused by Rossby waves (Cipollini et al. 2001). This first pattern accounts for 69% of the total covariance at intra-annual time scales, whereas the second pattern (not shown) explains less than 10% of the covariance.

Figure 9 shows the time evolution of the first mode of the SLA and CHL intra-annual anomalies. The figure only shows the period from 2002 to 2006 to facilitate the observation of short-term variability. The correlation between the first expansion coefficients for SLA and CHL intra-annual anomalies is 0.40 (significant at 99% level). Although the compression may play an important role on the enhancement of CHL observed by the satellite, the relatively modest value of correlation indicates that processes other than mechanical accumulation take place. For example, biological processes related to food web dynamics which are not coupled with the SLA should be important.

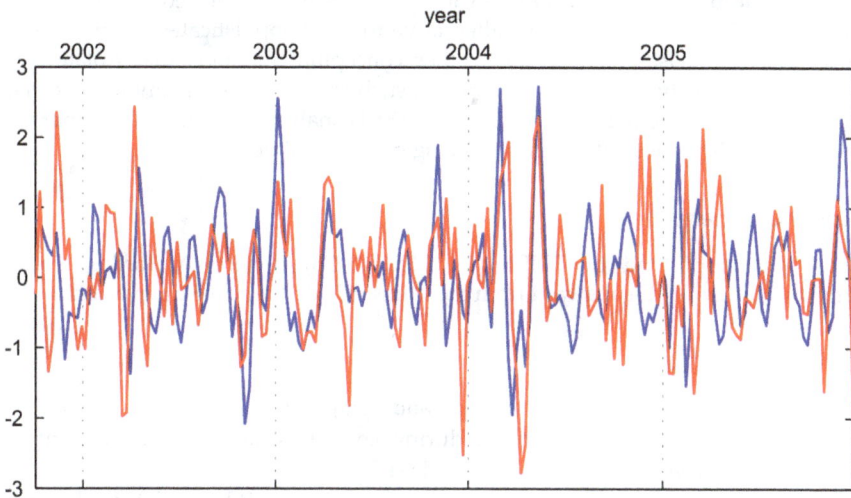

Fig. 9. Time evolution of normalized first expansions coefficients of SLA (blue line) and CHL (red line) intra-annual anomalies.

5. Conclusion

This study analyzes the basin scale physical-biological coupling in the Mediterranean Sea at inter-annual, seasonal and intra-annual time scales based on 12 years of concurrent satellite SLA and CHL data. Not surprisingly, the long-term (inter-annual and seasonal) variability of SLA and CHL is negatively correlated in most oceanic areas of the Mediterranean Sea. This is the typical behavior of temperate regions associated with the availability of nutrients in the mixed layer: summer stratification blocks upward entrainment of nutrients from deep layers and winter mixing brings nutrients to the surface (Cushing, 1959). In the coastal regions, particularly in those areas influenced by major rivers, the biological response is controlled by supply of nutrients of continental origin. However, other biological responses to the thermocline oscillations are also observed. The inter-annual SLA and CHL covary in areas dominated by mesoscale eddies, such as the Levantine basin and the southern part of the Western basin. Cyclonic eddies can enhance primary production by upwelling of nutrient rich water (McGillicuddy et al., 1998; Oschlies & Garçon, 1998). In the intra-annual variability, the coupling between SLA and CHL is exerted through a mechanical compression mechanism, which concentrates nutrients and phytoplankton cells into the surface layer. Nevertheless, the overall influence of the mesoscale eddies and the compression mechanism in the enhancement of phytoplankton in the Mediterranean Sea deserves further study.

The SVD analysis to link SLA and CHL is a quick, easily accessible and powerful method for assessing the ocean physical-biological coupling. Our results demonstrate its strength over the direct correlation. The correlation map indicates the spatial covariability of SLA and CHL but cannot provide any details about their temporal variability. SVD analysis extracts the dominant temporal and spatial components of covariability between SLA and CHL into a series of orthogonal functions or statistical modes, and their time evolution or expansion coefficients. In addition, the SVD modes can be related to different coupling mechanism. This methodology represents a simple alternative to more sophisticated coupled physical-biological ocean models. There are also other conceptual methods that isolate coupled modes of variability between spatial time series, such as joint EOFs or canonical correlation analysis. According to Bretherton et al. (1992), the SVD analysis is simpler and superior than these other methods in most situations involving geophysical fields.

6. Acknowledgment

This work was supported by EHRE (CTM2009-08270) project. A. Jordi's work was supported by a Ramón y Cajal award from the Ministerio de Ciencia e Innovación.

7. References

Abbott, M.R. & Zion, P.M. (1987). Spatial and temporal variability of phytoplankton pigment off northern California during coastal ocean dynamics experiment 1. *Journal of Geophysical Research*, 92 (1), 745–1,755.

Antoine, D.; Morel, A.; Gordon, H.R.; Banzon, V.F. & Evans, R.H. (2005). Bridging ocean color observations of the 1980s and 2000s in search of long-term trends. *Journal of Geophysical Research*, 110, C06009, doi:10.1029/2004JC002620.

Antoine, D; Morel, A. & André, J. M. (1995). Algal pigment distribution and primary production in the Eastern Mediterranean as derived from coastal zone color scanner observations. *Journal of Geophysical Research*, 100, 16193–16209.

Arnone, R.A. & La Violette, P.E. (1986). Satellite definition of the bio-optical and thermal variation of coastal eddies associated with the African Current. *Journal of Geophysical Research*, 91, 2351– 2364.

Bailey, S. W. & Werdell, P. J. (2006). A multi-sensor approach for the on-orbit validation of ocean color satellite data products. *Remote Sensing of the Environment*, 102, 12–23.

Barale, V.; Jaquet, J.M. & Ndiaye, M. (2008). Algal blooming patterns and anomalies in the Mediterranean Sea as derived from the SeaWiFS data set (1998-2003). *Remote Sensing of Environment*, 112 (8), 3300-3313.

Barber, R.T. & Hilting, A.K. (2000). Achievements in biological oceanography, *50 Years of Ocean Discovery: National Science Foundation 1950-2000*, pp. 11–21, National Academies Press, Washington, USA.

Behrenfeld, M.J. & Falkowski, P.G. (1997). Photosynthetic rates derived from satellite-based chlorophyll concentration. *Limnology and Oceanography*, 42 (1), 1– 20.

Behrenfeld, M.J.; Randerson, J.T.; McClain, C.R.; Feldman, G.C.; Los, S.O.; Tucker, C.J.; Falkowski, P.G.; Field, C.B.; Frouin, R.; Esaias, W.E. & others (2001). Biospheric primary production during an ENSO transition. *Science*, 291 (2), 594–2,597.

Bosc, E.; Bricaud, A. & Antoine, D. (2004). Seasonal and interannual variability in algal biomass and primary production in the Mediterranean Sea, as derived from 4 years of SeaWiFS observations. *Global Biogeochemical Cycles*, 18 (1), doi:10.1029/2003GB002034.

Bretherton, C.S.; Smith, C. & Wallace, J.M. (1992). An Intercomparison of Methods for Finding Coupled Patterns in Climate Data. *Journal of Climate*, 5 (6), 541-560.

Bricaud, A.; Bosc, E. & Antoine, D. (2002). Algal biomass and sea surface temperature in the Mediterranean Basin - Intercomparison of data from various satellite sensors, and implications for primary production estimates. *Remote Sensing of Environment*, 81 (2-3), 163-178.

Cazenave, A.; Cabanes, C.; Dominh, K. & Mangiarotti, S. (2001). Recent sea level change in the Mediterranean sea revealed by Topex/Poseidon satellite altimetry. *Geophysical Research Letters*, 28 (8), 1607-1610.

Chauhan, P.; Mohan, M.; Sarangi, R. K.; Kumari, B.; Nayak, S. & Matondkar, S.G.P. (2002). Surface chlorophyll a estimation in the Arabian Sea using IRS-P4 Ocean Colour Monitor (OCM) satellite data. *International Journal of Remote Sensing*, 23, 1663-1676.

Cipollini, P.; Cromwell, D.; Challenor, P.G. & Raffaglio, S. (2001). Rossby waves detected in global ocean colour data. *Geophysical Research Letters*, 28 (2), 323-326.

Cushing, D.H. (1959). The seasonal variation in oceanic production as a problem in population dynamics. *ICES Journal of Marine Science*, 24, 455– 464.

D'Ortenzio, F.; Marullo, S.; Ragni, M.; d'Alcala, M.R. & Santoleri, R. (2002). Validation of empirical SeaWiFS algorithms for chlorophyll-alpha retrieval in the Mediterranean Sea - A case study for oligotrophic seas. *Remote Sensing of Environment*, 82 (1), 79-94.

D'Ortenzio, F.; Ragni, M.; Marullo, S. & d'Alcala, M.R. (2003). Did biological activity in the Ionian Sea change after the Eastern Mediterranean Transient? Results from the analysis of remote sensing observations. *Journal of Geophysical Research*, 108 (C9), doi:10.1029/2002jc001556.

Esaias, W.E.; Abbott, M.R.; Barton, I.; Brown, O.B.; Campbell, J.W.; Carder, K.L.; Clark, D.K.; Evans, R.H.; Hoge, F.E.; Gordon, H.R.; Balch, W.M.; Letelier, R. & Minnett, P.J. (1998). An overview of MODIS capabilities for ocean science observations. *IEEE Transactions on Geoscience and Remote Sensing*, 36, 1250–1265.

Fu, L.-L.; Christensen, E.; Lefebvre; M. & Menard, Y. (1994). TOPEX/POSEIDON mission overview. *Journal of Geophysical Research*, 99, 24369-24382.

Gordon, H.R. (1997). Atmospheric correction of ocean color imagery in the Earth observing system era. *Journal of Geophysical Research*, 102, 17081–17106.

Gordon, H.R. (1998). In-orbit calibration strategy for ocean color sensors. *Remote Sensing of the Environment*, 63, 265– 278.

Gordon, H.R.; Clark, D.K.; Brown, J.W.; Brown, O.B.; Evans, R.H. & Broenkow, W.W. (1983). Phytoplankton pigment concentrations in the Middle Atlantic Bight: comparison of ship determinationos and CZCS estimates. *Applied Optics*, 22, 20-36

Hooker, S.B. & McClain, C.R. (2000). The calibration and validation of SeaWiFS data. *Progress in Oceanography*, 45, 427–465.

Jaquet, J.M.; Tassan, S.; Barale, V. & Sarbaji, M. (1999). Bathymetric and bottom effects on CZCS chlorophyll-like pigment estimation: data from the Kerkennah shelf (Tunisia). *International Journal of Remote Sensing*, 20(7), 1343−1362.

Jordi, A. & Wang, D.-P. (2009). Mean dynamic topography and eddy kinetic energy in the Mediterranean Sea: Comparison between altimetry and a 1/16 degree ocean circulation model. *Ocean Modelling*, 29 (2), 137-146.

Jordi, A.; Basterretxea, G. & Anglès, S. (2009). Influence of ocean circulation on phytoplankton biomass distribution in the Balearic Sea: Study based on Sea-viewing Wide Field-of-view Sensor and altimetry satellite data. *Journal of Geophysical Research*, 114 (C11005), doi:10.1029/2009JC005301.

Kishino, M.; Ishizaka, J.; Saitoh, S.; Senga, Y. & Utashima, M. (1997). Verification plan of OCTS atmospheric correction and phytoplankton pigment by moored optical buoy system. *Journal of Geophysical Research*, 102, 17197-17207.

Larnicol, G.; Ayoub, N. & Le Traon, P.Y. (2002). Major changes in Mediterranean Sea level variability from 7 years of TOPEX/Poseidon and ERS-1/2 data. *Journal of Marine Systems*, 33, 63-89.

Le Traon, P.Y. & Dibarboure, G. (1999). Mesoscale mapping capabilities of multiple-satellite altimeter missions. *Journal of Atmospheric and Oceanic Technology*, 16, 1208–1223.

Le Traon, P.Y.; Nadal, F. & Ducet, N. (1998). An improved mapping method of multisatellite altimeter data. *Journal of Atmospheric and Oceanic Technology*, 15 (2), 522-534.

Lehahn, Y.; d'Ovidio, F.; Lévy, M. & Heifetz, E. (2007). Stirring of the northeast Atlantic spring bloom: A Lagrangian analysis based on multisatellite data. *Journal of Geophysical Research*, 112 (8) doi:10.1029/2006JC003927.

Leterme, S.C. & Pingree, R.D. (2008). The Gulf Stream rings and North Atlantic eddy structures from remote sensing (Altimeter and SeaWiFS). *Journal of Marine Systems*, 69 (3-4) 177-190.

Maritorena, S.; d'Andon, O.H.F.; Mangin, A. & Siegel, D.A. (2010). Merged satellite ocean color data products using a bio-optical model: Characteristics, benefits and issues. *Remote Sensing of the Environment*, 114, 1791–1804.

McClain, C.R. (2009). A decade of satellite ocean color observations. *Annual Review of Marine Sciences*, 1, 19–42.

McGillicuddy, D.J.; Robinson, A.R.; Siegel, D.A.; Jannasch, H.W.; Johnson, R.; Dickeys, T.; McNeil, J.; Michaels, A.F. & Knap, A.H. (1998). Influence of mesoscale eddies on new production in the Sargasso Sea. *Nature*, 394 (6690), 263-266.

Millot, C.; Taupier-Letage, I. (2005). Additional evidence of LIW entrainment across the Algerian subbasin by mesoscale eddies and not by a permanent westward flow. *Progress in Oceanography*, 66(2-4), 231-250.

Morel, A. & Berthon, J.F. (1989). Surface pigments, algal biomass profiles, and potential production of the euphotic layer: Relationship reinvestigated in view of remote-sensing applications. *Limnology and Oceanography*, 34(8), 1545– 1562.

Morel, A.; Huot, Y.; Gentili, B.; Werdell, P.J.; Hooker, S.B. & Franz, B.A. (2007). Examining the consistency of products derived from various ocean color sensors in open ocean (case 1) waters in the perspective of a multisensor approach. *Remote Sensing of the Environment*, 111, 69–88.

Oschlies, A. & Garçon, V. (1998). Eddy-induced enhancement of primary production in a model of the north Atlantic Ocean. *Nature*, 394 (6690), 266-269.

Pascual, A.; Pujol, M.I.; Larnicol, G.; Le Traon, P.Y. & Rio, M.H. (2007). Mesoscale mapping capabilities of multi-satellite altimeter missions: First results with real data in the Mediterranean Sea. *Journal of Marine Systems*, 65(1-4), 190-211.

Platt, T. & Sathyendranath, S. (1988). Oceanic primary production: estimation by remote sensing at local and regional scales. *Science*, 241, 1613– 1620.

Pujol, M.I. & Larnicol, G. (2005). Mediterranean Sea eddy kinetic energy variability from 11 years of altimetric data. *Journal of Marine Systems*, 58 (3-4), 121-142.

Rast, M.; Bézy, J.L. & Bruzzi, S. (1999). The ESA Medium Resolution Imaging Spectrometer MERIS – A review of the instrument and its mission. *International Journal of Remote Sensing*, 20, 1681–1702.

Santoleri, R.; Banzon, V.; Marullo, S.; Napolitano, E.; D'Ortenzio, F. & Evans, R. (2003). Year-to-year variability of the phytoplankton bloom in the southern Adriatic Sea (1998-2000): Sea-viewing Wide Field-of-view Sensor observations and modeling study. *Journal of Geophysical Research*, 108(C9), doi:10.1029/2002JC001636.

Siegel, D.A.; Court, D.B.; Menzies, D.W.; Peterson, P.; Maritorena, S. & Nelson, N.B. (2008). Satellite and in situ observations of the bio-optical signatures of two mesoscale eddies in the Sargasso Sea. *Deep-Sea Research Part II*, 55 (1), 218–1,230.

Siegel, D.A.; Fields, E. & McGillicuddy, D.J. (1999). Mesoscale motions, satellite altimetry and new production in the Sargasso Sea. *Journal of Geophysical Research*, 104 (13), 359–13,379.

Stammer, D. (1997). Steric and wind-induced changes in TOPEX/POSEIDON large-scale sea surface topography observations. *Journal of Geophysical Research*, 102 (9), 20987-21009.

Taupier-Letage, I.; Puillat, I.; Millot, C. & Raimbault, P. (2003). Biological response to mesoscale eddies in the Algerian Basin. *Journal of Geophysical Research*, 108 (C8), doi:10.1029/1999JC000117.

Volpe, G.; Santoleri, R.; Vellucci, V.; d'Alcala, M.R.; Marullo, S. & D'Ortenzio, F. (2007). The colour of the Mediterranean Sea: Global versus regional bio-optical algorithms evaluation and implication for satellite chlorophyll estimates. *Remote Sensing of the Environment*, 107(4), 625 – 638.

Wilson, C. & Adamec, D. (2001). Correlations between surface chlorophyll and sea surface height in the tropical Pacific during the 1997-1999 El Nino-Southern Oscillation event. *Journal of Geophysical Research*, 106 (12), 31175-31188.

Wilson, C. & Adamec, D. (2002). A global view of bio-physical coupling from SeaWiFS and TOPEX satellite data, 1997-2001. *Geophysical Research Letters*, 29 (8), 10.1029/2001GL014063.

Wunsch, C. (1996). *The Ocean Circulation Inverse Problem*, Cambridge University Press, ISBN 0-521-48090-6, New York, USA.

Yong, S.-S.; Shim, H.-S.; Heo, H.-P.; Cho, Y.-M.; Oh, K.-H.; Woo, S.-H. & Paik, H.-Y. (1999). The ground checkout test of OSMI on KOMPSAT-1. *Journal of Korean Society of Remote Sensing*, 15, 297–305.*

6

Ocean Color Remote Sensing of Phytoplankton Functional Types

Tiffany A.H. Moisan[1], Shubha Sathyendranath[2] and Heather A. Bouman[3]
[1]*NASA Goddard Space Flight Center, Wallops Island,*
[2]*Plymouth Marine Laboratory, Bedford Institute of Oceanography,*
Prospect Place, The Hoe Plymouth,
[3]*Department of Earth Sciences, University of Oxford, South Parks Road, Oxford,*
[1]*USA*
[2]*Canada,*
[3]*UK*

1. Introduction

Interest in phytoplankton diversity has increased in recent years due to its possible role in regulating climate by production and consumption of greenhouse gases. For example, gases can diffuse across the air-sea interface, many of which are synthesized and emitted by certain phytoplankton species or groups. It has been suggested that these variations play an important role in moderating our climate through backscattering of solar radiation and within cloud formation. Climate will ultimately control fundamental environmental conditions that regulate algal growth, including water temperature, nutrients, and light and thus can be expected to result in changes in the species composition, trophic structure and function of marine ecosystems. In the past several decades, the scientific community has witnessed changes in phytoplankton distribution.

The marine phytoplankton community is diverse and includes on the order of tens of thousands of phytoplankton species (Jeffrey & Vesk 1997). On regional scales, phytoplankton biogeography is controlled by the physical, chemical, and meteorological characteristics that force ecosystem dynamics. There is a renewed impetus for new technologies to provide information about the phytoplankton community composition over global scales. Real-time, large-scale taxonomic information, if available, could open up new possibilities and approaches geared toward monitoring highly-dynamic oceanic processes and phenomena such as algal blooms (including harmful algal blooms), frontal structures, eddies, and episodic events (storms, river outflow, and wind mixing). Phytoplankton diversity information provides a valuable quantitative database for structuring sophisticated predictive models that includes taxonomic phytoplankton community information such as size spectra, probability distribution of taxa, and upper trophic level estimations including fisheries productivity (Cheson & Case 1986, De Angelis & Waterhouse 1987). There have been several reviews and books written on phytoplankton community structure, dynamics, and biogeochemistry as measured by ocean color (Mitchell 1994, Martin 2004, Mueller et al. 2004, Miller et al. 2005, Richardson and LeDrew 2006, Longhurst 2007, Robinson 2010).

Fig. 1. Distribution of *Nitzschia americana* (solid square), *Nitzschia bicapitata* (open circle) and *Nitzschia pseudonana* (open triangle) from shipboard observations. Modified from Hasle (1976). These data were compiled over several years in order to understand their global distribution.

Approaches have been proposed for remote sensing detection of phytoplankton functional types (PFTs) and may be classified as direct (defined as they exploit optical signatures of phytoplankton that may be detected by sensors on satellite platforms) or indirect (e.g. exploit relationships between chlorophyll-a concentration and functional types). Both approaches have their advantages and limitations. Direct methods are limited to few phytoplankton groups that possess optical signatures distinct from other constituents in seawater. Indirect methods exploit well-established algorithms for retrieval of chlorophyll-a concentrations from ocean color satellite products in combination with phytoplankton community structure, which is validated by *in situ* observations. Optical sensors have ushered in a new capability of remotely sensing the ocean synoptically compared with *in situ* data collected on board ships, which may take years to compile but is is capable of exploring ancillary properties inaccessible to remote sensing (Hasle 1976, Figure 1). Early shipboard studies have demonstrated the complexity of the marine ecosystem and how distribution, function, and physiology are linked closely to the biochemistry of carbon, nitrogen, sulfur, and other elements in the sea. This chapter discusses how information on PFTs may be derived from a combination of satellite data and *in situ* observations. It is anticipated that developing an understanding of the large-scale distributions and variability in PFTs will contribute to various ecological problems of the day of paramount economic importance, such as providing an ecological basis for fisheries fluctuations and occurrence of harmful algal blooms. Future satellite sensors with improved spectral resolution or temporal resolution holds promise for improving the accuracy of retrieval of PFTs and understanding their role in the ecology of the ocean.

Global patterns of diversity for the majority of marine organisms are virtually uncharacterized (Worm et al. 2005). For planktonic organisms, which form the base of the marine food web, global ocean geographic distribution of diversity has only been characterized for planktonic foraminifera (Rutherford et al. 1999). Marine biodiversity patterns show a worldwide consistency despite differences in environmental conditions of various oceanographic regions (Irigolen et al. 2004). Marine phytoplankton diversity, similar to terrestrial vegetation, is described as a uni-modal function of phytoplankton biomass with an occurrence of maximum diversity at intermediate levels of phytoplankton biomass (sub-bloom concentrations) and low diversity during massive blooms (Irigolen et al. 2004). Similar to terrestrial vegetation, phytoplankton diversity is a unimodal function of phytoplankton biomass, with maximal diversity at intermediate levels of phytoplankton biomass and minimum diversity during massive blooms.

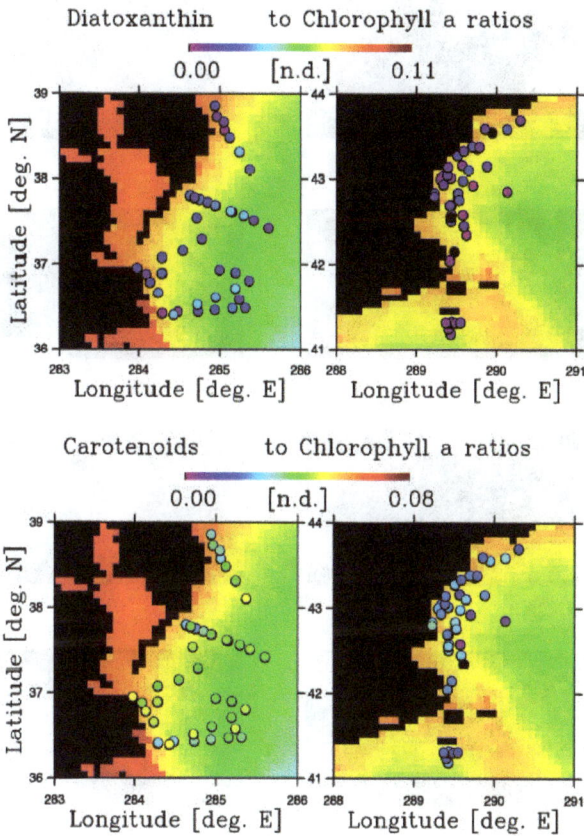

Fig. 2. Ratios of diatoxanthin and carotenoids (to chlorophyll-a) near the Delmarva Peninsula against the SeaWiFS 10 year average. Differences between the values and the SeaWiFS average emphasize that the phytoplankton community is structured differently from the bulk community (e.g. chlorophyll-a).

Moreover, the diversity–productivity relationship of oceanic plankton is similar to the classical relationship observed in early shipboard collections of phytoplankton. The results of PFT distribution return a significantly different pattern compared with the 10-year SeaWiFS average, underscoring that underlying biological processes produce a complex ecological food web matrix in the ocean (Figure 2).

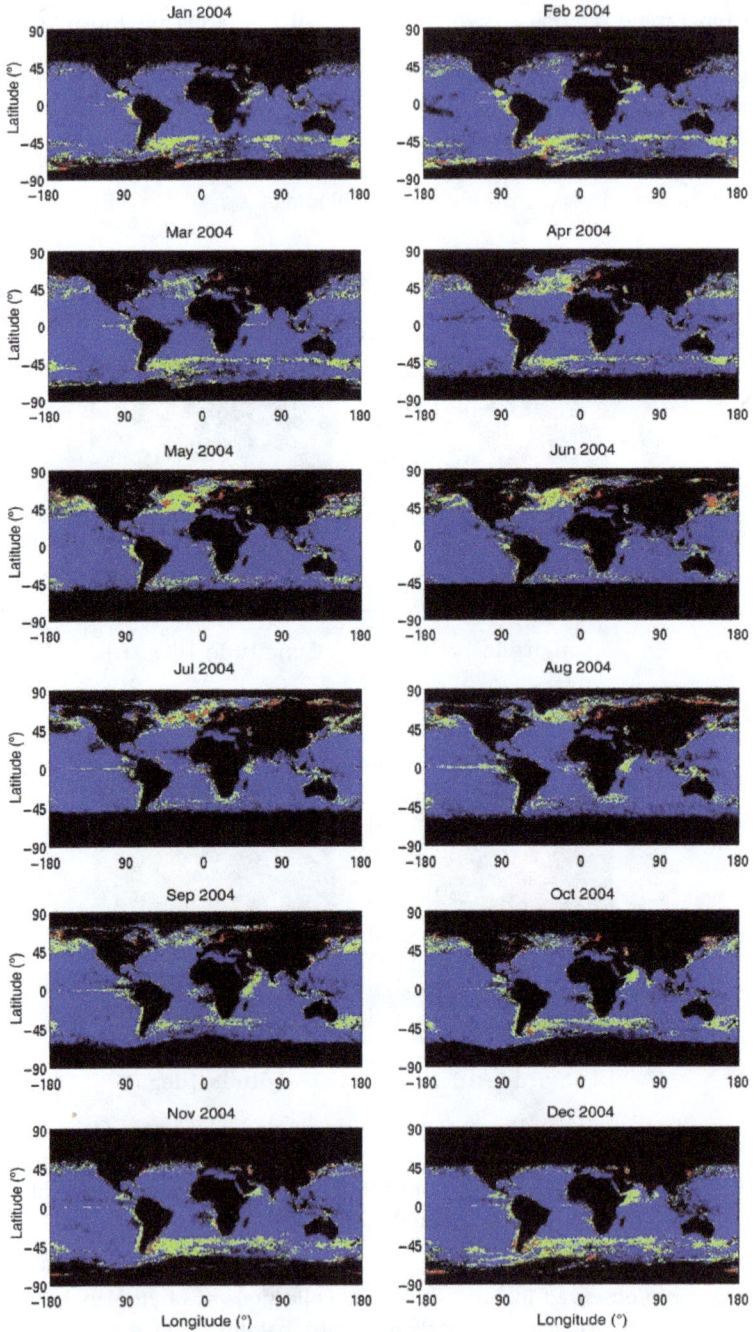

Fig. 3. Monthly SeaWiFS data for 2004 interpreted for phytoplankton, size classes: pico (blue); nano (green); micro (red, Aiken et al. 2009).

Our limited understanding of marine ecosystem responses to various physico-chemical climate drivers, which include genetic and phenotypic adaptation to the unprecedented pace of climate change call for novel methods to study the problem, and remote sensing is one of the potential avenues (Falkowski et al. 2004, Chavez et al. 1999). Understanding the cyclical relationship of phytoplankton groups and how the production of greenhouse gases relates to temperature and productivity provides a better understanding of their coupling to climate (Falkowski et al. 2004). Changes in phytoplankton community structure could potentially provide an early warning for climate-driven perturbations in marine ecosystems (Hallegraeff 2010). Establishing the biogeography of several morphologically and genetically distinct marine species is a complex target for which remote sensing may provide at best, a partial solution.

Historically, satellite sensors have provided oceanographers with bulk phytoplankton pigment concentrations (e.g. chlorophyll-*a*) over global scales with synoptic resolution (McClain, 2009). Coastal Zone Color Scanner (CZCS) was the beginning of the era of satellite remote sensing of ocean color. It was a relatively simple sensor and was successful at the retrieval of chlorophyll-a concentrations in open-ocean waters. Models were then developed that utilized satellite-derived chlorophyll-a data along with information on photosynthetically available radiation and temperature at the sea surface to compute water column primary production by remote sensing. However, CZCS performed poorly in coastal waters. SeaWiFS and MODIS, the successor to the CZCS, had better radiometric precision and more wavebands and was designed to address the limitations of CZCS, in particular for applications in coastal waters. Satellites with even higher spectral resolution followed (e.g. MERIS). Though the newer generations of satellites were designed for improving performance in coastal waters, there has been increasing recognition of their value in the detection of phytoplankton functional types as well. These satellite missions have provided knowledge and understanding of a variety of events including frontal features and episodic blooms on a global basis. The scientific community has now moved onto predicting ecologically significant characteristics of the food web such as size structure as interpreted for phytoplankton size classes including picoplankton to microplankton (Aiken et al. 2008). Ocean Color has been recognized as an essential climate variable by the Global Climate Observation System (GCOS). As the global time series of ocean color data grows (CZCS, MODIS, & SeaWiFS), there is an increase in satellite products available to study decadal-scale variations in phytoplankton distribution and primary production.

Understanding the spatial and temporal distribution of PFTs will allow us to improve our knowledge of biologically-mediated fluxes of elements between the upper ocean and the ocean interior (Falkowski & Raven, 1997). The performance of biogeochemical models in the ocean has improved substantially as a result of incorporating PFTs into ecosystem models. The spatial variability and concentration of various PFTs are critical to improving primary-productivity estimates, and understanding the feedbacks of climate change. The ability to observe PFTs on a global scale that relate to key biogeochemical processes such as nitrogen fixation, silicification, and calcification is valuable to studies of marine elemental cycles. Despite the paucity of data on functional groups, our understanding of ecosystem linkages is improving as we accrue larger amounts of data. Major divisions of phytoplankton taxonomic groups such as diatoms, coccolithophores, dinoflagellates, chlorophytes, and cyanobacteria are often separated into distinct functional groups, as these taxonomic groups have unique biogeochemical signatures

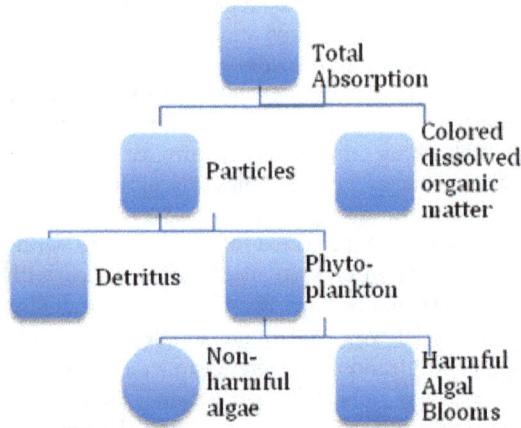

Fig. 4. Schematic diagram showing sources of variation in optical characteristics which influence ocean color. (Keller et al. 1989)

Aiken et al. (2009) shows how different size classes of phytoplankton are distributed globally, and size is often related to function (Nair et al. 2008). Nonetheless, there are limitations to discriminating phytoplankton taxonomic composition or even functional groups by remote sensing, due to our inability to discriminate various phytoplankton types optically. Remote-sensing reflectance (Rrs (λ)) is influenced by absorption and backscattering by seawater, phytoplankton, colored dissolved organic matter, detrital matter and other suspended material (Garver & Siegel 1997). We show the various optical pools that contribute to absorption, which contributes to the spectral variability of Rrs (λ, Figure 4). Remote sensing of particular types of phytoplankton is only possible if optical characteristics are identified for those types that can be used to distinguish them from all other types of material in the water.

New technological developments and improved scientific knowledge have allowed for the development of several approaches for detecting phytoplankton biomass and some functional groups of phytoplankton including coccolithophores (Balch et al. 1991, Balch et al. 1996) and *Trichodesmium* (Subramanian et al. 1994, 1999 a,b, Hu et al. 2010). More recently, algorithms have been developed to distinguish additional phytoplankton groups and size classes (Sathyendranath et al. 2004, Alvain et al. 2005, Moisan et al. 2011ab). High Performance Liquid Chromatography (HPLC) is the *in situ* method of choice to serve as ground truth for satellite products due to its accuracy and rapid processing. A relatively new analysis tool called CHEMical TAXonomy (CHEMTAX) has been developed that is designed to yield information on the phytoplankton composition using HPLC data (Mackey et al.1996), but it has to be recognized that HPLC models have their limitations (e.g. Latasa 2007).

2. "History of terminology for phytoplankton functional types": Ecological, biogeochemical, and optical definitions

The Ocean Color community has utilized a reductionist approach to describe phytoplankton taxonomic composition and refers to optical signatures of certain phytoplankton groups/species or PFTs. In this chapter, we will refer to Phytoplankton Functional Types (PFTs) in order to utilize current terminology. Please take into account that we may be defining groups within different levels of complexity that relate to phylogeny or bio-optically similarities etc. Taxonomic characterization has utilized different terminology over the years and we delve into the history of its classification because it may define many different levels of complexity.

Present studies on diversity are not too far removed from the original ideas invoked by G.E. Hutchinson's "Paradox of the Plankton" (Hutchinson, 1961) and J. H. Connell's "Intermediate Disturbance Hypothesis" (Connell, 1978). Ocean variability, through eddies, storms, seasonal cycles, and El Niño, have been documented to play a role in controlling diversity. How this level of stochastic variability manifests itself spatially and temporally due to climate change is unknown. Falkowski & Oliver (2007) suggest that resource competition theory is adequate to describe global distribution of marine eukaryotic phytoplankton taxa during equator-to-pole and continent-to-land thermal gradients. These results are supportive in light of describing phytoplankton beyond a bulk approach of utilizing only chlorophyll a.

A functional group is used as an ecological term to define a group of organisms that have similar biogeochemical processes e.g. nitrogen fixation, silicifiers, and calcifiers (Fauchald & Jumars 1979, Reynolds et al. 2002, Nair et al. 2008). Functional groups in phytoplankton are defined as groups of organisms related through common biogeochemical processes but are not necessarily phylogenetically related (Iglesias-Rodriguez et al. 2002). Homologously similar sets of organisms are referred to as "functional groups" or "biogeochemical guilds" following Totterdell et al. (1993). For marine phytoplankton, some attempts have been made to organize species within certain classes of phytoplankton according to their ecological and biogeochemical function or habitat distribution perspective (Iglesias-Rodríguez et al. 2002, Smayda & Reynolds, 2003, Vila & Masó, 2005). Thus, terminology has finally evolved to the usage of Phytoplankton Functional Type (PFT) or Phytoplankton Functional group but these classifications are not necessarily straightforward and include a wide range of different taxonomic classification approaches such as size, biogeochemistry, or traditional taxonomic hierarchies (see discussion by Nair et al. 2008).

3. On the road to predicting remote sensing reflectance: Fundamental basis between pigments and in vivo absorption

Algorithm development for remote sensing is focused on estimating the quantities of various optically-active constituents in the ocean. Various pigments present in phytoplankton play a role in determining the total absorption coefficient of phytoplankton, which is a key determinant of the spectral variability in remote-sensing reflectance. We utilize a reconstruction model to demonstrate the spectral variability of photoprotective and photosynthetic pigments (Sathyendranath et al. 1987, Bidigare 1990). There are differences in the weight-specific absorption spectra of various *in vitro* (extracted) phytoplankton pigments (Figure 5). The comparison of maximal peaks of mass-specific absorption spectra and center wavelengths on ocean color satellite platforms proves that coupling taxonomic and optical properties is quite challenging (Table 1). Future sensors with increased temporal and spectral resolution will provide more spectral information for development of algorithms. Future development of hyperspectral sensors would cover pigment-specific peaks at relatively high spatial and temporal resolution, which would allow retrieval of photosynthetic and photo-protective pigments.

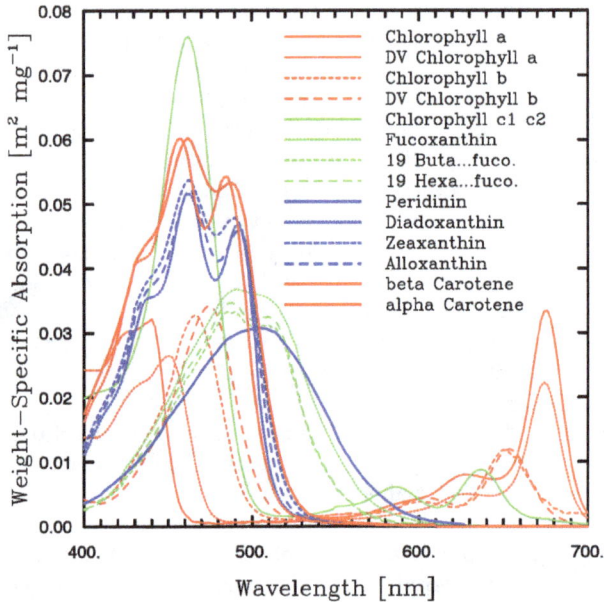

Fig. 5. Weight-specific in vitro absorption spectra of various pigments, $a_i^*(\lambda)$, derived from measuring the absorption spectra of individual pigments in solvent and shifting the maxima of the spectra according to Bidigare et al. (1990). Data obtained courtesy of A. Bricaud (Bricaud et al. 2004).

SeaWiFS	MODIS	MERIS	OCM2	OCTS	CZCS
412	412	413	415	412	443
443	443	443	442	443	520
490	469	490	491	490	550
510	488	510	512	516	670
555	531	560	557	565	
670	547	620	620	667	
	555	665			
	645	681			
	667	709			
	678				

Table 1. Past and present Ocean Color Sensors and their respective center wavelengths (nm).

The multi-spectral radiance measurements from the satellite sensor can be corrected for the influence of the atmosphere to yield remote-sensing reflectance. It has been demonstrated using various types of models (e.g. Morel & Prieur 1977; Sathyendranath & Platt 1988) that remote sensing reflectance, R_{rs} (λ), at a particular wavelength (λ) increases with the back-scattering coefficient of light at a wavelength, and decreases with the absorption coefficient:

$$R_{rs} (\lambda) \text{ is proportional to } b_b (\lambda) / [a (\lambda) + b_b(\lambda)]$$

where $b_b(\lambda)$ is backscattering coefficient and $a(\lambda)$ is the total absorption coefficient. The total absorption coefficient can be partitioned into its components:

$$a(\lambda) = a_w(\lambda) + a_{ph}(\lambda) + a_g(\lambda) + a_d(\lambda)$$

where the subscripts w, ph, g, and d relate to seawater, phytoplankton, gelbstoff and detritus, respectively. A similar equation can be wiritten for backscatter. The absorption coefficient of phytoplankton can be further broken down into contributions from various pigments to the total phytoplankton absorption.

$$a(\lambda) = \sum_{i=1}^{n} C_i a^*_{ci} (\lambda)$$

where c_i is the concentration of the individual pigments derived from HPLC analysis and $a_i^*(\lambda)$ is the absorption coefficient for the phytoplankton pigment.

As the pigment composition of phytoplankton changes with changes in community structure, the absorption spectra will be modified accordingly. Furthermore, it has been well demonstrated that the absorption spectra of phytoplankton are also modified by changes in the size structure of the community (Duysens 1956; Morel and Bricaud 1981; Sathyendranath et al. 1987; Moisan and Mitchell 1999). Therefore, pigment packaging significantly affect a*ph (λ, Figure 6). To the extent that size and function are related, size-dependent changes may also be related to changes in functional types. Remote sensing of PFTs must rely on such deviations in the spectral signatures of phytoplankton associated with changes in the phytoplankton community structure. Note that such approaches can be adapted to account for the contribution of mycosporine-like amino acids to absorption in the UV regions, which have some but limited taxonomic value (Moisan et al. 2011a, b).

4. Algorithm development of ocean color phytoplankton functional types

PFT algorithms have been developed to map the distribution of numerically- and ecologically-important organisms that demonstrate an anomalous relationship to a satellite-derived reflectance product, such as chlorophyll-a. Generally, algorithm development has capitalized on the backscattering or absorption characteristics of a particular phytoplankton group, or utilized sophisticated modeling efforts. In earlier years, a particular species/group was targeted, whereas more recently mathematically sophisticated algorithms have produced multiple products relating to phytoplankton community structure. Both approaches provide valuable data products, which enhance our understanding of the spatial and temporal variability in ecological structure and function. We present a generalized description of present algorithms for *in situ* data and satellite-derived data that describe phytoplankton taxonomic composition.

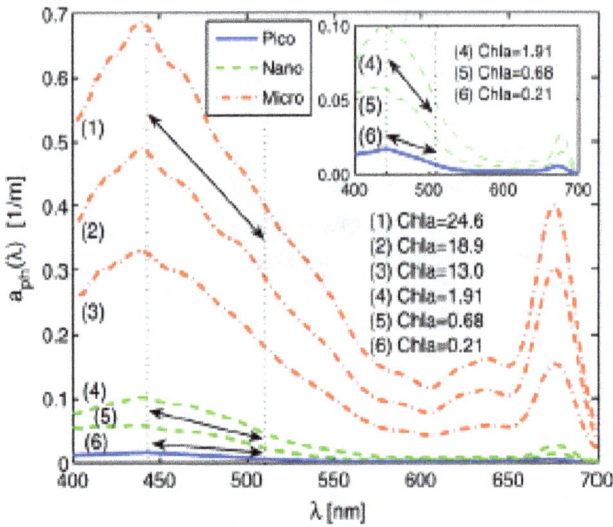

Fig. 6. Phytoplankton absorption spectra for a range of Chla (24.6, 18.9, 13.0, 1.91, 0.68, 0.21 mg m−3) and taxonomic size classes (pico, nano and micro) with decreasing slope from high to low aph (λ) and Chla; inset spectra of pico and nanoplankton at expanded range (Hirata et al 2008).

	Chlorophylls							Carotenes		Xanthophylls											
	Chl-*a*	Chl-*b*	DVChl-*a*	DVChl-*b*	Chlc1	Chlc2	Chlc3	β,ε-car	β,β-car	Allo	19_But	Diadino	Dino	Fuco	19_Hex	Lut	Neo	Per	Pras	Viola	Zea
Prokaryota																					
Synechococcus	–							–	–												–
Prochlorococcus			–	–				–	–												–
Eukaryota																					–
Rhodophyta	–							–													
Cryptophyta	–					–		–		•											
Chlorophyceae	–	–						•								•	–			–	
Prasinophyceae	–	–						•						•		–			–	–	
Eulgenophyta	–	–						–				–						•			
Eustigmatophyta	–					•		–												–	•
Bacillariophyta	–			–		•		–						–							
Dinophyta	–					•		–				–	•					–			
Prymnesiophyta	–			–	–	•		•			–			–							
Chrysophyceae	–			–		•		–			–			–							
Raphidophyceae	–			–		–		–			–			–							

Table 2. Distribution of pigments across the divisions and classes of the algae that are conventionally measured by high performance liquid chromatography. ● = major pigment (>10%); •=minor pigment (≤10%) of the total chorophylls or carotenoids. (Modified from Jeffrey & Vesk 1997)

5. Pigments as a diagnostic tool for identifying phytoplankton functional types

Because conventional light microscopy is labor-intensive and biased towards the larger phytoplankton, biological oceanographers have sought alternative approaches to derive information on community structure of the entire phytoplankton population. The most common method to achieve this is by measuring the phytoplankton pigment composition. Algal pigments have been routinely used as chemotaxonomic markers in studies of phytoplankton ecology and biogeochemistry. Chlorophyll-a, in its monovinyl and divinyl forms, is found within all photosynthetic microalgae and cyanobacteria, and is used as a universal proxy of autotrophic biomass. However, there also exist other chlorophylls and carotenoids that are routinely measured by High Performance Liquid Chromatography (HPLC) and may serve as class-specific markers.

It should be noted that most algal pigments are found in more than one class, while some are not necessarily present in every member of the same class. Given the complexity of the distribution of pigments between and within various phytoplankton taxa, a number of statistical and mathematical methods have been developed to partition the bulk pigment biomass into the various phytoplankton groups. In addition to using pigments as class-specific markers, pigments indices of phytoplankton classes have been developed based on certain diagnostic pigments that tend to dominate in a particular size class. In this section we briefly describe these approaches. For a more detailed review of the use of pigments as taxonomic markers see Wright & Jeffrey (2005).

Pigments in relation to taxa

The increased use of HPLC pigment analyses in ecological, remote sensing and biogeochemical studies of marine systems has created an impetus to try to extract from pigment data not only the bulk chlorophyll-a biomass of the entire population, but also the distribution of phytoplankton groups from the class to the genus level. The quantitative use of pigment markers to assess the relative contribution of taxa to chlorophyll-a biomass is still a relatively new and developing area of research. The task of estimating the contribution of various taxa to the pigment signature of natural samples is complex, given that many pigments are found in more than one algal class. The distribution of phytoplankton pigments across the various phytoplankton classes is illustrated in Table 1.

There are three basic approaches used to extract information on algal taxonomic composition based on pigments: multiple linear regression (Gieskes & Kraay 1983, 1986), inverse methods (Bidigare & Ondrusek 1996, Everitt et al. 1990, Letelier et al. 1993, Vidussi et al 2001) and matrix factorization analysis (Mackey et al. 1996, Wright & Jeffrey 2006).

Multiple linear regression analysis

The principal motivation behind the use of multiple linear regression studies of phytoplankton pigments was to establish a relationship between diagnostic pigments & the ubiquitous light harvesting pigment, chlorophyll-*a*, a universal proxy of phytoplankton biomass. One of the aims of the approach was to account for the fraction of phytoplankton cell biomass that may not be detected by conventional microscopy counts. Using multiple linear regression of diagnostic pigments against chlorophyll-a concentration, Gieskes & Kraay (1983) revealed that cryptophytes accounted for about half of the chlorophyll-a standing stock during a spring bloom study of the central North Sea, although these cells were not detected by microscope counts. The evidence was compelling, given that cryptophytes have a unique 'pigment fingerprint' (alloxanthin), and this carotenoid was the most abundant accessory pigment in the field samples. The explanation for the discrepancy between the pigment data and the cell counts was that small flagellates and photosynthetic ciliates that harbour alloxanthin are poorly preserved in Lugol's solution, resulting in underestimated cell abundances whereas the larger microphytoplankton (diatoms and dinoflagellates) are better preserved and more easily to identified. The authors also commented on the difficulty in assigning other pigments, such as fucoxanthin, to a particular class, given it is also found in other algal classes, such as prymnesiophytes and chrysophytes.

There are caveats to using the multiple linear regression approach to examine shifts in phytoplankton community structure (Gieskes et al. 1988). In the study by Gieskes et al. (1988), a clear influence of light adaptation on pigment composition was apparent in surface and deep populations, which were characterized by different ratios and intracellular concentrations of diagnostic pigments. As a result, poor taxon specificity of pigment markers such as fucoxanthin may mislead authors to conclude that microscopic counts provide superior quantitative estimates of taxon-specific biomass compared to pigment data.

Inverse methods

In order to estimate the contribution of various algal taxa to chlorophyll-a, inverse methods have been adopted to studies of phytoplankton pigments. The inversion method uses ratios of chlorophyll-a to accessory pigments based on literature values for representative taxa, (Everitt et al. 1990, Letelier et al. 1993, Bidigare et al. 1996). From these data a series of simultaneous equations for each algal group is derived to determine their contribution to total chlorophyll-a biomass. The method also addresses the problem of shared pigments between classes by subtracting out the contribution by other taxonomic groups for a particular pigment marker. The culture-based pigment: chlorophyll-a seed values are then modified by matrix inversion to the field data to find the least squares best solution. Constraints are also applied to the pigment ratios to avoid negative contributions by taxa.

The inverse method was used by Everitt et al. (1990) to demonstrate that nanophytoplankton species dominated the waters of the western Equatorial Pacific, and showed significant variation in community structure. Seasonal changes in the composition of phytoplankton groups (*Prochlorococcus*, other cyanobacteria, prymnesiophytes and chrysophytes) that comprise the deep chlorophyll maximum were examined for the Station ALOHA time series (Letelier et al. 1993). This approach has also been adopted by Vidussi et al. (2001) to assess the shift in community structure during a spring bloom in the northwestern Mediterranean Sea.

Matrix factorization of HPLC pigments (CHEMTAX)

CHEMTAX (CHEMical TAXonomy) is a MATLAB program that estimates the relative contributions of different phytoplankton taxa to the bulk chlorophyll-a concentration of a given sample (Mackey et al. 1996, Wright et al. 2000). Over the years this method has been used to separate out the phytoplankton community into taxonomic groups to at least the class, and in some cases, the genus or species level. To implement this algorithm an input pigment matrix is constructed based on knowledge of the kinds of taxa likely to be present in the study area and information on the cell-specific pigment composition and concentrations of these groups from culture studies. Unlike the multiple linear regression or inversion methods, which use only one or two diagnostic pigment markers for each taxa in the analyses, the CHEMTAX pigment matrixes make use of a wider range of accessory pigments. However, some criticisms within the community have occurred regarding its application because it assumes constant pigment ratios (Latasa 2007).

Given that the growth conditions of microalgae are known to influence the pigment composition of phytoplankton. The principal factors that led to chromatic adaptation of microalgae include the quality and spectral quantity of irradiance and nutrient status. In addition to the effect of taxa, growth conditions of microalgae are known to influence the pigment composition of phytoplankton. The principal factors that lead to chromatic adaptation of microalgae include the quality and spectral quantity of irradiance and nutrient status. Variability in pigment composition caused by these factors is seen both in the change in the intracellular concentration of the cells as well as the ratios of accessory pigments (such as the relative concentrations of photosynthetic or photoprotective pigments). Therefore, similar to inversion methods, it is recommended that sample be divided according to light regime before running the matrix factorization.

Other inverse methods focus on the absorption term related to phytoplankton as has been estimated using a wide range of algorithms. An alternative approach proposed by Bidigare et al. (1989 & 1990) utilizes the absorption signatures of the various pigments to reconstruct the unpackaged absorption spectra of marine phytoplankton. Using the pigment outputs for various algal groups from CHEMTAX, an unpackaged absorption spectra for a particular phytoplankton group can be constructed when regarding the absorption spectra solely as a function of the chlorophyll-a concentration and the chlorophyll-a specific absorption spectra. Several inverse modeling capabilities are available that can make use of these absorption formulations for other pigments. Moisan et al. (2011ab) have developed a matrix inverse modeling technique that produces photoprotective and photosynthetic pigments with relatively accurate results ($r^2 > 0.80$) using HPLC pigments as validation (Figure 7).

Modeling pigments in relation to cell size fractions

Fortuitously, the major taxa of marine phytoplankton tend to fall within the three size classes originally proposed by Sieburth (1979), the micro- (>20µm), nano- (2-20µm) and picophytoplankton (<2µm). Thus pigment markers of particular taxonomic groups can provide insight into the relative contribution of a particular size class to the pigment biomass. The use of pigment markers to derive a size index of phytoplankton populations was proposed by Claustre (1994). The approach used seven diagnostic pigments to obtain an index of the relative contribution of microphytoplankton to pigment biomass integrated over the watercolumn. The two marker pigments ascribed to the microphytoplankton size class are fucoxanthin, associated with the diatom fraction, and peridinin, representing the dinoflagellate fraction. The index also used markers pigments that tend be dominant accessory pigments in natural populations: 19'-hex, 19'-but, alloxanthin, zeaxanthin and chlorophyll-b. Thus by summing the watercolumn-integrated concentration of diagnostic pigments associated with the microphytoplankton fraction, and dividing by the sum of all water-column integrated accessory pigments, an index of the relative contribution of microphytoplankton to total integrated pigment biomass can be calculated according to the equation:

$$F_p = \frac{(\sum fuco + \sum perid)}{\sum fuco + \sum perid + \sum 19'hex + \sum 19'but + \sum allox + \sum zeax + \sum chlb}.$$

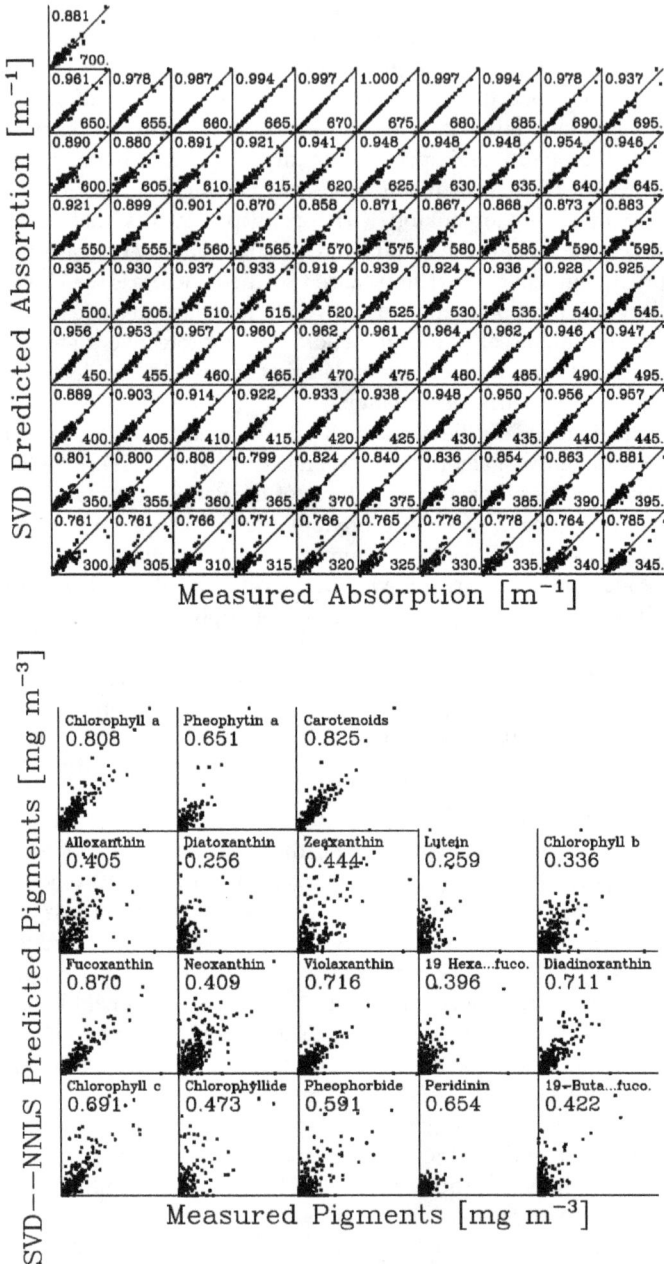

Fig. 7. (top) Predicted absorption versus measured absorption (Algorithm II) for values every 10 nm ranging from 300 nm to 700 nm using matrix inversion methods using a single value decomposition. (bottom) Predicted pigments versus measured HPLC pigments for use in photosynthesis and photoprotection pigments. See Moisan et al. (2011 ab for more details).

A F_p value equal to one means that fucoxanthin and peridinin made up the entire integrated diagnostic pigment concentration, whereas a value of zero means that neither indicator pigment of microphytoplankton were found in the water column.

The size-based approach was refined by Vidussi et al. (2001) to obtain a pigment-derived indices of the contribution of the three size classes of phytoplankton (micro- (>20mm), nano- (2-20mm) and picophytoplankton (<2mm) to integrate pigment biomass. To determine the relative contribution of the three size classes to chlorophyll-a biomass, Uitz et al. (2006) combined the multiple linear regression analysis of Gieskes et al. (1983, 1988) with the size-specific index of Vidussi et al. (2001) to derive estimates of all three size fractions and used not only to derive an index of the presence of the three size classes, but also to estimate their relative contribution to total chlorophyll concentration, similar to the inversion methods and CHEMTAX. It is important to note that in this approach integrated rather than discrete pigment concentrations are used. The approach to using size as a description of the phytoplankton community is becoming a well-accepted PFT, which provides an important link between the fields of remote sensing and marine ecology.

6. Modeling trophic structure through correspondence between taxa, size and function

The relationship between phytoplankton size and biogeochemical function has been well established. Diatoms (silicifiers) tend to be large, and their negative buoyancy leads to their significant contribution to export production (Sarmiento & Gruber 2004). Picophytoplankton, on the other hand, are more important in microbial food webs were rapid recycling of organic matter in the surface ocean leads to a reduction in biogenic export to the deep ocean. Recent molecular studies on the picocyanobacterium *Prochlorococcus*, which dominates the chlorophyll-a biomass in the oligotrophic gyres, have revealed that many strains are unable to utilize nitrate as a nutrient source, and consequently can be considered obligate participants in regenerative production.

The use of pigment markers to provide insight on the trophic structure of marine ecosystems was examined by Claustre (1994). He proposes that the contribution of microphytoplankton (in particular diatoms) to the integrated concentration of diagnostic pigments, which can be considered to be a indicator of the *f*-ratio and hence new production. In addition, information on the contribution of various phytoplankton taxa to chlorophyll-a standing stock has led to the generation of taxon-specific maps of primary production using global relationships on the relationship between surface chlorophyll-a concentration and the ratio of integrated diagnostic pigments to integrated chlorophyll-a concentration (Uitz et al. 2006). Although the simple pigment indices invoked by Claustre (1994) and Uitz et al. (2006) cannot resolve instances where other taxa contribute to the diagnostic pigment markers used in their analysis, it provides a useful first attempt at obtaining information on the global distribution of phytoplankton size classes (Uitz et al. 2006), the environmental factors that control their biogeography (Bouman et al. 2003) and their relative contribution to marine productivity (Bouman et al. 2005, Uitz et al. 2006).

7. Societal benefits of ocean color approaches to algorithm development for key phytoplankton functional types: Fisheries and habs applications

A fundamental goal of phytoplankton biogeography is to describe how PFTs are distributed spatially and temporally and how these patterns relate to processes that control primary production/new production. Several theories have been postulated about what governs biological diversity, which stabilizes community dynamics. Future satellites with increased spectral resolution will allow for algorithms of different proxies for diversity whether the products are pigments, size spectra, or biogeochemical indicators. Such phytoplankton diversity indicators will lead to a better understanding of the response of marine ecosystems to climate and human activities (Platt et al. 2008, 2009). To date, the scientific community has been able to describe phytoplankton community structure based on broad size classes, pigmentation, probability of occurrence, or some other index of presence or absence. We have focused on two successful applications including fisheries management and harmful algal blooms.

a. Fisheries applications

Satellite ocean color application on broad scales allow for PFTs, namely diatoms, to be put into an ecological context regarding El Niño events in the Humboldt ecosystem (Jackson et al. 2011, Figure 8). A locally tuned algorithm for detecting diatom distribution allowed for critical observations in shifts in the size structure of the phytoplankton community, which provide a food source for anchovies. Changes in carbon structure will affect food stress on the fish populations and appears to influence landings in the following year. Mapping the distribution and abundance of phytoplankton using remotely sensed data assists in the creation of targeted and quantitative fishing strategies. Satellites can provide strategies and tools together with modeling and other satellite. Both ocean color and other satellite data, coupled with models, can provide resource managers the information of food web components (other than the chlorophyll-a product), which may be used to direct shipboard sampling on regional scales.

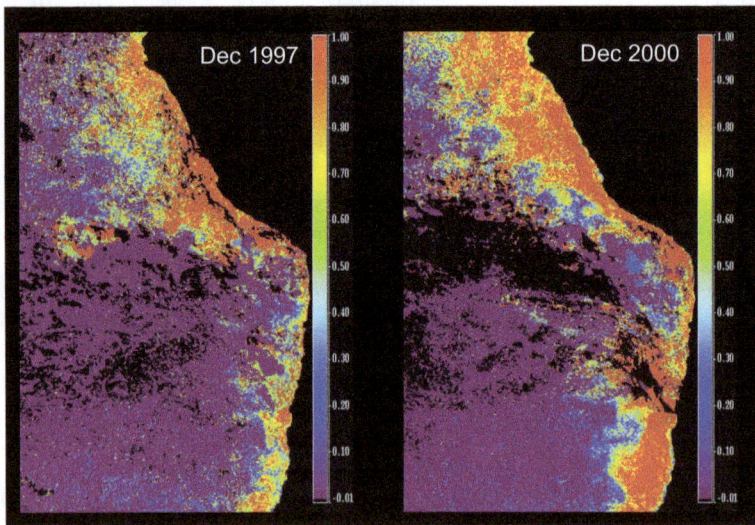

Fig. 8. Climate-induced changes in the spatial distribution of diatoms in the Humboldt system. The figure shows a marked reduction in the presence of diatoms during the 1997 El Niño event compared with the control period (December 2000), which represents typical oceanic conditions for the study region. The color bar represents the fraction of pixels identified as diatoms. Modified from Jackson et al. (2011).

b. Monitoring and tracking through the identification of harmful algal blooms

Ocean Color satellites have had a valuable application for many marine biological processes such as red tides and harmful algal blooms (HABs, reviewed in Stumpf & Tomlinson 2005). HABs are not only harmful to the marine ecosystem but substantive economic losses occur in coastal regions worldwide and the need for remote detection of blooms is critical (Anderson et al. 1997, Schofield et al. 1999, Sellner et al. 2003). HAB events threaten human health, living marine resources, and ecosystem health. Their occurrence has increased in frequency, duration, and severity over the last several decades (Figure 9). Current monitoring efforts of toxic cells and toxin levels in shellfish tissue are relatively slow and lack the synoptic coverage of satellites (Schofield et al. 1999). Nevertheless, HAB monitoring can occur at unprecedented scales with detailed shipboard and drifters to predict arrival, transport time, and possibly toxicity of the bloom (Figure 9, Schofield et al. 1999). Knowledge such as this allows the public to know location, timing of the bloom and close any fisheries that are impacted economically. An excellent example of this type work has been conducted on the mesoscale blooms of *Karenia brevis* in the Gulf of Mexico with ocean color derived chlorophyll a and solar stimulated fluorescence (Stumpf et al. 2003, Hu et al. 2005, Figure 9). Most HAB studies have focused on monitoring with chlorophyll-a as a biomass indicator, however, photo-protective pigments such UV mycosporine-like amino acids and gyroxanthin from other non-toxic organisms (Kahru & Mitchell 1998, Moisan et al. 2011). Several countries all over the world have successfully monitored HABs over broad spatial scales with supplemented satellite imagery and aircraft monitoring in the Gulf of Mexico that allows for tracking of the bloom by satellite and drifters to notify the public.

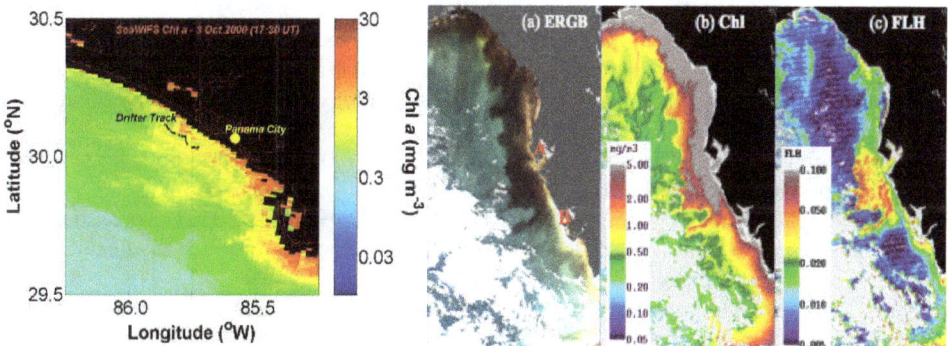

Fig. 9. Left figure: Harmful algal bloom image in the Gulf of Mexico while using drifters and ocean color satellite imagery that allows for tracking for responses to the public (Lohrenz et al., unpub.). Right figure: Fluorescence Line Height is a better indicator than the satellite-based Chl for chlrophyll-a biomass in CDOM-rich waters. This image is for a *Karenia brevis* bloom (modified from Hu et al. 2005, Courtesy of Dr. Chuanmin Hu). Please note that images from different years.

8. Future directions of algorithm development using ocean color

Over the past few decades, the dominant phytoplankton biomass product, chlorophyll-a, has been observed over a long time period on global scales. Recent investigations using ecological provinces have indicated additional regional variability and have provided information based on marine ecology and biogeochemistry in the broader context of climate change (Falkowski et al. 2004). Although validated satellite data is not a substitute for ship-based sampling, *in situ* information about the physical/biological regimes is still required to produce sound ecological products. Hence, the future lies in the combined utilization of *in situ* data, remote sensing, and modeling. The remote sensing of PFTs in the ocean will bring about a greater understanding of how phytoplankton community structure affects climate with biologically-produced greenhouse gases. In contrast, PFTs will contribute to an overall understanding of global marine biodiversity and that knowledge will provide insight on the relationship between ecosystem stability and ocean biogeochemistry.

9. Acknowledgements

We would like to thank Rich Landa and Rachel Steinhardt for their excellent assistance. Many thanks to Chuanmin Hu, and Steve Lohrenz. We thank our reviewers for their excellent and generous comments on our earlier work. Our work was partially funded by the National Atmospheric and Space Administration's Biodiversity Program 05-TEB/05-0016, the National Oceanic and Atmospheric Administration (NOAA) and the National Environmental Research Council (NERC) UK.

10. References

Aiken, J., Pradhan, Y., Barlow, R., Lavender, S., Poulton, A., Holligan, P., & Hardman-Mountford, N. (2009). Phytoplankton pigments and functional types in the Atlantic Ocean: A decadal assessment, 1995-2005. *Deep Sea Res* 56: 899-917

Alvain, S., Moulin, C., & Dandonneau, Y. (2005). Remote sensing of phytoplankton groups in case 1 waters from global SeaWiFS imagery. *Deep Sea Res* 52: 1989–2004

Anderson, D.M. & Garrison, D.J. (editors) (1997). The ecology and oceanography of harmful algal blooms. *Limnol Oceanogr* 42: 1009–1305

Balch, W., Holligan, PM., Ackleson, SG., & Voss, KJ. (1991). Biological and optical properties of mesoscale coccolithophore blooms. *Limnol Oceanogr* 36: 629–643

Balch, W., Kilpatrick, K.A., Holligan, P.M., & Trees, C. (1996). The 1991 coccolithophore bloom in the central north Atlantic I—Optical properties and factors affecting their distribution. *Limnol Oceanogr* 41: 1669 – 1683

Bidigare, R.R. (1989). Potential effects of UV-B radiation on marine organisms of the southern-ocean-Distributions of phytoplankton and krill during austral spring photochemistry and photoplankton and krill during austral spring. *Photochem Photobiol* 50: 469-477.

Bidigare, R.R., Ondrusek, M.E., Morrow J.H. and Kiefer D.A. (1990) In-vivo absorption properties of algal pigments. Proc. SPIE 1302: 290-302.

Bidigare, R.R., & Ondrusek, M.E. (1996). Spatial and temporal variability of phytoplankton pigment distributions in the central equatorial Pacific Ocean. *Deep Sea Res.* II 43: 809-833

Bouman, H.A., Platt, T., Sathyendranath, S., Li W.K.W., Stuart V. & Fuentes-Yaco C. (2003). Temperature as indicator of optical properties and community structure of marine phytoplankton: implications for remote sensing. *Mar Ecol Prog Ser* 258: 19-30

Bouman, H.A., Platt, T., Sathyendranath S. & Stuart V. (2005). Dependence of light-saturated photosynthesis on temperature and community structure. *Deep Sea Res.* I 52: 1284-1299

Bricaud A.H., Claustre J. Ras, and Oubelkheir K. (2004) Natural variability of phytoplanktonic absorption in oceanic waters: Influence of the size structure of algal populations. *J Geophys Res. 109:* C11010.

Chavez F.P., Strutton, P.G., Friederich, G. E., Feely, R. A., Feldman, G. C., Foley, D. G., & McPhaden, M. J. (1999) Biological and Chemical Response of the Equatorial Pacific Ocean to the 1997-98 El Niño. *Science* 286: 2126-2131

Chesson, P.L., & Case, T.J. (1986). Nonequilibrium community theories: chance, variability, history, and coexistence. In J. Diamond and T. Case (eds), "Community Ecology," Harper and Row, pp. 229-239

Claustre, H. (1994). The trophic status of various oceanic provinces as revealed by phytoplankton pigment signatures. *Limnol Oceanogr* 39 (5): 1206-1210

Connell, J.H. (1978). Diversity in rain forests and coral reefs. *Science* 199: 1302-1310

DeAngelis, D. & Waterhouse, J.C. (1987). Equilibrium and nonequilibrium concepts in ecological models. *Ecol Monogr* 57: 1-21

Duysens, L.M.N. (1956). The flattening effect of the absorption spectra of suspensions as compared to that of solutions. *Biochem Biophys* Acta 19: 1-12

Everitt, D.A., Wright, S.W., Volkman, J.K., Thomas, D.P., Lindstrom, E.J. (1990). Phytoplankton community compositions in the western equatorial Pacific determined form chlorophyll and carotenoid pigment distributions. *Deep Sea Res.* 37: 975-997

Falkowski, P.G., Katz, M.E., Knoll, A.H., Quigg, A., Raven J.A., & Schofield O. (2004). The Evolution of Modern Eukaryotic Phytoplankton. *Science* 305: 354-360

Falkowski, P.G., & Oliver, M.J. (2007). Mix and match: How climate selects phytoplankton. *Nat Rev Microbiol* 5: 813-819

Falkowski, P.G., & Raven, J.A. (1997). Aquatic Photosynthesis. Blackwell, Oxford, UK.

Fauchald, K., & Jumars, P.A. (1979). The diet of worms: a study of polychaete feeding guilds. - *Annu Rev Oceanogr Mar Biol* 17: 193-284

Garver, S. & Siegel, D.A. (1997). Inherent optical property inversion of ocean color spectra and its biogeochemical interpretation. I. Time series from the Sargasso Sea, *J Geophys Res* 102: 18607-18625

Gieskes, W.W.C., & Kraay, G.W. (1983). Dominance of Cryptophyceae during the phytoplankton spring bloom in the central North Sea detected by HPLC analysis of pigments. *Mar. Biol.* 75: 179-185

Gieskes, W.W. & Kraay, G.W. (1986). Floristic and physiological differences between the shallow and the deep nanophytoplankton community in the euphotic zone of the open tropical Atlantic revealed by HPLC analysis of pigments. *Mar. Biol.* 91: 567-576

Gieskes, W.W.C., Kraay, G.W., Nontji, A., Setiapermana & Sutomo, D. (1988). Monsoonal alternation of a mixed and a layered structure in the phytoplankton of the euphotic zone of the Banda Sea (Indonesia): A mathematical analysis of algal pigment fingerprints. *Neth J Sea Res* 22: 123-137

Hallegraeff, G. M. 2010. Ocean climate change, phy- toplankton community responses, and harmful algal blooms: A formidable predictive challenge. Journal of Phycologia 46: 220-235.

Hasle, G.R. (1976). The biogeography of some marine planktonic diatoms. *Deep Sea Res* 23: 319-338

Hirata, T., J. Aiken, Hardman-Mountford N., Smyth T.J., & Barlow R.G. (2008) An absorption model to determine phytoplankton size classes from satellite ocean colour. Remote Sensing of the Environment 112: 3153-3159.

Hu, C., Muller-Karger, FE., Taylor, C., Carder, K.L., Kelble, C., Johns, E., & Heil, C. (2005). Red tide detection and tracing using MODIS fluorescence data: A regional example in SW Florida coastal waters. *Remote Sens Environ* 97: 311-321

Hu, C., Cannizzaro, J., Carder, KL., Muller-Karger, F.E. & Hardy R. (2010). Remote detection of *Trichodesmium* blooms in optically complex coastal waters: Examples with MODIS full-spectral data. *Remote Sens. Environ.* 114:2048-2058

Hutchinson, G.E. (1961). The paradox of the plankton. *American Naturalist* 95: 137–147

Iglesias-Rodríguez MD, CW Brown, SC Doney, J Kleypas, D Kolber, Z Kolber, PK Hayes and PG Falkowski. (2002). Representing key phytoplankton functional groups in ocean carbon cycle models: Coccolithophorids, *Global Biogeochem. Cycles*, 16(4), 1100, doi: 10.1029/2001GB001454, 200

Irigolen, X., Hulsman, J., & Harris, R.P. (2004). Global biodiversity patterns of marine phytoplankton and zooplankton. *Nature* 429: 863-7

Jackson, T., Bouman, H.A., Sathyendranath, S., & E. Devred. (2011). Regional-scale changes in diatom distribution in the Humboldt upwelling system as revealed by remote sensing: implications for fisheries. ICES Journal of Mrine Science 68: 729-736.

Jeffrey, S.W. & Vesk, M. (1997). Introduction to marine phytoplankton and their pigment signatures. In Jeffrey SW, Mantoura RFC and Wright SW (eds), Phytoplankton Pigments in Oceanography: A Guide to Advanced Methods. SCOR-UNESCO, Paris, pp. 37–8

Kahru, M. & Mitchell, B.G. (1998). Spectral reflectance and absorption of a massive red tide off Southern California. *Journal of Geophysical Research* 103: 21601–21609

Keller, M.D., Bellows, W.K. & Guillard R.R.L. (1989). Dimethyl sulfide production in marine phytoplankton. In: Biogenic Sulfur in the Environment. Symp. Ser. 393 (Saltzman, E.S. and Cooper, W.J., Eds.), pp. 167–182. American Chemical Society, Washington, DC.

Latasa, M. (2007). Improving estimates of phytoplankton class abundances using CHEMTAX. *Mar Ecol Prog Ser* 329: 13-21

Letelier, R.M., Bidigare, R.R., Hebel, D.V., Ondrusek, M., Winn, C.D. & Karl, D.M. (1993). Temporal variability of phytoplankton community structure based on pigment analysis. *Limnol Oceanogr* 38: 1420-1437

Longhurst, A.R. (2007). *Ecological geography of the sea* (2). Academic Press. 542 pp.

MacKey, M.D., MacKey, D.J., Higgins, H.W. & S.W. Wright. (1996). CHEMTAX—a program for estimating class abundances from chemical markers: application to HPLC measurements of phytoplankton. *Mar. Ecol. Prog. Ser.* 144: 265–283

Martin, S. (2004). *An Introduction to Ocean Remote Sensing.* Cambridge, UK: Cambridge University Press. 426 pp.

McClain, C. (2009). A decade of satellite ocean color observations. *Annu Rev Mar Sci* 1: 19–42.

Miller, R.L., Del Castillo, C.E., & McKee, B.A. (2005). *Remote Sensing of Coastal Aquatic Environments: Technologies, Techniques and Applications: Vol.7. Remote Sensing and Digital Image Processing.* Dordrecht, The Netherlands: Springer. 345 pp.

Mitchell BG. 1994. Coastal zone color scanner retrospective. J. Geophys. Res. 99: 7291-92.

Moisan, T.A., & B.G. Mitchell. (1999). Photophysiological acclimation of *Phaeocystis antarctica* Karsten under light limitation. *Limnol and Oceanogr* 44: 247-258

Moisan, T.A.H., Moisan, J.R., & M. Linkswiler. (2011). Retrieving phytoplankton pigments through a mathematical inversion method. Submitted to J Cont Res

Moisan, J.R., Moisan, T.A.H. & Linkswiler M.A. (2011). Estimating Phytoplankton Pigment Concentrations from Phytoplankton Absorption Spectra. J. Geophys. Res. 116: 0148-0227

Morel, A. & L. Prieur. (1977). Analysis of variation in ocean color. *Limnol. Oceanogr.* 22: 709-722

Morel, A. & Bricaud, A. (1981). Theoretical results concerning light absorption in a discrete medium, and application to specific absorption of phytoplankton. *Deep Sea Res* Part A. Oceanographic Research Papers 28 (11): 1375-1393

Mueller, J.L., Fargion, G.S., & McClain, C.R. (2004). *Ocean optics protocols for satellite ocean color sensor validation, revision 5: Biogeochemical and bio-optical measurements and data analysis protocols: Vol. 5; Vol. 211621. NASA technical memorandum.* Goddard Space Flight Center. 36 pp.

Nair, A., Sathyendranath, S., Platt, T., Morales, J., Stuart, V., Forget, M-H.E., Devred, E. and Bouman H. (2008), Remote sensing of phytoplankton functional types, Remote Sens. Environ 112: 3366 – 3375.

Platt, T. & Sathyendranath, S. (2008). Ecological indicators for the pelagic zone of the ocean from remote sensing. *Rem Sens Environ* 112: 3426–3436

Platt, T., White, I.I.I., GN, Zhai L. et al. 2009. The phenology of phytoplankton blooms: Ecosystem indicators from remote sensing. *Ecological Modelling* 21: 3057–3069

Reynolds, C.S., Huszar, V., & Kruk, C. 2002. Towards a functional classification of the freshwater phytoplankton. *J Plankt Res* 24: 417-428

Richardson, L.L., LeDrew, E. (2006). *Remote sensing of aquatic coastal ecosystem processes: science and management applications: Vol. 9. Remote sensing and digital image processing.* Dordrecht, The Netherlands: Springer. 324 pp.

Robinson, I.S. (2010). *Understanding the Oceans from Space: The Unique Applications of Satellite Oceanography.* Springer. 638 pp.

Rutherford, E.S., Rose, K.A., Mills, E.L., Forney, J.L., Mayer, C.M, & Rudstam, L.G. (1999). Individual-based model simulations of a zebra musssel (*Dreissena polymorpha*) induced energy shunt on walleye (*Stizostedion vitreum*) and yellow perch (*Perca flavescens*) populations in Oneida Lake, New York. *Can. J Fish Aquat Sci* 56: 2148-2160

Sarmiento, J.L. & N. Gruber. (2004). Ocean Biogeochemical Dynamics. Princeton Univ. Press, Princeton, N. J.

Sathyendranath, S., and Platt, T. (1988). The Spectral Irradiance Field at the Surface and in the Interior of the Ocean: A Model for Applications in Oceanography and Remote Sensing, *J. Geophys. Res., 93:* 9270–9280.

Sathyendranath S., Lazzara, L., & Prieur, L. (1987). Variations in the Spectral Values of Specific Absorption of Phytoplankton. *Limnol Oceanogr* 32 (2): 403-415

Sathyendranath, S., Watts, L., & Devred, E. (2004). Discrimination of diatoms from other phytoplankton using ocean-colour data. *Mar. Ecol. Prog. Ser.* 272: 59–68

Schofield, O., Bergmann, T., Grzymski, J. & Glenn, S. (1999). Spectral fluorescence and inherent optical properties during upwelling events off the coast of New Jersey. *SPIE Ocean Optics* XIV 3:60–67

Sieburth, J. MCN. 1979. *Sea microbes.* New York: Oxford University Press 491 pp.

Sellner, S.G., Sellner, K.G., & Brownlee, E.F. (2003). Effects of Barley Straw (*Hordeum vulgare*) on Freshwater and Brackish Phytoplankton and Cyanobacteria. *J Applied Phycology* 15: 525-531

Smayda, T.J. & Reynolds, C.S. (2003). Strategies of marine dinoflagellate survival and some rules of assembly. *J Sea Res* 49: 95–106

Stumpf, R.P., Culver, M.E., Tester, P.A., Tomlinson, M., Kirkpatrick, G.J. & B.A. Pederson. (2003). Monitoring *Karenia brevis* blooms in the Gulf of Mexico using satellite ocean color imagery and other data. *Harmful Algae News* 35: 1–14

Stumpf, R.P. & Tomlinson, M.C. (2005). Remote sensing of harmful algal blooms. In Remote Sensing of Coastal Aquatic Environments, ed. RL Miller, CE Del Castillo, BA McKee, pp. 277–96. AH Dordrecht, The Netherlands: Springer. 347 pp.

Subramanian, A., & Carpenter, E.J. (1994). An empirically derived protocol for the detection of blooms of the marine cyanobacterium *Trichodesmium*using CZCS imagery. *Int. J. Remote Sensing* 158: 1559–1569

Subramanian, A., Carpenter, E.J., Karentz, P.G. & Falkowski, PG. (1999a). Optical properties of the marine diazotrophic cyanobacteria *Trichodesmium* spp. I. Absorption and spectral photosynthetic characteristics. *Limnol. Oceanogr.* 44: 608-617

Subramanian, A., Carpenter, E.J. & Falkowski, P.G. (1999b) Optical properties of the marine diazotrophic cyanobacteria *Trichodesmium* spp. II. Reflectance model for remote sensing. Limnol. Oceanogr. 44: 618-627

Totterdell, I.J., Armstrong, R.A., & Drange, H. (1993).Trophic resolution. In Evans, G. T. and Fasham, M. J. R. (eds), Towards a Model of Ocean Biogeochemical Processes. NATO ASI, Vol. I 10. Springer-Verlag, Berlin, pp. 71–92

Uitz, J., Claustre, H., Morel, A., Hooker, & S.B. (2006). Vertical distribution of phytoplankton communities in open ocean: An assessment based on surface chlorophyll. *J. Geophys. Res.* 111:C08005

Vidussi, F., Claustre, H., Manca, B.B., Luchetta, A., & Marty, J.C. (2001). Phytoplankton pigment distribution in relation to upper thermocline circulation in the eastern Mediterranean Sea during winter. *J Geophys Res* 106 (C9): 19,939–19,956

Vila, M., Giacobbe, M.G., MasÓ M., Gangemi E., Penna A., Sampedro N., Azzaro F., Camp J., & Galluzzi L. (2005). A comparative study on recurrent blooms of Alexandrium minutum in two Mediterranean coastal areas. *Harmful algae* 4: 673-695

Worm, B., Sandow, M., & Oschlies, A. (2005). Global patterns of predator diversity in the open oceans. *Science* 309: 1365–1369

Wright, S.W. & RL van den Enden. (2000). Phytoplankton community structure and stocks in the East Antarctic marginal ice zone (BROKE survey, January-March 1996) determined by CHEMTAX analysis of HPLC pigment signatures. *Deep Sea Research Part II: Topical Studies in Oceanography* 47 (12-13): 2363-240

Wright, S.W. & Jeffrey, S.W. (2006). Pigment markers for phytoplankton production. *Handbook Env Chem* 2N: 71-104

Part 3

Fires

7

Advances in Remote Sensing of Post-Fire Vegetation Recovery Monitoring – A Review

Ioannis Gitas[1], George Mitri[2],
Sander Veraverbeke[3,4] and Anastasia Polychronaki[1]
*[1]Laboratory of Forest Management and Remote Sensing,
Aristotle University of Thessaloniki, Thessaloniki,
[2]Biodiversity Program, Institute of the Environment,
University of Balamand and Department of Environmental Sciences,
Faculty of Science, University of Balamand,
[3]Department of Geography, Ghent University, Ghent
[4]Jet Propulsion Laboratory, California Institute of Technology,Pasadena, CA,
[1]Greece
[2]Lebanon
[3]Belgium
[4]USA*

1. Introduction

Accurate information relating to the impact of fire on the environment and the way it is distributed throughout the burned area is a key factor in quantifying the impact of fires on landscapes (van Wagtendonk et al. 2004), selecting and prioritizing treatments applied on site (Patterson and Yool 1998), planning and monitoring restoration and recovery activities (Jakubauskas 1988; Jakubauskas et al. 1990; Gitas 1999) and, finally, providing baseline information for future monitoring (Brewer et al. 2005).

In order to assess economic losses and ecological effects, post-fire impact assessment requires precise information on extent, type and severity of fire (short-term impact assessment) as well as on forest regeneration and vegetation recovery (long-term impact assessment). Assessing the short-term impact is related to the study of fire behaviour, fire suppression and fire effects while the long-term impact assessment of fires is needed in order to establish post-fire monitoring management and introduce restoration and recovery activities.

As fire sizes increase and time becomes a constraining factor, traditional methods to assess post-fire impact on vegetation have become costly and labour-intensive (Bertolette and Spotskey 2001; Mitri and Gitas 2008). Given the extremely broad spatial expanse and often limited accessibility of the areas affected by fire, satellite remote sensing is an essential technology for gathering post-fire related information in a cost-effective and time-saving manner (Smith and Woodgate 1985; Chuvieco and Congalton 1988; Jakubauskas et al. 1990; White et al. 1996; Patterson and Yool 1998; Beaty and Taylor 2001; Escuin et al. 2002).

In addition, the development of high spatial and spectral resolution remote sensing instruments, both airborne and spaceborne, as well as advanced image analysis techniques have provided an opportunity to evaluate patterns of forest regeneration and vegetation recovery after wildfire.

The aim of this chapter is to review the role of Remote Sensing (RS) in post-fire monitoring of vegetation recovery. More specifically, traditional and advanced methods and techniques that have been so far employed to monitor vegetation regrowth after fire by RS will be reviewed and future trends will be identified.

More specifically, Part 2 deals with the ecological framework of the effects of fire on the ecosystem, Part 3 describes the methods and techniques that have so far been employed to estimate forest regeneration and vegetation recovery by means of field survey and by RS, Part 4 focuses on the advances in RS of post-fire vegetation monitoring, Part 5 emphasises on future trends in RS of post-fire monitoring, and Part 6 outlines the main conclusions of the chapter.

2. Ecological framework

Fire is an integral part of many ecosystems (Trabaud 1994). However, in recent decades the general trend in the number of fires and in the surface burned has increased spectacularly. This increase can be attributed to: (a) land-use changes (Rego 1992; García-Ruiz et al. 1996), and (b) climatic warming (Maheras 1988; Torn and Fried 1992; Amanatidis et al. 1993; Piñol et al. 1998; EPA 2001).

The **ecological effects** of forest fires are very diverse. This is not only because of the complexity of plant communities and the interface of disturbances such as grazing and cutting with burning, but also because of the different responses to the type, duration and intensity of fire, the season in which it occurs and its frequency (Le Houerou 1987).

The effect of fire on forested ecosystems can range from disastrous to beneficial. Harmful effects include changes in the physical, chemical and biological properties of soils; benefits are the removal of accumulated fuels, an increase in water yield, the control of insects and diseases, the preparation of seedbeds, and the release of seeds from serotinous cones (Kozlowski 2002).

Specifically, the main effects of fire on soils are the loss of nutrients during burning and the increased risk of erosion after burning. The latter is in fact related to the regeneration traits of the previous vegetation and to environmental conditions (Pausas et al. 1999). Large fires that produce a greater number of intensely burned patches can favour the colonization of invasive, fire tolerant species at the expense of rare/endemic species that are less tolerant to post-fire conditions. Thus biodiversity is also affected by fire (Dafis 1990). Climate change predictions and repercussions of forest fires on erosion, water yield and desertification further add to these threats (Moreno et al. 1990).

The effects of forest fires on vegetation are the most evident due to plants vulnerability to burning leading to permanent changes in the composition of the vegetation community, decreased vegetation cover, biomass loss and the alteration of landscape patterns (Perez-Cabello et al. 2009). However, forest fire is the major force in the biological evolution of biota such as in the Mediterranean (Naveh 1991). For instance, most Mediterranean plant species exhibit effective regeneration mechanisms for overcoming the immediate effects of fire (Mooney and Hobbs 1986). Plant species mechanisms can be passive (e.g. thick insulating bark), or active (e.g. re-sprouting from underground storage organs and seedlings from fire protected seeds), which may lead to a rapid process of vegetation cover with similar characteristics to the previous communities (Perez-Cabello et al. 2009). However, the response of vegetation to fire is very complex and it is not easy to generalize because of the large number of factors that can affect the regeneration process (e.g. topographic-climatic influences, plant composition, topographic parameters, soil characteristics, etc.).

Natural regeneration (the regrowth of lost or destroyed parts or organs) of disturbed mature forests to a pre-disturbance condition is often slow, unpredictable, and fraught with difficulties (Kozlowski 2002). The natural regeneration of disturbed forest stands typically occurs in four sequential stages:

1. A stand initiation and regeneration stage: interactions among propagules (including seeds in seed banks and those dispersed into a site as well as sprouting or layering of residual trees) and soil and climatic conditions.
2. A thinning or stem exclusion stage: canopy closes and mortality of trees accelerates, competition for resources (light, water, mineral nutrients), leaf area index reaches its maximum.
3. A transition or understory regeneration stage: death of some overstorey trees, resulting in the formation of gaps in the canopy and the reintroduction of understorey vegetation.
4. A steady-state or old-growth stage: the continuation of a series of successional stages beginning in the previous stage and which may culminate in an old-growth climax forest (Oliver 1981; Oliver and Larson 1996; Kozlowski 2002).

Plant species react to fire through different morphological and physiological traits (Perez-Cabello et al. 2009). Some can survive fires due to protected plant tissues sprouting (e.g. underground storage organs) and high growth rates after fire and others can rapidly establish seedlings (Buhk et al. 2007). Resprouting ability is a very common survival strategy. Post-fire buds respond by producing new shoots (Miller 2000) and this engenders a rapid return to pre-fire conditions. Differently, seedling establishment may originate from on-site seeds or from off-site seed sources if favourable environmental conditions appear following fire (Baeza and Roy 2008).

The total recovery of a burned area includes different aspects such as revegetation, fauna recuperation, biodiversity, landscape aesthetics, 'natural' runoff rates and sediment yield (Inbar et al. 1998). Opinions about the natural state (the target value of post-fire recovery) differ among disciplines. Seen from a soil and water conservation point of view, a return to the original vegetation cover is sufficient, while ecologists consider recovery as a return in the richness of the original species. From a silvicultural perspective, it is important to consider both the quantity and quality of fire-induced tree regeneration (Gould et al. 2002).

The assessment of the ecological effects of fires on biodiversity, soil degradation and on the cycling of carbon and nitrogen requires not only a detailed and accurate mapping of the burned areas but also an accurate mapping of the type and severity of fire and of post-fire forest regeneration (for example pine regeneration) and vegetation recovery (for example shrub recovery) (Le Houerou 1987; Jabukauskas et al. 1990; Naveh 1991).

3. Post-fire monitoring using remote sensing

This part of the chapter consists of three different sections. Section (3.1) provides information on field based post-fire vegetation monitoring, section (3.2) discusses the airborne and spaceborne sensors used in post-fire monitoring, and section (3.3) introduces the relative satellite image analysis techniques. The latter, provides a thoroughly description of well-known methods employed in post-fire monitoring and discuses studies related to the estimation of post-fire albedo and Land Surface Temperature. Subsection (3.3.6) deals with the use of SAR data to monitor the post-fire impact on forests. A summary table (Table 2) is also included.

3.1 Field based post-fire monitoring

Traditional methods of recording post-fire impact on vegetation include extensive field work or observations from an airborne platform, followed by the initial mapping (manually) of resource damage into predetermined classes (Bertolette and Spotskey 2001). As fire sizes increase and time becomes a constraining factor, traditional methods have become costly and labour-intensive.

Most studies are carried out in the first years after fire, and are mainly focused on seedling germination and on the survival and the restoration of plant cover (Table 1). Several years after the fire, measurements usually focus more on the characteristics of, for example, the trees, namely height, canopy width, basal diameter and volume. It has been realised that long-term monitoring is often required in order to evaluate the resilience of the different ecosystems towards forest fires. Normally, in long-term post-fire vegetation monitoring sampling units are established right after the fire event and the studied variables are monitored for several consecutive years (Calvo et al. 1998; Tarrega et al. 2001).

The most common sampling technique for monitoring plant populations is the use of permanent square plots. Taking into account both species characteristics and the extent of the study area, the size, the shape and the number of different plots are determined (Arianoutsou 1984; Clemente et al. 1996; Calvo et al. 2002; Cruz et al. 2003; Mitri and Gitas 2010). Accordingly, and in order to facilitate the objective collection of data on the ground, a number of field variables have been measured. In addition, there is a wide variety of sampling protocols depending on the type of survey (Daskalakou and Thanos 2004) that have been developed.

The analysis of post-fire vegetation recovery and monitoring can be either structural (involving variables such as cover and spatial heterogeneity), which is based on visual or floristic assessments (species composition, richness, community diversity, etc.), or a combination of the above (Pausas et al. 1999; Eshel et al. 2000; Kazanis and Arianoutsou 2004). In addition, protocols used can be distinguished into two general categories, that of plots (Thanos et al. 1996; Daskalakou and Thanos 1997; Tsitsoni 1997; Martínez-Sánchez et al. 1999; De Luis et al. 2001; Bailey and Covington 2002; Kennard et al. 2002) and that of transects (Ne'eman et al. 1999; Pausas et al. 1999; Caturla et al. 2000; Wahren et al. 2001; Gould et al. 2002).

Pausas et al. (1999) investigated the effect of different environmental conditions (climatic zones, aspect, and lithology) on the recovery process. They found large spatial variation in plant recovery in different localities with the same aspect, and for different aspects in the same locality. The recovery rate was different for different years due to changing climatic conditions and was higher on north-facing slopes, which are wetter than south-facing slopes. Belda and Meliá (2000) investigated the influence of climate on the natural post-fire regeneration of the burned area and found that regeneration followed an exponential curve, which was stronger in wet zones and had high correlation coefficients. Tsitsoni (1997) found that a high value of organic matter of the burned soil was a favourable factor for natural regeneration, as well as the position on the hillside, with lower regeneration indice values for the upper position and increasing values when descending the slope. De Luís et al. (2001) investigated the combined effect of fire and rainstorm on short-term ecosystem response by simulating high intensity rainfall on burned field plots in a Mediterranean shrubland. They found that one-year seedling survival was lower in the plots affected by rainfall simulation than in the control plots. A higher mortality rate, as a consequence of rainfall, was observed in the most abundant species. Pre-fire conditions may also affect the heterogeneity of burn severity, creating a wide range of local and landscape effects (Ne'eman et al. 1999). The propagation of fires and recolonisation processes are events that depend on the spatial organization of vegetation (Mouillet et al. 2001).

Variables assessed in the field	Reference
Different environmental conditions (climatic zones, high intensity rainfall, aspect, and lithology)	(Pausas et al. 1999; Belda and Meliá 2000; De Luis et al. 2001)
Organic matter of the burned soil	(Tsitsoni 1997)
Pre-fire conditions (i.e. spatial organisation of vegetation)	(Calvo et al. 1994; Ne'eman et al. 1999; Mouillet et al. 2001)
Resprouting vigour (plant's anatomical features, and the characteristics of the individual before disturbance), intensity of disturbance, and the environmental conditions after disturbance	(Lloret and Vilà 1997; Díaz-Delgado and Pons 2001)
Seed banks and germination - Seed number contained in soil samples - Cone opening and seed dispersal	(Thanos and Georghiou 1988; Skordilis and Thanos 1995; Daskalakou and Thanos 1996; Ferrandis et al. 1996; Herranz et al. 1999; Keeley 2000)
Seedling germination, survival and growth in a plot and along a transect on which samples are taken of plant cover, tree cover, tree characteristics or the floristic composition and cover per species	(Thanos et al. 1996; Daskalakou and Thanos 1997; Tsitsoni 1997; Martínez-Sánchez et al. 1999; Ne'eman et al. 1999; Pausas et al. 1999; Caturla et al. 2000; De Luis et al. 2001; Wahren et al. 2001; Bailey and Covington 2002; Gould et al. 2002; Kennard et al. 2002)
Post-fire structural dynamics	(Calvo et al. 1991)
Species richness patterns and vegetation diversity -density of seedlings – abundance of different age classes	(Pausas et al. 2003; Perula et al. 2003; Kavgaci et al. 2010)
Permanent plot collection per plant community type and data reduction - Visual cover of the plant species	(Santalla et al. 2002; Clemente et al. 2009)
Non-native species cover correlated with high native species richness	(Hunter et al. 2006)
Field spectrometry	(Broge and Leblanc 2000; Thenkabail et al. 2000; Thenkabail et al. 2002; Mitri and Gitas 2010)

Table 1. Field variables assessed to determine post-fire monitoring

Vegetative resprouting is possible when buds survive the fire to resprout. Some species regenerate by both sexual and vegetative reproduction. Resprouting vigour depends upon the plant's anatomical features, the characteristics of the individual before disturbance (plant size, number of shoots and physiological status of the plant), intensity of disturbance, and the environmental conditions after disturbance (Lloret and Vilà 1997). Areas where sprouting species are available usually show higher recovery rates than areas with mainly obligated seeders (Díaz-Delgado and Pons 2001). Seed banks are important in the dynamics of many plant communities as they provide an immediate source of propagules for recruitment after disturbance.

Measurements may be focused on the monitoring of seedling germination, survival and growth (height) in a plot (Tsitsoni 1997; Martínez-Sánchez et al. 1999; De Luis et al. 2001; Bailey and Covington 2002; Kennard et al. 2002), or may be along a transect on which samples are taken of plant cover (Caturla et al. 2000), tree cover (Ne'eman et al. 1999), tree characteristics (Ne'eman et al. 1999; Gould et al. 2002) or the floristic composition and cover per species (Pausas et al. 1999; Wahren et al. 2001).

Field spectrometry measurements were employed to investigate the spectral properties of plants, vegetation recovery, and naturally regenerating forest (Broge and Leblanc 2000; Thenkabail et al. 2000; Thenkabail et al. 2002; Mitri and Gitas 2010). More general methods to estimate vegetation abundance (Bonham 1989), recovery and forest regeneration in the field were adopted.

3.2 Airborne and spaceborne sensors

In comparison with extensive and labour-intensive field campaigns, remote sensing offers a time- and cost-effective alternative for mapping post-fire vegetation over large areas. Ground truthing based on limited sample sets is, however, always suggested for calibration and validation purposes (Shaw et al. 1998; Mitri and Gitas 2010). Airborne platforms provide a first option to acquire remotely sensed imagery. Stueve et al. (2009) used aerial photography in combination with KH-4B (Key Hole 4B) imagery from the CORONA mission to detect post-fire tree establishment at an alpine treeline ecotone, whereas Amiro et al. (1999) employed airborne measurements to monitor the post-fire energy balance of boreal forest. Peterson and Stow (2003) applied Spectral Mixture Analysis on Airborne Data and Acquisition and Registration (ADAR) data for the mapping of post-fire chaparral regrowth in Southern California. Airborne imagery allows detailed spatial information (Bobbe et al. 2001). However, despite the increasing availability of digital aerial images, these data are rather seldom used. This is explained by the fact that airborne measurements usually cover relatively small areas. As a result, many photographs are required to cover large burned areas, which subsequently require correction and mosaicking (Gitas et al. 2009).

In addition to aerial photographs, spaceborne sensors have shown big potential for assessing post-fire regrowth effects. Table 2 lists a multitude of studies focusing on the use of remote sensing for assessing post-fire vegetation recovery. Satellite sensors are characterized by their technical specifications. These technical specifications determine the sensor's capabilities with regards to the monitoring of post-fire recovery trajectories. In this context, the term resolution is of paramount importance. Resolution is the character of data that limits the user's ability to detect and identify an object of feature within the data (Bobbe et al. 2001). Resolution is fourfold; difference is made between spatial, temporal, spectral and radiometric resolution:

- Spatial resolution is a measure of fineness of spatial detail and it determines the smallest object that can be identified in the data. For digital sensors, spatial resolution is expressed as pixel size.
- Temporal resolution refers to the sensor's revisiting time, i.e. the time period in which the same area is successively sensed. Temporal resolution depends upon orbital characteristics.
- Spectral resolution is a measure of the specific wavelength intervals in which the sensor records. It is important to distinguish between broadband sensors, in which reflectance values are averaged over relatively wide bandpasses, and hyperspectral sensors, which are characterized by the acquisition over many fine wavebands.
- Radiometric resolution is a measure of the sensor's ability to distinguish between two features of similar reflectance.

Sensor design is a determinant factor when choosing the appropriate image analysis technique. In this context, it is important to trade-off between spatial and temporal resolution. High to moderate spatial resolution (e.g. smaller than 30 m) acquire only a few images a year, while low spatial resolution sensors (e.g. larger than 250 m) are characterized by daily image acquisition (Veraverbeke et al. 2011a).

Most studies listed in Table 2 are based on optical satellite sensors. These images generally are subject to preprocessing prior to the analysis. In order to obtain geometrically registered top-of-canopy (TOC) reflectance values the imagery usually requires geometric, radiometric, atmospheric and topographic corrections (e.g. Veraverbeke et al. 2010a). After the abrupt changes caused by the fire, the more gradual vegetation recovery leads to alterations in radiometric response at landscape scale. These changes are governed by: (i) disappearance of the charcoal/ash, (ii) changes in the proportion of bare soils and (iii) an increase in vegetative cover. So far, many studies focused on the red-near infrared (R-NIR) bi-spectral space to discriminate between bare soils and vegetated areas because vegetation recovery results in higher NIR reflectance values and lower R reflectance values due to augmented chlorophyll absorption.

According to the references included in Table 2, applications conducted for post-fire monitoring can be sorted based on the characteristics of the sensor that was used:

- High resolution sensors: 6 studies were based on airborne imagery, 1 on Quickbird imagery and one on KH-4B imagery.
- Moderate resolutions sensors: the majority of the studies applied Landsat sensors: 27 studies used Thematic Mapper (TM) images, 10 Enhanced Thematic Mapper plus (ETM+) images and 5 Multispectral Scanner images (MSS). In addition, 3 authors applied Satellite Pour l'Observation de la Terra (SPOT) Multispectral (XS) data, 2 Synthetic Aperture Radar (SAR), 1 Earth Observing-1 (EO1) Hyperion and 1 ICESAT Geoscience Laser Altimeter System (GLAS).
- Low resolution sensors: although these sensors are characterized by their low spatial resolution, they have the advantage of repeated temporal sampling with high temporal frequency. Nine studies used Terra Moderate Resolution Imaging Spectroradiometer (MODIS) time series, 5 SPOT Vegetation (VGT) and 5 National Oceanic and Atmospheric Administration (NOAA) Advanced Very High Resolution (AVHRR) Data.
- SAR: three studies used multi-temporal ERS images (C band), in one study the potential of ALOS PALSAR (L band) was investigated for post-fire monitoring, while multiple-polarization aircraft L-band was used to monitor burn recovery in a coastal marsh.

Reference	Ecological parameter(s) measured	Technique	Ecosystem	Field data	Remotely sensed data
Alcaraz-Segura et al. 2010	Greening	NDVI	Boreal forest	/	AVHRR
Amiro et al. 1999	Sensible heat (H), latent heat (LE), CO_2 flux, surface radiometric temperature net radiation	Surface energy balance	Boreal forest	/	BOREAS Twin Otter airborne data
Belda and Melia 2000	/	NDVI	Mediterranean ecosystem	/	TM
Bisson et al. 2008	/	NDVI	Mediterranean ecosystem	Selected plots	TM, ETM+
Bourgeau-Chavez et al 2007	Soil moisture variations in fire disturbed areas	Regression models	Boreal forest	5 10X10 sample plots, randomly selected along 200m long transcects	ERS 2
Carranza et al. 2001	/	NDVI and landscape structure	Mediterranean ecosystem	33 line transect plots	TM
Clemente et al. 2009	Fractional vegetation cover	NDVI and other indices	Mediterranean ecosystem	/	TM, ETM+
Cuevas-Gonzalez et al. 2008	fAPAR	NDVI transformation	Boreal forest	/	MODIS
Cuevas-Gonzalez et al. 2009	/	NDVI and other indices	Boreal forest	/	MODIS
Diaz-Delgado and Pons 2001	/	NDVI and control plot selection	Mediterranean ecosystem	/	MSS
Diaz-Delgado et al. 2002	/	NDVI and control plot selection	Mediterranean ecosystem	/	MSS
Diaz-Delgado et al. 2003	/	NDVI and control plot selection	Mediterranean ecosystem	/	MSS, TM

Reference	Ecological parameter(s) measured	Technique	Ecosystem	Field data	Remotely sensed data
Goetz et al. 2006	/	NDVI	Boreal forest	11 sites	AVHRR
Goetz et al. 2010	Tree height	LIDAR	Boreal forest	/	GLAS, TM, MODIS
Gouveia et al. 2010	/	NDVI	Mediterranean ecosystem	/	VGT
Hall et al. 1991	Land cover classification including regeneration classes	Supervised classification	Boreal forest	/	MSS
Henry and Hope 1998	/	NDVI and other indices	Mediterranean ecosystem	36 line transect plots with plant community statistics	XS
Hernandez-Clemente et al. 2009	/	NDVI	Mediterranean ecosystem	/	TM, ETM+
Hicke et al. 2003	NPP	NDVI transformation	Boreal forest	/	AVHRR
Hope et al. 2007	/	NDVI	Mediterranean ecosystem	/	TM, ETM+
Idris et al. 2005	/	NDVI and control plot selection	Tropical and boreal forest	/	AVHRR
Jabukauskas et al. 1990	Land cover classification including regeneration classes	Supervised minimum distance-to-mean classification	Temperate coniferous forest	32 linguistic sample points	TM
Jacobson 2010	/	NDVI and other indices	Woodland community	/	XS
Kasischke et al 2007	Relations between soil moisture patterns and post-fire tree recruitment	Empirical relations	Boreal forest	2 test sites	ERS 1 and 2

Reference	Ecological parameter(s) measured	Technique	Ecosystem	Field data	Remotely sensed data
Kasischke et al 2011	Biomass estimation of regenerating forests	Empirical relations	Boreal forest	/	ALOS PALSAR
Lhermitte et al. 2010	/	NDVI and control plot selection	Savannah ecosystem	/	VGT
Lhermitte et al. 2011	/	NDVI and control plot selection	Savannah ecosystem	/	VGT
Li et al. 2008	/	NDVI, EVI and control plot selection	Temperate forest	/	MODIS
Lozano et al. 2010	/	NDVI and other indices, landscape structure	Mediterranean ecosystem	/	TM, ETM+
Lyons et al. 2008	/	Albedo	Boreal forest	/	MODIS
Malak and Pausas 2006	/	NDVI	Mediterranean ecosystem	/	TM
Marchetti et al. 1995	/	Infrared index	Mediterranean ecosystem	/	TM
McMichael et al. 2004	LAI	NDVI transformation	Mediterranean ecosystem	62 sample points and field spectroscopy	TM, ETM+
Minchella et al. 2009	Vegetation recovery	Simulations with scattering model	Mediterranean ecosystem	/	ERS
Mitchell and Yuan 2010	/	NDVI	Temperate forest	/	TM
Mitri and Gitas 2010	Land cover classification including regeneration classes	Object-based classification	Mediterranean ecosystem	/	Hyperion
Palandjian et al. 2009	Classification of 4 regeneration classes	Density slicing on NDVI data	Mediterranean ecosystem	Flux tower	Quickbird

Reference	Ecological parameter(s) measured	Technique	Ecosystem	Field data	Remotely sensed data
Peterson and Stow 2003	Fractional vegetation cover	SMA	Mediterranean ecosystem	/	TM, ADAR
Ramsey et al. 1999	Time-since-burn	Regression analysis	Marsh ecosystem	/	Aircraft SAR
Randerson et al. 2006	/	Albedo	Boreal forest	93 line transect points and field spectroscopy	MODIS
Riaño et al. 2002	/	NDVI and control plot selection	Mediterranean ecosystem	/	AVIRIS
Ricotta et al. 1998	/	NDVI and landscape structure	Mediterranean ecosystem	/	TM
Roder et al. 2008	Fractional vegetation cover	SMA	Mediterranean ecosystem	Ground-based NDVI	MSS, TM, ETM+
Sankey et al. 2008	Fractional vegetation cover	SMA	Sagebrush community	Field spectroscopy	XS, aerial photographs
Schroeder and Pereira 2002	/	Landscape structure	Boreal forest	/	TM
Segah et al. 2010	/	NDVI	Tropical forest	60 sample points	TM, VGT
Shaw et al. 1998	/	NDVI and other indices	Boreal forest	/	Field spectroscopy
Steyaert et al. 1997	Land cover classification including regeneration classes	Unsupervised cluster classification	Boreal forest	/	AVHRR, TM
Stueve et al. 2009	Tree establishment classification	Supervised minimum distance-to-mean binary classification	Alpine treeline border	Line transect plots	Airborne and KH-4B imagery

Reference	Ecological parameter(s) measured	Technique	Ecosystem	Field data	Remotely sensed data
Telesca and Lasaponara 2006	/	Detrended fluctuation analysis on NDVI time series	Mediterranean ecosystem	78 line transect plots and field spectroscopy	VGT
Van Leeuwen 2008	Phenological metrics	NDVI	Temperate ecosystem	/	MODIS
Van Leeuwen et al. 2010	Phenological metrics	NDVI and control plot selection	Temperate and Mediterranean ecosystems	78 line transect plots and field spectroscopy	MODIS
Veraverbeke et al. 2012	Fractional vegetation cover	NDVI and other indices	Mediterranean ecosystem	/	TM
Veraverbeke et al. 2012	/	NDVI, albedo, LST and control plot selection	Mediterranean ecosystem	/	MODIS
Veraverbeke et al. 2012	Fractional vegetation cover	SMA	Mediterranean ecosystem	19 line transect points	TM
Vicente-Serrano et al. 2008	/	NDVI and other indices	Mediterranean ecosystem	/	TM/ETM+
Viedma et al. 1997	Regeneration rate	NDVI	Mediterranean ecosystem	/	TM
Vila and Barbosa 2010	Fractional vegetation cover	NDVI and other indices, SMA	Mediterranean ecosystem	Field data	TM, ETM+
White et al. 1996	/	NDVI	Boreal forest	/	TM
Wittenberg et al. 2007	/	EVI	Mediterranean ecosystem	/	TM, ETM+

(ADAR: Airborne Data and Acquisition and Registration, AVHRR: Advanced Very High Resolution Radiometer, AVIRIS: Airborne Visible/Infrared Imaging Spectroradiometer, ETM+: Enhanced Thematic Mapper plus, EVI: Enhanced Vegetation Index, fAPAR: fraction of Absorbed Photosynthetically Active Radiation, GLAS: Geoscience Laser Altimeter System, KH: Key Hole, LAI: Leaf Area Index, LST: Land Surface Temperature, LIDAR, Light Detection and Ranging)MSS: Multispectral Scanner, MODIS: Moderate Resolution Imaging Spectroradiometer, NDVI: Normalized Difference Vegetation Index, NPP: Net Primary Productivity, RADAR: Radio Detection and Ranging, SAR: Synthetic Aperture Radar, SMA: Spectral Mixture Analysis, TM: Thematic Mapper, XS: Multispectral VGT: Vegetation

Table 2. Examples of post-fire vegetation recovery studies using remotely sensed data

A review of the studies presented in Table 2, shows that current research mainly focused on multispectral Landsat, MODIS, VGT and AVHRR data. The moderate and low resolution scales are relatively well documented. For the coarse spatial resolution this usually goes hand in hand with time series analysis. The overview also highlights a need for further exploration of high spatial resolution and high spectral resolution, i.e. hyperspectral, data. High resolution data (e.g. Quickbird, IKONOS) could help in the recognition of individual plants, whereas hyperspectral data have the potential to discriminate between different vegetation species (Asner and Lobell 2000; Mitri and Gitas 2010; Somers et al. 2010). Besides optical imagery, Synthetic Aperture Radar (SAR) data is also worth a more in-depth evaluation.

From Table 2 it can also be concluded that vegetation recovery studies have been carried out in a number of different ecosystems including:

- Mediterranean ecosystems: 26 studies were conducted in the Mediterranean basin, whereas 5 papers concentrate on the Mediterranean ecozone of California (USA).
- Boreal forests: 20 study areas were carried out in boreal forests of North American and the Eurasian boreal zone.
- Temperate forests: 5 studies focused on temperate forests..
- Other types of forests: 2 studies were conducted in tropical forests, 2 in savannah ecosystems, 1 along an alpine treeline border, 1 in a sagebrush community and 1 in a marsh ecosystem.

Post-fire recovery rates depend on fire severity (Diaz-Delgado et al. 2003), soil properties (Bisson et al. 2008), post-fire meteorological conditions (Henry and Hope 1998; van Leeuwen et al. 2010) and ecotype (Viedma et al. 1997; Veraverbeke et al. 2010b; Lhermitte et al. 2011; Veraverbeke et al. 2011b). In fire-adapted sclerophyllous shrub lands, for example, recovery only takes a few years (Viedma et al. 1997; Pausas and Verdu 2005) whereas in boreal forests recovery lasts several decades (Nepstad et al. 1999). The summary above clearly shows that recovery research so far focused on boreal and Mediterranean ecosystems. Table 2 also reveals that only 26 % of the papers included in the list were supported by ground truth. This highlights a need to conduct research on the remote sensing of post-fire vegetation recovery supported by field campaigns, while other ecosystems such as tropical forest and savannah ecosystems urgently require a knowledge gain.

3.3 Image analysis techniques

Several image analysis techniques are employed in the remote sensing of post-fire recovery. Most of the traditional approaches have an origin outside fire applications but their methodology is easily adjustable for recovery studies. The most important traditional methods are image classification, Vegetation Indices (VIs) and Spectral Mixture Analysis (SMA). They are thoroughly discussed below. In addition, one specific technique applicable for ecological disturbances, i.e. control plot selection, is also incorporated. Finally, some less frequently used approaches are shortly described.

3.3.1 Image classification

Since long, multispectral image classification is known to be a powerful technique to translate remotely sensed data into ecologically relevant cover classes. Both supervised (Jakubauskas et al. 1990; Hall et al. 1991; Stueve et al. 2009; Mitri and Gitas 2010) and unsupervised (Steyaert et al. 1997) techniques have been applied in post-fire recovery studies. Most applications rely on pixel-based classifiers such as the minimum-distance-to-mean (Jakubauskas et al. 1990; Stueve et al. 2009) and maximum likelihood classifiers (Hall et al. 1991). While the majority of the studies focused on four or more cover classes (Jabukauskas et al. 1990; Hall et al. 1991; Steyaert et al. 1997; Mitri and Gitas 2010), the study of Stueve et al. (2009) was restricted to a binary classification of tree establishment along an alpine treeline border. A major problem associated with pixel-based classifications is the occurrence of salt-and-pepper artifacts. As a solution, object-based classification schemes include both spectral and contextual information (Wicks et al. 2002) (See more in Section 4).

Apart from multispectral classification approaches Palandjian et al. (2009) applied density slicing on post-fire Normalized Difference Vegetation Index (NDVI) data. As such, they discriminated four different regeneration classes. Generally spoken, relatively few studies applied image classification to monitor post-fire recovery. This is mainly explained by the fact that the spatial resolution of most popular satellite systems (e.g. Landsat) exceeds the size of individual regenerating plants. As a result, it is very difficult in these applications to find pure training data because most image pixels are mixed. This incites a need to explore the potential of high resolution imagery (e.g. Quickbird, IKONOS) to classify individual plants. This would also open new research pathways to study the small-scale spatial patterns of post-fire vegetation recovery.

3.3.2 VIs

By far the most widely used remote sensing technique to assess post-fire recovery is the NDVI (Tucker 1979) because of its strongly established relationship with above-ground biomass in a wide range of ecosystems (Carlson and Ripley 1997; Henry and Hope 1998; Cuevas-Gonzalez et al. 2009). The post-fire environment typically consists of a mixture of vegetation and substrate. Theoretically, Soil Adjusted Vegetation Indices (SAVIs) (Huete 1988; Baret and Guyot 1991; Qi et al. 1994) are better suited for these mixed environments. Relatively few studies have assessed the correlation between field estimates of vegetative cover and VIs. Clemente et al. (2009) contrasted the NDVI with the SAVI (Huete 1988), Transformed SAVI (TSAVI), (Baret and Guyot 1991) and Modified SAVI (MSAVI) (Qi et al. 1994) for estimating post-fire vegetation regrowth 7 and 12 years after a fire in Spain. The NDVI was stronger related to field estimates of vegetation cover than any other index. Vila and Barbosa (2010) drew more or less the same conclusion. They also found that the NDVI was most accurately related to field data eight years after a fire in Italy. Van Leeuwen et al. (2010) also retrieved high correlations between NDVI and field data of recovery. Veravrbeke et al. (2012b) comprehensively evaluated thirteen R-NIR vegetation indices for assessing post-fire vegetation recovery. They found that the NDVI indeed obtained the best correlations with line transect field data and the failure of the SAVIs was due to their inability to account for variations in background brightness. This approves the use of the NDVI as an appropriate recovery measure, however, it should be noted that the potential of spectral indices with a SWIR or MIR spectral band has not been fully explored yet. These spectral regions have proven to be very effective in discriminating soil and vegetation (Drake et al. 1999; Asner and Lobell 2000). In the context of post-fire recovery, VIs including a SWIR or MIR band have shown prospect in the studies of Marchetti et al. (1995), Cuevas-Gonzalez et al. (2009) and Jacobson (2010).

Several studies used NDVI data as a linkage to more ecologically relevant parameters such as fractional vegetation cover (Clemente et al. 2009; Vila and Barbosa 2010; Veraverbeke et al. 2012b), fraction of Absorbed Photosynthetically Active Radiation (fAPAR), (Cuevas-Gonzalez et al. 2008), Net Primary Production (NPP), (Hicke et al. 2003) and Leaf Area Index (LAI), (McMichael et al. 2004). Usually, these index transforms are based on thorough field calibration. Only few studies incorporated a substantial number of field plots to calibrate and validate the VI approach (Shaw et al. 1998; Bisson et al. 2008; Clemente et al. 2009; Jacobson 2010; Segah et al. 2010; van Leeuwen et al. 2010; Vila and Barbosa 2010; Veraverbeke et al. 2011b Hernandez-Clemente et al. 2009). Figure 1 presents an example of the relationship between the NDVI and field estimates of vegetative cover which was used to model post-fire vegetation cover in the burned area. The majority of the authors, however, use the NDVI as a well-accepted methodology without additional field efforts. In a mono-temporal context NDVI were related to climatic variables (Belda and Meliá 2000), topographic parameters (Mitchell and Yuan 2010) and fire severity (White et al. 1996; Mitchell and Yuan 2010).

Other studies employed multiple images to construct recovery trajectories (Viedma et al. 1997; Henry and Hope 1998; Ricotta et al. 1998; Carranza et al. 2001; Díaz-Delgado and Pons 2001; Diaz-Delgado et al. 2002; Riaño et al. 2002; Malak and Pausas 2006; Wittenberg et al. 2007; Bisson et al. 2008; Vicente-Serrano et al. 2008; Clemente et al. 2009; Hernandez-Clemente et al. 2009). Where Landsat-based studies allow only a few cloud-free images a year (Ju and Roy 2008), satellite sensors with high temporal frequency permit the construction of continuous time series. More recently, several authors have explored this data type for assessing post-fire effects (Idris et al. 2005; Goetz et al. 2006; Telesca and Lasaponara 2006; Li et al. 2008; van Leeuwen 2008; Alcaraz-Segura et al. 2010; Gouveia et al. 2010; Lhermitte et al. 2010; Segah et al. 2010; van Leeuwen et al. 2010; Lhermitte et al. 2011; Veraverbeke et al. 2012a; Veraverbeke et al. 2012c). Thanks to this it is possible to discriminate between regeneration patters and seasonal fluctuations (Veraverbeke et al. 2010b; Lhermitte et al. 2011; Veraverbeke et al. 2012b). A major advantage of multi-temporal data is that regression fits between time since fire and NDVI data give reliable recovery rate estimates (Viedma et al. 1997; Díaz-Delgado and Pons 2001; Gouveia et al. 2010). Extension of these regeneration rates allows prediction on the future state of biomass. This is of major interest for decision makers in rangeland management. NDVI data also served as the preferred data source for the control plot selection procedure (Díaz-Delgado and Pons 2001; Diaz-Delgado et al. 2002; Lhermitte et al. 2010; Lhermitte et al. 2011) and landscape ecological applications (Ricotta et al. 1998; Carranza et al. 2001).

3.3.3 Spectral mixture analysis

The post-fire environment typically consists of a mixture of vegetation and substrate. Thus, monitoring post-fire regeneration processes essentially poses a sub-pixel issue at the resolution of most operational satellite systems such as Landsat. A number of image analysis techniques accommodating mixing problems exist (Atkinson et al. 1997; Arai 2008) with SMA being the most common technique utilized in many applications (Roberts et al. 1998; Asner and Lobell 2000; Riaño et al. 2002; Roder et al. 2008; Somers et al. 2010). SMA effectively addresses this issue by quantifying the sub-pixel fraction of cover of different endmembers, which are assumed to represent the spectral variability among the dominant terrain features. A major advantage of SMA is its ability to detect low cover fractions, something which remains difficult with the traditional vegetation indices (VIs) approach (Henry and Hope 1998; Elmore et al. 2000; Rogan and Franklin 2001). Moreover, SMA

A.

B.

$y = 1.39x - 0.21 \quad R^2 = 0.68$

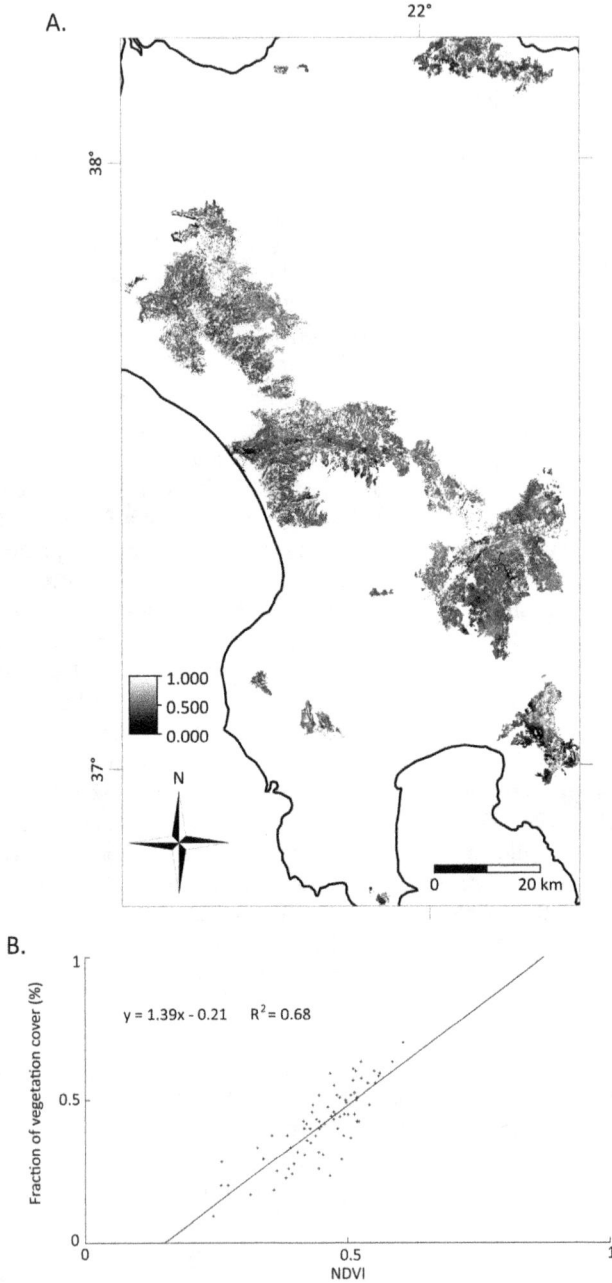

Fig. 1. Fractional vegetation cover map (A) three years after the large Peloponnese (Greece) fires based on the regression fit between the Landsat Normalized Difference Vegetation Index (NDVI) and line transect field ratings of vegetation cover (B) (Veraverbeke et al. 2012b).

directly results in quantitative abundance maps, without the need of an initial calibration based on field data as with VIs (Somers et al. 2010; Vila and Barbosa 2010). With regards to post-fire effects, rather few studies employed SMA to monitor post-fire vegetation responses (Riaño et al. 2002; Peterson and Stow 2003; Roder et al. 2008; Sankey et al. 2008; Vila and Barbosa 2010; Veraverbeke et al. 2012a). Although results of these studies were consistent, most of them were all restricted to simple linear SMA models in which only one spectrum was allowed for each endmember. As a consequence, the performance of these SMA models often appeared to be suboptimal (Roder et al. 2008; Vila and Barbosa 2010) because these models did not incorporate the natural variability in scene conditions of terrain features inherent in remote sensing data (Asner 1998). To overcome this variability effect Peterson and Stow (2003) applied multiple endmember SMA (MESMA), (Roberts et al. 1998). MESMA incorporates natural variability by allowing multiple endmembers for each constituting terrain feature. These endmember sets represent the within-class variability (Somers et al. 2010) and MESMA models search for the most optimal endmember combination by reducing the residual error when estimating fractional covers (Asner and Lobell 2000). Rogge et al. (2006) and Veraverbeke et al. (2012a), however, clearly demonstrated that reducing the residual error by applying MESMA not always results in the selection of the most appropriate endmember spectrum. An initial segmentation of the area prior to the unmixing process in order to retain areas which reveal a high similarity in the spectral properties of a certain endmember has been presented as a sound and computationally efficient solution to address this issue (Rogge et al. 2006; Veraverbeke et al. 2012a).

A possible amelioration in post-fire vegetation mapping using SMA could be the inclusion of SWIR and MIR spectral regions in the unmixing process. These spectral regions have proven to be very effective in discriminating soil and vegetation (Drake et al. 1999; Asner and Lobell 2000). Carreiras et al. (2006) demonstrated that adding the SWIR-MIR Landsat bands resulted in better estimates of tree canopy cover in Mediterranean shrublands. Additionally, enhancing the spectral resolution by employing hyperspectral data would increase the amount of spectral detail which would benefit the differentiation between spectra (Mitri and Gitas 2008). By including more and other spectral wavebands the unmixing model could gain discriminative power. Potentially, this would make it even possible to distinguish between non-photosynthetic vegetation and substrate (Asner and Lobell 2000; Somers et al. 2010), which appeared to be impossible in current applications.

3.3.4 Control plot selection

A major difficulty in post-fire time series analysis is that the analysis can be hampered by phenological effects, both due to the differences in acquisition data and due to inter-annual meteorological variability (Díaz-Delgado and Pons 2001). To deal with these phenological effects Diaz-Delgado and Pons (2001) proposed to compare vegetation regrowth in a burned area with unburned reference plots within the same image. As such, external and phenological variations are minimized among the compared areas. Several authors have successfully adapted the reference plot approach (Diaz-Delgado et al. 2002; Diaz-Delgado et al. 2003; Idris et al. 2005; Li et al. 2008; van Leeuwen et al. 2010). The reference plot selection procedure has, however, two main difficulties. Firstly, large scale application remains constrained due to the necessity of profound field knowledge to select relevant control plots.

Secondly, the reference plot approach fails to describe within-burn heterogeneity as it uses mean values per fire plot. To solve these problems, Lhermitte et al. (2010) proposed a pixel-based control plot selection method which follows the same reasoning with respect to the minimization of phenological effects by comparison with image-based control plots. The difference with the reference plot procedure, however, is situated in the fact that the pixel-based method assigns a unique unburned control pixel to each burned pixel. This control pixel selection is based on the similarity between the time series of the burned pixel and the time series of its surrounding unburned pixels for a pre-fire year (Lhermitte et al. 2010). The method allows the quantification of the heterogeneity within a fire plot since each fire pixel is considered independently as a focal study pixel and a control pixel is selected from a contextual neighbourhood around the focal pixel. This approach has been used in studies assessing the temporal dimension of fire impact and subsequent recovery (Veraverbeke et al. 2010b; Lhermitte et al. 2011; Veraverbeke et al. 2011a; Veraverbeke et al. 2012c). Figure 2 presents an example of the principle of the control plot selection procedure on a NDVI time series. With the exception of Veraverbeke et al. (2012c) who also incorporated Land Surface Temperature (LST) and albedo data in the procedure, so far, the control plot selection procedure has only been applied on NDVI data. Nevertheless, in theory, the control plot selection procedure allows any kind of remotely sensed data as input. Moreover, the procedure has the potential to provide valuable reference information other disturbance in which external forces abruptly remove the vegetation (e.g. volcanic eruptions, landslides, hurricanes, tsunamis, etc.).

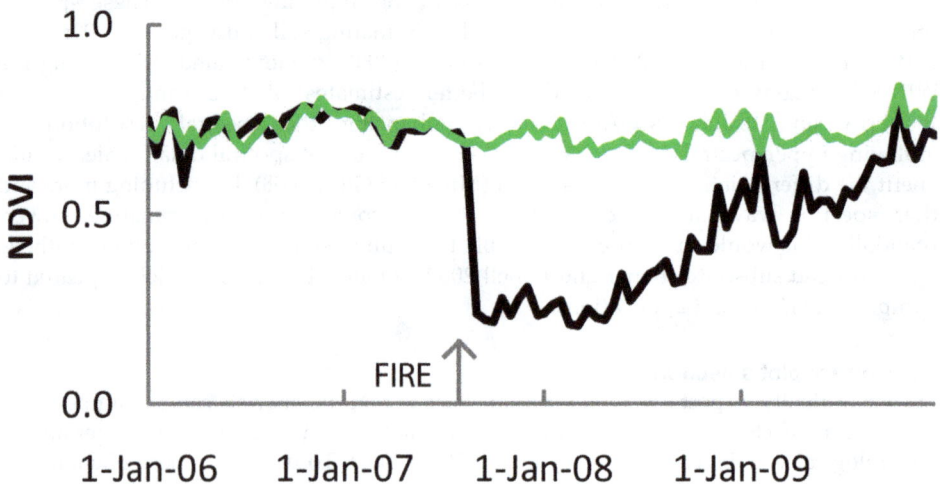

Fig. 2. Example of Normalized Difference Vegetation Index (NDVI) time series of a burned pixel (black line) and its corresponding control pixel (green line). The control pixel mimics how the burned pixel would have behaved without fire occurrence (after Veraverbeke et al. (2012c)).

Despite of the merits of the control pixel selection procedure as presented by Lhermitte et al. (2010) and Veraverbeke et al. (2010b) two constraints remain. Firstly, due to the necessity to search in larger windows for pixels in the middle of the burn the performance of the procedure is likely to be better near the contours of the burn perimeter. On one hand this is inevitable as the potentially most similar neighbour pixels are burned. On the other hand one could argue that this phenomenon incites to make the control pixel selection settings dependent on the distance to the fire perimeter. The procedure is also affected by a second constraint, i.e. the heterogeneity of the unburned landscape matrix. It is obvious that the procedure will be more optimal in highly homogeneous landscapes, even for large search windows. In contrast, in highly heterogeneous mixtures of different land cover types the procedure will potentially fail to retrieve similar pixels for small window sizes. It is a hard task to uncouple and quantify the effects of both constraints. Solutions to this have the potential to further improve the selection procedure.

3.3.5 Post-fire albedo and land surface temperature

Besides the use of optical data and its derivatives (e.g. VIs) some authors focused on the recovery of remotely sensed bioclimatic variables such as albedo and Land Surface Temperature (LST) (Amiro et al. 1999; Lyons and Randerson 2008; Veraverbeke et al. (2012c)). The increase in post-fire LST progressively weakens over time (Veraverbeke et al. (2012c)), whereas Lyons et al. (2008) and Veraverbeke et al. (2012c)) observed that albedo quickly recovers after an initial post-fire drop. The albedo even exceeds pre-fire values when char materials are removed and vegetation starts to regenerate (Veraverbeke et al. (2012c)). Thus, where the immediate fire effect results in an increased absorption of radiative energy, the long-term effect generally is an increased albedo (Amiro et al. 2006; Randerson et al. 2006). The quantification of these effects, together with an accurate estimation of the amount of greenhouse gasses emitted by the fire and the subsequent post-fire carbon sequestration of regenerating vegetation, are necessary for a holistic comprehension of the effect of wildfires on regional and global climate. In this context, Randerson et al. (2006) comprehensively demonstrated that, although the first post-fire year resulted in a net warming, the long-term balance was negative. As such they concluded that an increasing fire activity in the boreal region would not necessarily lead to a net climate warming. Remotely sensed proxies of albedo and LST can also be used to estimate the spatio-temporal behaviour of several radiative budget parameters of paramount biophysical importance such as sensible and latent heat fluxes (Bastiaanssen et al. 1998; Roerink et al. 2000). The immediate post-fire surface warming and its ecological consequences as well as the long-term post-fire temporal development of heat fluxes could form a relatively unexplored and captivating research topic.

3.3.6 Post-fire monitoring using SAR

Synthetic-aperture radar (SAR) data has been extensively used for various ecological processes (Kasischke et al. 1997) and have been especially useful in areas characterized by frequent cloud conditions such as the tropics and in the remote locations of the boreal forests. However, the application of SAR data in monitoring vegetation regrowth has been rather limited, while in most of the studies empirical relationships between field measurements and the backscatter values have been investigated.

Ramsey et al (1999) investigated the use of multiple-polarization aircraft L-band to monitor burn recovery in a coastal marsh. The authors found a significant relationship between VH-polarization and time-since-burn. In addition, Ramsey et al 1999 examined the same relationship with scaled SAR returns. Scaled by control data [e.g. VH(burn)-VH(control)], all three polarizations (VV, HH, VH) regressions were found significant, with 83% of the time-since-burn explained by the VH variable.

Kasischke et all (2011) investigated the utility of L-band ALOS PALSAR data (23.6 cm wavelength) for estimating low aboveground biomass in a fire-distributed black spruce forests in interior Alaska nearly 20 years after the fire events. Field measurements were analyzed against the radar backscatter coefficient. Significant linear correlations were found between the log of the aboveground biomass and σ^o (L-HH) and σ^o (L-HV), with the highest correlation found when soil moisture was high. Kasischke et all 2011 concluded that using spaceborne SAR systems to monitor forest regrowth will not only require collection of biomass data to establish the relationship between biomass and backscatter, but may also require developing methods to account for variations in soil moisture.

Kasischke et al (2007) based on the findings of Bourgeau-Chavez et al (2007) explored the relations between soil moisture patterns and post-fire tree recruitment in fire-disturbed black spruce forests in Interior Alaska using ERS-1 and ERS-2 C band (5.7 cm wavelength). Both Kasischke et al (2007) and Bourgeau-Chavez et al (2007) found high correlations between ERS SAR backscatter and measured soil moisture in the burned areas. Furthermore, Kasischke et al (2007) found that the measured levels of tree recruitment are related to the levels of soil moisture: aspen seedlings were able to germinate and grow within the severely burned areas because of adequate soil moisture was present during the growing season. In contrast, low aspen recruitment at a severely burned area was attributed to lower soil moisture.

Minchella et al. (2009) used multitemporal ERS SAR images to monitor the vegetation recovery in a Mediterranean burned area. Following a qualitative approach (analysis of the multitemporal backscattering signatures) they observed that, due to the increase in soil moisture in the backscattering, the measurements, taken throughout at least one year, of the similarity between the backscattering of the burned area and the backscattering of a bare soil around or inside the burned area, may provide a SAR-based index for the vegetation recovery in the burned area. In addition, Minchella et all 2009 used a microwave scattering model. A minimization of the distance between simulated results and measured data has been carried out using the re-growth rate as the key variable. Results showed that the retrieved values were in agreement with in-situ measurements.

Tanase et al 2011 analyzed SAR metrics from burned forested areas in Spain and Alaska. SAR dataset that were used consisted of ERS (C band), TerraSAR-X (X-band), Environmental Satellite (Envisat) Advanced SAR (ASAR) (C-band) and ALOS PALSAR (L-band) images. The authors concluded that for Mediterranean forests, the L-band HV-polarized SAR backscatter allowed the best differentiation of regrowth phases whereas at X- and C-band the HV-polarized backscatter was less sensitive to modification in forest structure due to the rapid saturation of the signal. For boreal forest four different regrowth phases were separated. Co-polarized repeat-pass coherence presented weak sensitivity to the different forest regrowth phases. Separation was possible only for the most recently affected sites (<15 year since disturbance) regardless of the radar frequency.

4. Advances in remote sensing of post-fire monitoring

Given that older generation sensors have many known limitations with respect to their suitability for studying complex biophysical characteristics (De Jong et al. 2000, Steininger 2000, Sampson et al. 2001, Salas et al. 2002), the need to benefit from new generation of high spatial and spectral resolution sensors as well as active sensors is of critical importance. Accurate quantification of vegetation regeneration could be essential for biodiversity assessment, land cover characterization and biomass modelling (Blackburn and Milton 1995). High spectral resolution facilitates the identification of features while high spatial resolution permits accurate location of features (Gross and Scott 1998). Additionally, advanced multispectral sensors also allow significantly improved signal to noise ratios (Levesque and King 2003).

The development of new hyperspectral remote sensing instruments, both airborne and spaceborne, has provided an opportunity to study vegetation recovery after wildfire (Riaño et al. 2002). A number of recent studies have indicated the advantages of using discrete narrowband data from specific portions of the spectrum, rather than broadband data, to obtain the most sensitive quantitative or qualitative information on vegetation characteristics.

Mitri and Gitas (2010) mapped post-fire vegetation recovery using EO-1 Hyperion imagery and OBIA and an overall accuracy of 75.81 % was reported (Figure 3). Object-oriented image analysis has been developed to overcome the limitations and weaknesses of traditional image processing methods for feature extraction from high resolution images (Mitri and Gitas 2004; Mitri and Gitas 2010). The basic difference, especially when compared with pixel-based procedures is that image object analysis does not classify single pixels but rather image objects that have been extracted in a previous image segmentation step (Baatz and Schape 1999). The concept here is that the information that is necessary to interpret an image is not represented in a single pixel, but in image objects. Object-based classification involves three main steps, namely, image segmentation, object training, and object classification. "Ground truth" information using field spectroradiometry instruments is equally important for validation of representative image wavebands to be used in object-based classification. According to Wicks et al. (2002), object-based classification may result in an increased accuracy and more realistic presentation of the environment.

Furthermore, many applications of remote sensing require high spatial resolution data for a correct determination of small objects. For instance, high spatial resolution imagery can be used before, during, and after a fire to measure fuel potential, access, progress, extent, as well as damage and financial loss. High spatial resolution multispectral data such as QuickBird (60 centimetres in panchromatic and 2.4 m in multispectral) can identify not only individual tree crowns, but often also the type of tree, estimate biomass, condition and age class (Wang et al. 2004, Palandjian et al. 2009).

LIDAR data have been used extensively for estimating various forest attributes such as canopy height, biomass, basal area and LAI forest variables as diameter at breast height, volume and density (Dubayah and Drake 2000; Lim et al. 2003a; Bortolot and Wynne 2005) and individual tree heights. The limited use of LIDAR data for monitoring vegetation regrowth can be attributed to the limited existence of spaceborne LIDAR Also, the operation of airborne LIDAR can be hampered by weather conditions, it is cost prohibited and can cover only limited areas.

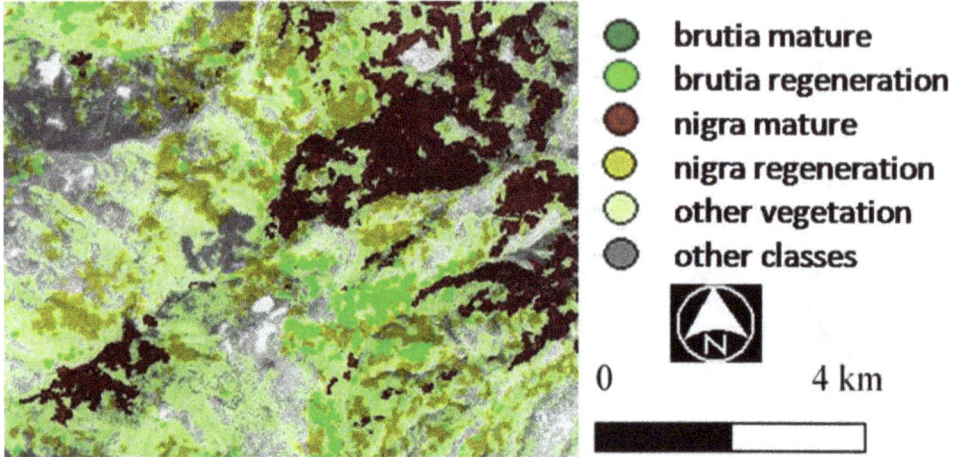

Fig. 3. Example of post-fire recovery map obtained after applying object-based classification in a Mediterranean ecosystem (Mitri and Gitas 2010).

However, it is worth mentioning the work of Wulder et al (2009) who used the integration of Landsat TM/ETM+ imagery and profiling LIDAR transects to characterize post-fire conditions on boreal forest in Canada shortly after the fire event. The main goals of their research were to evaluate whether LIDAR can be used to detect changes in vertical forest structural characteristics associated with the wildfire and find relationships between the vertical information extracted from the LIDAR datasets and horizontal information such as the indices NBR, dNBR and RdNBR extracted from the Landsat datasets in pre- and post-fire conditions. The authors had available LIDAR data collected along the same transect pre- and post-fire and could identify differences in forest structure before and after the fire, but those differences in structure were more related to post fire effects in dense forest than open or sparse forests. However, no significant correlation was found between the Landsat measures of post-fire effect and the LIDAR -derived measures of pre-fire forest structure. Kim et al (2009) used LIDAR intensity values in conjunction with field measurements to distinguish between live and dead standing tree biomass in a mixed coniferous forest in USA. Result of their regression analysis showed that low intensity returns from LIDAR were associated with dead tree biomass. It was suggested that knowing the background value of dead biomass in the forest (from field observations) one can estimate the additional contribution of dead standing tree biomass associated with a fire.

In addition to the airborne LIDAR applications, there has been a study using the satellite-based LIDAR instrument Geoscience Laser Altimeter System (GLAS) onboard ICESAT. Goetz et al (2010) used the LIDAR data acquired from GLAS to derive canopy structure information in burned areas and associated forest regrowth across Alaska. The LIDAR data were stratified by age class (using fire perimeters) and vegetation type (using the EVI calculated from MODIS NBAR). Data were also analyzed in relation to the burn severity. Tree height estimates derived from GLAS data were then compared to field measurements. The results showed that the GLAS data have the utility for inferring height properties of vegetation in fire disturbed areas of boreal forest, but it was mentioned that the data require both careful screening and some knowledge of the study area.

Furthermore, even though Polarimetric Interferometric SAR (PolInSAR) data has not been used so far for monitoring vegetation regrowth in fire affected areas, PolInSAR proved capable in estimating tree height (Papathanasiou and Cloude, 2001). The height estimation with PolInSAR has been successfully demonstrated on various forest types, ranging from boreal to tropical forests, and achieves high accuracies, (Praks et al. 2007; Garestier et al. 2008; Hajnsek et al. 2009). The operation of single-pass interferometric systems (TanDEM-X has been operating since May 2010) in space opens the door to a new set of unique applications (Krieger et al. 2010), one of them being the monitoring of forest biomass after fire.

5. Future investigation

Given that older generation sensors have many known limitations with respect to their suitability for studying complex biophysical characteristics, the need to benefit from new generation of high spatial and spectral resolution sensors as well as active sensors is of critical importance. The development of new remote sensing instruments, both airborne and spaceborne, has provided an opportunity to advance studies and researches on vegetation recovery after wildfire. Future research related to advances in remote sensing in post-fire monitoring is expected to focus on the following:

- Ecosystems with high environmental importance such as tropical forest and savannah ecosystems urgently require a knowledge gain.
- The need to conduct research on the remote sensing of post-fire vegetation recovery supported by field campaigns.
- The potential of spectral indices with a SWIR or MIR spectral band.
- Given that older generation sensors have many known limitations with respect to their suitability for studying complex biophysical characteristics, the need to benefit from new generation of high spatial and spectral resolution sensors and active sensors with different characteristics is of critical importance. Also the use of combined data acquired from more than one sensor.
- New research pathways to study the small-scale spatial patterns of post-fire vegetation recovery are required (e.g. the need to explore the potential of high resolution imagery such as QuickBird and IKONOS to classify individual plants).
- Investigation of the potential of Polarimetric Interferometric SAR data (TanDEM-X) to estimate post-fire biomass.
- Exploitation of advanced image analysis techniques in order to develop automated and transferable procedures.

6. Conclusions

Based on a review of the literature, a number of conclusions can be drawn:
- The role of remote sensing is increasingly becoming very important in post-fire monitoring.
- Most research has so far been carried out in areas covered by Mediterranean forests and shrublands.
- Current research is mainly based on the employment of Landsat, MODIS, VGT and AVHRR data. The extensive use of moderate and low resolution imagery is related to time series analysis.

- Vegetation indices and SMA are the main techniques employed so far in post-fire monitoring while the NDVI is the most commonly used index due to its strong relationship with above-ground biomass in a wide range of ecosystems. Only a small number of studies employ image classification due to the fact that the spatial resolution of the most commonly used satellite sensors exceeds the size of individual regenerating plants
- A number of developments including: the increase in the number of sensors with different characteristics suitable for post-fire monitoring (e.g. LIDAR, hyperspectral), the improved access to and availability of satellite data and derived products, and the development of new methods and advanced digital image analysis techniques (e.g. OBIA, Control plot selection) are expected to move forward research and establish RS an operational tool for post-fire monitoring.

7. Acknowledgments

This publication was financially supported by the Aristotle University of Thessaloniki (AUTh) Research Committee. The review was also supported by the Lebanese National Council for Scientific Research (CNRS) and part of the work was carried out at the Jet Propulsion Laboratory, California Institute of Technology, under a contract with the National Aeronautics and Space Administration.

8. References

Alcaraz-Segura, D., Chuvieco, E., Epstein, H., Kasischke, E. and Trishchenko, A., 2010. Debating the greening vs. browning of the North American boreal forest: differences between satellite datasets. Global Change Biology 16: 760-770.

Amanatidis, G., Paliatsos, A., Repapis, C. and Bartzis, J., 1993. Decreasing precipitation trend in the Marathon area, Greece. International Journal of Climatology 13: 191-201.

Amiro, B., MacPherson, J. and Desjardins, R., 1999. BOREAS flight measurements of forest-fire effects on carbon dioxide and energy fluxes. Agricultural and Forest Meteorology 96: 199-208.

Amiro, B., Orchansky, A., Barr, A., Black, T., Chambers, S., Chapin, F., Goulden, M., Litvak, M., Liu, H., McCaughey, J., McMillan, A. and Randerson, J., 2006. The effect of post-fire stand age on the boreal forest energy balance. Agricultural and Forest Meteorology 140: 41-50.

Arai, K., 2008. Nonlinear mixture model of mixed pixels in remote sensing satellite images based on Monte Carlo simulation. Advances in Space Research 41: 1725-1743.

Arianoutsou, M., 1984. Post-fire successional recovery of a phryganic (East Mediterranean) ecosystem. Acta Oecol. (Oecol. Plant.) 5(19): 287- 394.

Asner, G., 1998. Biophysical and biochemical sources of variability in canopy reflectance. Remote Sensing of Environment 64: 234-253.

Asner, G. and Lobell, D., 2000. A biogeophysical approach for automated SWIR unmixing of soils and vegetation. Remote Sensing of Environment 74: 99-112.

Atkinson, P., Cutler, M. and Lewis, H., 1997. Mapping sub-pixel proportional land-cover with AVHRR imagery. International Journal of Remote Sensing 18: 917-935.

Attema, E. P. W. and Ulaby, F. T., 1978. Vegetation modeled as a water cloud. Radio Science 13(2): 357-364.

Baatz, M. and Schape, A., 1999. Object-Oriented and Multi-Scale Image Analysis in Semantic Networks. In: 2nd International Symposium on Operationalization of Remote Sensing: 15-20.

Baeza, M. and Roy, J., 2008. Germination of an obligate seeder (Ulex parviflorus) and consequences for wildfire management. Forest ecology management 256: 685-693.

Bailey, J. and Covington, W., 2002. Evaluating ponderosa pine regeneration rates following ecological restoration treatments in northern Arizona, USA. Forest ecology and management 155.

Baret, F. and Guyot, G., 1991. Potentials and limits of vegetation indices for LAI and APAR assessment. Remote Sensing of Environment 35: 161-173.

Bastiaanssen, W., Menenti, M., Feddes, R. and Holtslag, A., 1998. A remote sensing surface energy balance algorithm for land (SEBAL) 1. Formulation. Journal of Hydrology 212-213: 198-212.

Beaty, R. and Taylor, A., 2001. Spatial and temporal variation of fire regimes in a mixed conifer forest landscape, southern Cascades, California, USA. Journal of Biogeography 28: 955-966.

Belda, F. and Meliá, J., 2000. Relationships between climatic parameters and forest vegetation: application to burned area in Alicante (Spain) Forest ecology and management 135 195-204.

Bertolette, D. and Spotskey, D., 2001. Remotely sensed burn severity mapping. In: H. David Harmon, Michigan, Crossing Boundaries in Park Management: Proceedings of the 11th Conference on research and Resource Management in Parks and on Public Lands, The George Wright Society: 44-51.

Bisson, M., Fornaciai, A., Coli, A., Mazzarini, F. and Pareschi, M., 2008. The Vegetation Resilience After Fire (VRAF) index: development, implementation and an illustration from central Italy. International Journal of Applied Earth Observation and Geoinformation 10: 312-329.

Bobbe, T., Lachowski, H., Maus, P., Greer, J. and Dull, C., 2001. A primer on mapping vegetation using remote sensing. International Journal of Wildland Fire 10: 277-287.

Bonham, C., 1989. Measurements for terrestrial vegetation. Wiley, New York, NY, USA.

Bortolot, Z. J. and Wynne, R. H., 2005. Estimating forest biomass using small footprint LIDAR data: An individual tree-based approach that incorporates training data. ISPRS Journal of Photogrammetry & Remote Sensing 59(6): 342- 360.

Bourgeau-Chavez, L. L., Kasischke, E. S., Riordan, K., Brunzell, S., Nolan, M., Hyer, E., Slawski, J., Medvecz, M., Walters, T. and Ames, S., 2007. Remote monitoring of spatial and temporal surface soil moisture in fire disturbed boreal forest ecosystems with ERS SAR imagery. International Journal of Remote Sensing 28(10): 2133 – 2162.

Brewer, C. K., Winne, J. C., Redmond, R. L., Opitz, D. W., Mangrich, M. V., 2005. Classifying and mapping wildfire severity: A comparison of methods. Photogrammetric Engineering and Remote Sensing 71: 1311-1320.

Broge, N. and Leblanc, E., 2000. Comparing prediction power and stability of broadband and hyperspectral vegetation indices for estimation of green leaf area index and canopy chlorophyll density. Remote Sensing of Environment 76: 156-172.

Buhk, C., Meyn, A. and Jentsch, A., 2007. The challenge if plant regeneration after fire in the Mediterranean Basin: scientific gaps in our knowledge on plant strategies and evolution of traits. Plant ecology 192: 1-19.

Calvo, L., Tarrega, R. and Luis, E., 1991. Regeneration in Quercus Pyrenaica Ecosystems After Surface Fires. International Journal of Wildland Fire 1: 205-210.

Calvo, L., Tarrega, R. and Luis, E., 1994. Comparative study of post-fire successional development in two communities with different maturity stage in Quercus pyrenaica climax. In: Viegas, D.X. (Ed.) Proceedings of the 2nd International Conference on Forest Fire Research : Coimbra: 951-960.

Calvo, L., Tarrega, R. and Luis, E., 1998. Twelve years of vegetation changes after fire in an Erica australis community. In: Trabaud, L. (Ed.) Fire Management and Landscape ecology. Fairfield, Whashington, International Association of Wildland Fire: 123-136.

Calvo, L., Tarrega, R. and Luis, E., 2002. The dynamics of Mediterranean shrubs species over 12 years following perturbations. Plant ecology 160: 25-42.

Carlson, T. and Ripley, T., 1997. On the relation between NDVI, fractional vegetation cover and leaf area index. Remote Sensing of Environment 62: 241-252.

Carranza, M., Ricotta, C., Napolitano, P., Massaro, E. and Blasi, C., 2001. Quantifying post-fire regrowth of remotely sensed Mediterranean vegetation with percolation-based methods. Plant Biosystems 135: 311-318.

Carreiras, J., Pereira, M. C. J. and Pereira, S. J., 2006. Estimation of tree canopy cover in evergreen oak woodlands using remote sensing. Forest Ecology and Management 23: 45-53.

Caturla, N., Raventós, J., Guàrdia, R. and Vallejo, R., 2000. Early post-fire regeneration dynamics of Brachypodium retusum Pers. (Beauv.) In old fields of the Valencia region (eastern Spain). Acta Oecologica 21: 1-12.

Chuvieco, E. and Congalton, R., 1988. Mapping and inventory of forest fires from digital processing of TM data. Geocarto International 4: 41–53.

Clemente, A., Rego, F. and Correia, A., 1996. Demographic patterns and productivity of post-fire regeneration in Portuguese Mediterranean maquis. Int. J. of Wildland Fire 6(1): 5-12.

Clemente, R., Navarro Cerrillo, R. and Gitas, I., 2009. Monitoring post-fire regeneration in Mediterranean ecosystems by employing multitemporal satellite imagery. International Journal of Wildland Fire 18: 648–658.

Cruz, A., Pérez, B. and Moreno, J. M., 2003. Resprouting of the Mediterranean type shrub Erica australis with modified lignotuber carbohydrate content. J. Ecol 91: 348-356.

Cuevas-Gonzalez, M., Gerard, F., Baltzer, H. and Riano, D., 2008. Studying the change in fAPAR after forest fire in Siberia using MODIS. International Journal of Remote Sensing 29: 6873-6892.

Cuevas-Gonzalez, M., Gerard, F., Baltzer, H. and Riano, D., 2009. Analysing forest recovery after wildfire disturbance in boreal Siberia using remotely sensed vegetation indices. Global Change Biology 15: 561-577.

Dafis, S., 1990. Prevention of fire and postfire rehabilitation using silvicultural methods. In: Workshop on forest fires in Greece. Department of Forestry and Natural Environment, Aristotelian University of Thessaloniki, Thessaloniki, Greece. 65-72.

Daskalakou, E. and Thanos, C., 1997. Post-fire establishment and survival of Aleppo pine seedlings. In: P. Balabanis, G. Eftichidis, R. Fantechi (Eds.) Forest fire risk and management European Commission, Directorate General XII, Science, Research and Development, Luxembourg. 357-368.

Daskalakou, E. N. and Thanos, C. A., 1996. Aleppo Pine (Pinus Halepensis) Postfire Regeneration: the Role of Canopy and Soil Seed Banks. International Journal of Wildland Fire 6: 59-66.

Daskalakou, E. N. and Thanos, C. A., 2004. Postfire regeneration of Aleppo pine - The temporal pattern of seedling recruitment. Plant ecology 171: 81-89.

De Luis, M., Garcia-Cano, M., Cortina, J., Raventos, J., Gonzales-Hidalgo, J. and Sanchez, J., 2001. Climatic trends, disturbances and short-term vegetation dynamics in a Mediterranean shrubland. Forest ecology and management: 25-37.

Diaz-Delgado, R., Lloret, F. and Pons, X., 2003. Influence of fire severity on plant regeneration by means of remote sensing. International Journal of Remote Sensing 24: 1751-1763.

Diaz-Delgado, R., Lloret, F., Pons, X. and Terradas, J., 2002. Satellite evidence of decreasing resilience in mediterranean plant communities after recurrent wildfires. Ecology 83: 2293-2303.

Díaz-Delgado, R. and Pons, X., 2001. Spatial patterns of forest fires in Catalonia (NE of Spain) along the period 1975-1995. Analysis of vegetation recovery after fire. Forest ecology and management 147: 67-74.

Drake, N., Mackin, S. and Settle, J., 1999. Mapping vegetation, soils and geology in semiarid shrublands using spectral matching and mixture modeling of SWIR AVIRIS imagery. Remote Sensing of Environment 68: 12-25.

Dubayah, R. O. and Drake, J. B., 2000. LIDAR remote sensing for forestry applications. Journal of Forestry 98: 44-46.

Elmore, A., Mustard, J., Manning, S. and Lobell, D., 2000. Quantifying vegetation change in semiarid environments: precision and accuracy of spectral mixture analysis and the normalized difference vegetation index. Remote Sensing of Environment 73: 87-102.

EPA, 2001. Inventory of U.S. Greenhouse Gas Emissions and Sinks: 1990 – 1999. In: Environmental Protection Agency, U.S. 75-85.

Escuin, S., Fernández-Rebollo, P. and Navarro, R. M., 2002. Aplicación de escenas Landsat a la asignación de grados de afectación producidos por incendios forestales. Revista de Teledetección 1: 25–36.

Eshel, A., Henig-Sever, N. and Ne'eman, G., 2000. Spatial variation of seedling distribution in an east Mediterranean pine woodland at the beginning of post-fire succession. Plant Ecology 148: 175-182.

Ferrandis, P., Herranz, J. M. and Martínez-Sánchez, J. J., 1996. The Role of Soil Seed Bank in the Early Stages of Plant Recovery After Fire in a Pinus Pinaster Forest in SE Spain. International Journal of Wildland Fire 6: 31-35.

García-Ruiz, J., Lasanta, T., Ruiz-Flano, P., Ortigosa, L., White, S., González, C. and Martí, C., 1996. Land-use changes and sustainable development in mountain areas: A case study in the Spanish Pyrenees. Landscape Ecology 11: 267-277.

Garestier, F., Dubois Fernandez, P. and Papathanassiou, K., 2008. Pine forest height inversion using single-pass X-band PolInSAR data. IEEE Trans. Geosci. Remote Sensing 46: 59–68.

Gitas, I., 1999. Geographical Information Systems and Remote Sensing in mapping and monitoring fire-altered forest landscapes. PhD dissertation. Department of Geography, University of Cambridge: 237.

Gitas, I., De Santis, A. and Mitri, G., 2009. Remote sensing of burn severity. In: Chuvieco, E., (Eds), Earth observation of wildland fires in mediterranean ecosystems. Springer-Verlag, Berlin: 129-148.

Goetz, S., Fiske, G. and Bunn, A., 2006. Using satellite time-series data sets to analyze fire disturbance and forest recovery across Canada. Remote Sensing of Environment 101: 352-365.

Goetz, S., Sun, M., Baccini, A. and Beck, P., 2010. Synergetic use of spaceborne LIDAR and optical imagery for assessing forest fire disturbance: an Alaskan case study. Journal of Geophysical Research 115(G00E07).

Gould, K., Fredericksen, T., Morales, F., Kennard, D., Putz, F., Mostacedo, B. and Toledo, M., 2002. Post-fire tree regeneration in lowland Bolivia: implications for fire management. Forest ecology and management 165: 225-234.

Gouveia, C., DaCamara, C. and Trigo, R., 2010. Post-fire vegetation recovery in Portugal based on spot/vegetation data. Natural Hazards and Earth System Sciences 10: 673-684.

Hajnsek, I., Kugler, F., Lee, S. and Papathanassiou, K., 2009. Tropical-forest-parameter estimation by means of Pol-InSAR: The INDREX-II campaign. IEEE Trans. Geosci. Remote Sensing 47: 481–493.

Hall, F., Botkin, D., Strebel, D., Woods, K. and Goetz, S., 1991. Large-scale patterns of forest succession as determined by remote sensing. Ecology 72: 628-640.

Henry, M. and Hope, S., 1998. Monitoring post-burn recovery of chaparral vegetation in southern California using multi-temporal satellite data. International Journal of Remote Sensing 19: 3097-3107.

Hernandez-Clemente, R., Navarro, R., Hernandez-Bermejo, J., Escuin, S. and Kasimis, N., 2009. Analysis of postfire vegetation dynamics of Mediterranean shrub species based on terrestrial and NDVI data. Environmental Management 43: 876-887.

Herranz, J., Ferrandis, P. and Martínez-Sánchez, J., 1999. Influence of heat on seed germination of nine woody Cistaceae species. International Journal of Wildland Fire 9: 173-182.

Hicke, J., Asner, G., Kasischke, E., French, N., Randerson, J., Collatz, J., Stocks, B., Tucker, C., Los, S. and Field, C., 2003. Postfire response of North American boreal forest net primary productivity analyzed with satellite observations. Global Change Biology 9: 1145-1157.

Hope, A., Tague, C. and Clark, R., 2007. Characterizing post-fire vegetation recovery of California chaparral using TM/ETM+ time-series data. International Journal of Remote Sensing 28: 1339-1354.

Huete, A., 1988. A Soil-Adjusted Vegetation Index (SAVI). Remote Sensing of Environment 25: 295–309.

Hunter, M. E., Omi, P. N., Martinson, E. J. and Chong, G. W., 2006. Establishment of non-native plant species after wildfires: effects of fuel treatments, abiotic and biotic factors, and post-fire grass seeding treatments. International Journal of Wildland Fire 15: 271-281.

Idris, M., Kuraji, K. and Suzuki, M., 2005. Evaluating vegetation recovery following large-scale forest fires in Borneo and northeastern China using multi-temporal NOAA/AVHRR images. Journal of Forest Research 10: 101-111.

Inbar, M., Tamir, M. and Wittenberg, L., 1998. Runoff and erosion processes after a forest fire in Mount Carmel, a Mediterranean area. Geomorphology 24: 17-33.

Jabukauskas, M., Lulla, K. and Mausel, P., 1990. Assessment of vegetation change in a fire-altered forest landscape. Photogrammetric Engineering and Remote Sensing 56: 371-377.

Jacobson, C., 2010. Use of linguistic estimates and vegetation indices to assess post-fire vegetation regrowth in woodland areas. International Journal of Wildland Fire 19: 94-103.

Jakubauskas, M., 1988. Postfire vegetation change detection using LANDSAT MSS and TM data. School of Graduate Studies. Terre Haute, Indiana, Indiana State University: 99.

Jakubauskas, M., Lulla, K. and Mausel, P., 1990. Assessment of vegetation change in a fire-altered forest landscape. Photogrammetric Engineering and Remote Sensing 56: 371-377.

Ju, J. and Roy, D., 2008. The availability of cloud-free Landsat ETM+ data over the conterminous United States and globally. Remote Sensing of Environment 112: 1196-1211.

Kasischke, E. S., Bourgeau-Chavez, L. L. and Johnstone, J. F., 2007. Assessing spatial and temporal variations in surface soil moisture in fire-disturbed black spruce forests in Interior Alaska using spaceborne synthetic aperture radar imagery – Implications for post-fire tree recruitment. Remote Sensing of Environment 108: 42–58.

Kasischke, E. S., Melack, J. M. and Dobson, M. C., 1997. The use of imaging radars for ecological applications-A review. Remote Sensing Environment 59: 141-156.

Kasischke, E. S., Tanase, M. A., Bourgeau-Chavez, L. L. and Borr, M., 2011. Soil moisture limitations on monitoring boreal forest regrowth using spaceborne L-band SAR data. Remote Sensing of Environment 115: 227–232.

Kavgaci, A., Carni, A., Basaran, S., Basaran, M. A., Kosir, P., Marinsek, A. and Silc, U., 2010. Long-term post-fire succession of Pinus brutia forest in the east Mediterranean. International Journal of Wildland Fire 19: 599-605.

Kazanis, D. and Arianoutsou, M., 2004. Long-term post-fire vegetation dynamics in Pinus halepensis forests of central Greece: a functional-group approach. Plant ecology 171: 101-121.

Keeley, J., 2000. Chaparral. In: M. Barbour and W. Billings, North American terrestrial vegetation, New York, Cambridge University Press: 204-253.

Kennard, D., Gould, K., Putz, F., Fredericksen, T. and Morales, F., 2002. Effect of disturbance intensity on regeneration meachanisms in a tropical dry forest. Forest ecology and management 162: 197-208.

Kim, Y., Yang, Z., Cohen, W. B., Pflugmacher, D., Lauver, C. L. and Vankat, J. L., 2009. Distinguishing between live and dead standing tree biomass on the North Rim of Grand Canyon National Park, USA using small-footprint LIDAR data. Remote Sensing of Environment 113: 2499–2510.

Kozlowski, T., 2002. Physiological ecology of natural regeneration of harvested and disturbed forest stands: implications for forest management. Forest ecology and management: 195-221.

Krieger, G., Papathanassiou, K. P., Younis, M., Moreira, A. and Hajnsek, I., 2010. Interferometric Synthetic Aperture Radar (SAR) Missions Employing Formation Flying A German Earth satellite system is designed to produce data on forest structure, biomass, tectonic shifts and glacier movements, and to advance understanding of Earth dynamics. Proceedings of the IEEE 98(5).

Le Houerou, H., 1987. Vegetation wildfires in the Mediterranean basin: evolution and trends. Ecologia Mediterranea 13: 13-24.

Lhermitte, S., Verbesselt, J., Verstraeten, W. and Coppin, P., 2010. A pixel based regeneration index using time series similarity and spatial context. Photogrammetric Engineering and Remote Sensing 76: 673-682.

Lhermitte, S., Verbesselt, J., Verstraeten, W. W., Veraverbeke, S. and Coppin, P., 2011. Assessing intra-annual vegetation regrowth after fire using the pixel based regeneration index. ISPRS Journal of Photogrammetry and Remote Sensing 66: 17-27.

Li, M., Qu, J. and Hao, X., 2008. Detecting vegetation change with satellite remote sensing over 2007 Georgia wildfire regions. Journal of Applied Remote Sensing 2(021505).

Lim, K., Treitz, P., Wulder, M., Stonge, B. and Flood, M., 2003a. LIDAR remote sensing of forest structure. Progress in Physical Geography 27(1): 88-106.

Lloret, F. and Vilà, M., 1997. Clearing of vegetation in Mediterranean garrigue: response after a wildfire. Forest ecology and management 93: 227-234.

Lyons, E. J., Y. and Randerson, J., 2008. Changes in surface albedo after fire in boreal forest ecosystems of interior Alaska assessed using MODIS satellite observations. Journal of Geophysical Research 113(G02012).

Maheras, P., 1988. Changes in precipitation conditions in the western Mediterranean over the last century. J. Clim 8: 179-189.

Malak, D. and Pausas, J., 2006. Fire regime and post-fire Normalized Difference Vegetation Index changes in the eastern Iberian peninsula. International Journal of Wildland Fire 15: 407–413.

Marchetti, M., Ricotta, C. and Volpe, F., 1995. A qualitative approach to the mapping of post-fire regrowth in Mediterranean vegetation with Landsat TM data. International Journal of Remote Sensing 16: 2487-2494.

Martínez-Sánchez, J., Ferrandis, P., De las Heras, J. and Herranz, J., 1999. Effect of burnt wood removal on the natural regeneration of Pinus halepensis after fire in a pine forest in Tus valley (SE Spain). Forest ecology and management: 1-10.

McMichael, C., Hope, A., Roberts, D. and Anaya, M., 2004. Post-fire recovery of leaf area index in California chaparral: a remote sensing-chronosequence approach. International Journal of Remote Sensing 25: 4743–4760.

Miller, M., 2000. Fire autoecology. In: J.K. Brown and J.K. Smith (Eds.), Wildland Fire in Ecosystems. Ogden, Uta: RMRS-GTR-42, USDA Forest Service, Rocky Mountain Research Station.

Minchella, A., Del Frate, F., Capogna, F., Anselmi, S. and Manes, F., 2009. Use of multitemporal SAR data for monitoring vegetation recovery of Mediterranean burned areas. Remote Sensing of Environment 113: 588-597.

Mitchell, M. and Yuan, F., 2010. Assessing forest fire and vegetation recovery in the Black Hills, South Dakota. GIScience and Remote Sensing 47: 276-299.

Mitri, G. and Gitas, I., 2004. A semi-automated object-oriented model for burned area mapping in the Mediterranean region using Landsat-TM imagery. International Journal of Wildland Fire 13(3): 367-376.

Mitri, G. and Gitas, I., 2008. Mapping the severity of fire using object-based classification of IKONOS imagery. International Journal of Wildland Fire 17: 431–442.

Mitri, G. and Gitas, I. Z., 2010. Mapping postfire vegetation recovery using EO-1 Hyperion imagery. IEEE Transactions on Geoscience and Remote Sensing 48: 1613-1618.

Mooney, R. and Hobbs, R., 1986. Resilience at the individual plant level. In: B.B. Dell, D. Lamont, and A. Hopkins (Eds.), Resilience in Mediterranean – type ecosystems: 65-82.

Moreno, J. M., Pineda, F. and Rivas-Martinez, S., 1990. Climate vegetation at the Eurosiberian- Mediterranean boundary in the Iberian Peninsula. Journal of Vegetation Science 1: 233-244.

Mouillet, F., Rambal, S. and Lavorel, S., 2001. A generic process-based SImulator for meditERRanean landscApes (SIERRA): design and validation exercises. Forest ecology and management: 75-97.

Naveh, Z., 1991. The role of fire in Mediterranean vegetation. Botanica Chronika (Greece) 10: 385-405.

Ne'eman, G., Fotheringham, C. J. and Keeley, J., 1999. Patch to landscape patterns in post fire recruitment of a serotinous conifer. Plant ecology 12: 235-242.

Nepstad, D., Verssimo, A., Alencar, A., Nobre, C., Lima, E., Lefebvre, P., Schlesinger, P., Potter, C., Moutinho, P., Mendoza, E., Cochrane, M. and Brooks, V., 1999. Large-scale impoverishment of Amazonian forest by logging and fire. Nature 398: 505-508.

Oliver, C., 1981. Forest development in North America following major disturbances. Forest ecology and management 3: 153-168.

Oliver, C. and Larson, B., 1996. Forest Stand Dynamics. Wiley, New York, NY, USA.

Palandjian, D., Gitas, I. and Wright, R., 2009. Burned area mapping and post-fire impact assessment in the Kassandra peninsula (Greece) using Landsat TM and Quickbird data. Geocarto International 24: 193-205.

Papathanassiou, K. P. and Cloude, S. R., 2001. Single-Baseline Polarimetric SAR Interferometry. IEEE Transactions on Geoscience and Remote Sensing 39(11).

Patterson, M. and Yool, S., 1998. Mapping fire-induced vegetation mortality using Landsat thematic mapper data: A comparison of linear transformation techniques. Remote Sensing of Environment 65: 132-142.

Pausas, J., Carbó, E., Caturla, R., Gil, J. and Vallejo, R., 1999. Post-fire regeneration patterns in the eastern Iberian Peninsula. Acta Oecologica 20: 499-508.

Pausas, J. and Verdu, M., 2005. Plant persistence traits in the fire-prone ecosystems of the Mediterranean basin: a phylogenetic approach. Oikos 109: 196-202.

Pausas, J. G., Ouadah, N., Ferran, A., Gimeno, T. and Vallejo, R., 2003. Fire severity and seedling establishment in Pinus halepensis woodlands, Eastern Iberian Peninsula. Plant Ecology 169: 205- 213.

Perez-Cabello, F., Echeverria, M., Ibarra, P. and De la Riva, J., 2009. Effects of fire on vegetation, soil and hydrogeomorphological behavior in Mediterranean ecosystems. In: Earth Observation of Wildland fires in Mediterranean ecosystems (Emilio Chuvieco Ed.) – Springer: 111-128.

Perula, V. G., Cerrillo, R. N., Reboloo, P. F. and Murillo, G. V., 2003. Postfire regeneration in Pinus pinea L. and Pinus pinaster Aiton in Andalucia (Spain). Env. Manage 31(1): 86-99.

Peterson, S. and Stow, D., 2003. Using multiple image endmember spectral mixture analysis to study chaparral regrowth in Southern California. International Journal of Remote Sensing 24: 4481-4504.

Piñol, J., Terradas, J. and Lloret, F., 1998. Climatic warming hazard, and wildfire occurrence in coastal eastern Spain. Climate Change 38: 345-357.

Praks, J., Kugler, F., Papathanassiou, K., Hajnsek, I. and Hallikainen, M., 2007. Tree height estimation for boreal forest by means of L- and X-band PoLInSAR and HUTSCAT profiling scatterometer. IEEE Geosci. Remote Sens. Lett 4: 466–470.

Qi, J., Chehbounidi, A., Huete, A., Kerr, Y. and Sorooshian, S., 1994. A modified soil adjusted vegetation index. Remote Sensing of Environment 48: 119–126.

Ramsey, E., Nelson, G., Sapkota, S., Laine, S., Verdi, J. and Krasznay, S., 1999. Using multiple-polarization L-band radar to monitor marsh burn recovery. IEEE Transactions on Geoscience and Remote Sensing 37: 636-639.

Randerson, J., Liu, H., Flanner, M., Chamber, S., Jin, Y., Hess, P., Pfister, G., Mack, M., Treseder, K., Welp, L., Chapin, F., Harden, J., Goulden, M., Lyons, E., Neff, J., Schuur, E. and Zender, C., 2006. The impact of boreal forest fire on climate warming. Science 314: 1130-1132.

Rego, F., 1992. Land use changes and wildfires. In: A. Teller, P. Mathy, J. Jeffers (Eds.) Response of forest fires to environmental change, Elsevier, London: 367-373.

Riaño, D., Chuvieco, E., Ustin, S., Zomer, R., Dennison, P., Roberts, D. and Salas, J., 2002. Assessment of vegetation regeneration after fire through multitemporal analysis of AVIRIS images in the Santa Monica Mountains. Remote Sensing of Environment 79: 60-71.

Ricotta, C., Avena, G., Olsen, E., Ramsey, R. and Winn, D., 1998. Monitoring the landscape stability of Mediterranean vegetation in relation to fire with a fractal algorithm. International Journal of Remote Sensing 19: 871-881.

Roberts, D., Gardner, M., Church, R., Ustin, S., Scheer, G. and Green, R., 1998. Mapping chaparral in the Santa Monica mountains using multiple endmember spectral mixture models. Remote Sensing of Environment 65: 267-279.

Roder, A., Hill, J., Duguy, B., Alloza, J. and Vallejo, R., 2008. Using long time series of Landsat data to monitor fire events and post-fire dynamics and identify driving factors. A case study in the Ayora region (eastern Spain). Remote Sensing of Environment 112: 259-273.

Roerink, G., Su, Z. and Menenti, M., 2000. S-SEBI: A simple remote sensing algorithm to estimate the surface energy balance. Physics and Chemistry of the Earth 25: 147-157.

Rogan, J. and Franklin, J., 2001. Mapping wildfire burn severity in southern California forests and shrublands using Enhanced Thematic Mapper imagery. Geocarto International 16: 1–11.

Rogge, D., Rivard, B., Zhang, J. and Feng, J., 2006. Iterative spectral unmixing for optimizing per-pixel endmember sets. IEEE Transactions on Geoscience and Remote Sensing 44: 3725–3736.

Sankey, T., Moffet, C. and Weber, K., 2008. Postfire recovery of sagebrush communities: assessment using SPOT-5 and very large-scale aerial imagery. Rangeland Ecology and Management 61: 598–604.

Santalla, S., Marcos, E., Valbuena, L., Calvo, L., Tárrega, R. and Luis, E., 2002. First years of regeneration in Quercus pyrenaica forest and Pinus pinaster stand after wildland fire. In: D. Viegas (Ed.) Forest Fire Research & Wildland Fire Safety, Millpress, Rotterdam.

Schroeder, D. and Perera, A., 2002. A comparison of large-scale spatial vegetation patterns following clearcuts and fires in Ontario's boreal forests. Forest Ecology and Management 159: 217-230.

Segah, H., Tani, H. and Hirano, T., 2010. Detection of fire impact and vegetation recovery over tropical peat swamp forest by satellite data and ground-based NDVI instrument. International Journal of Remote Sensing 31: 5297-5314.

Shaw, D., Malthus, D. and Kupiec, J., 1998. High-spectral resolution data for monitoring Scots pine (Pinus sylvestris L.) regeneration. International Journal of Remote Sensing 19: 2601-2608.

Skordilis, A. and Thanos, C., 1995. Seed stratification and germination strategy in the Mediterranean pines Pinus brutia and P. halepensis. Seed Science Research 5: 151-160.

Smith, R. and Woodgate, P., 1985. Appraisal of fire damage and inventory for timber salvage by remote sensing in mountain ash forests in Victoria. Australian Journal of Forestry 48(4): 252–263.

Somers, B., Verbesselt, J., Ampe, E., Sims, N., Verstraeten, W. W. and Coppin, P., 2010. Spectral mixture analysis to monitor defoliation in mixed-age Eucalyptus globulus Labill plantations in southern Australia using Landsat 5-TM and EO-A Hyperion data. International Journal of Applied Earth Observation and Geoinformation 12: 270–277.

Steyaert, L., Hall, F. and Loveland, T., 1997. Land cover mapping, fire regeneration, and scaling studies in the Canadian boreal forest with 1 km AVHRR and Landsat TM data. International Journal of Geophysical Research 102: 29581-29598.

Stueve, K., Cerney, D., Rochefort, R. and Kurt, L., 2009. Post-fire tree establishment patterns at the alpine treeline ecotone: Mount Rainier National Park, Washington, USA. Journal of Vegetation Science 20: 107–120.

Tanase, M., Riva, J.d.l., Santoro, M., Pérez-Cabello, F., Kasischke, E., 2011. Sensitivity of SAR data to post-fire forest regrowth in Mediterranean and boreal forests. Remote Sensing of Environment 115: 2075-2085.

Tarrega, R., Luis-Calabuig, E. and Valbuena, L., 2001. Eleven years of recovery dynamic after experimental burning and cutting in two Cistus communities. Acta Oecologica 22: 277-283.

Telesca, L. and Lasaponara, R., 2006. Pre- and post-fire behavioral trends revealed in satellite NDVI time series. Geophysical Research Letters 33(L14401).

Thanos, C., Daskalakou, E. and Nikolaidou, S., 1996. Early post-fire regeneration of a Pinus halepensis forest on Mount Parnis, Greece. Journal of Vegetation Science 7: 273-280.

Thanos, C. A. and Georghiou, K., 1988. Ecophysiology of fire-stimulated seed germination in Cistus incanus ssp creticus (L.) Heywood and C. salvifolius L. Plant, Cell and Environment 11: 841-849.

Thenkabail, P., Smith, R. and De-Pauw, E., 2000. Hyperspectral vegetation indices for determining agricultural crop characteristics. Remote Sensing of Environment 71: 158–182.

Thenkabail, P., Smith, R. and De-Pauw, E., 2002. Evaluation of narrowband and broadband vegetation indices for determining optimal hyperspectral wavebands for agricultural crop characterization. Photogrammetric Engineering and Remote Sensing 68: 607– 621.

Torn, M. and Fried, J., 1992. Predicting the impacts of global warming on wildland fire. Climate Change 21: 257-274.

Trabaud, L., 1994. Plant and fire variability relationships more specifically in the Mediterranean basin. In: 2nd International Conference in Forest Fire Research, Coimbra, Portugal: 53-58.

Tsitsoni, T., 1997. Conditions determining natural regeneration after wildfires in the Pinus halepensis (Miller, 1768) forests of Kassandra Peninsula (North Greece). Forest ecology and management 92: 199-208.

Tucker, C., 1979. Red and photographic infrared linear combinations for monitoring vegetation. Remote Sensing of Environment 8: 127-150.

van Leeuwen, W., 2008. Monitoring the effects of forest restoration treatments on post-fire vegetation recovery with MODIS multitemporal data. Sensors 8: 2017-2042.

van Leeuwen, W., Casady, G., Neary, D., Bautista, S., Alloza, J., Carmel, J., Wittenberg, L., Malkinson, D. and Orr, B., 2010. Monitoring post-wildfire vegetation response with remotely sensed time series data in Spain, USA and Israel. International Journal of Wildland Fire 19: 75-93.

van Wagtendonk, J., Root, R., Key, C. and Running, S., 2004. Comparison of AVIRIS and Landsat ETM+ detection capabilities for burn severity. International Journal of Wildland Fire 92: 397-408.

Veraverbeke, S., Verstraeten, W. W., Lhermitte, S. and Goossens, R., 2010a. Illumination effects on the differenced Normalized Burn Ratio's optimality for assessing fire severity. International Journal of Applied Earth Observation and Geoinformation 12: 60-70.

Veraverbeke, S., Lhermitte, S., Verstraeten, W. W. and Goossens, R., 2010b. The temporal dimension of differenced Normalized Burn Ratio (dNBR) fire/burn severity studies: the case of the large 2007 Peloponnese wildfires in Greece. Remote Sensing of Environment 114: 2548-2563.

Veraverbeke, S., Lhermitte, S., Verstraeten, W. W. and Goossens, R., 2011a. A time-integrated MODIS burn severity assessment using the multi-temporal differenced Normalized Burn Ratio (dNBRMT). International Journal of Applied Earth Observation and Geoinformation 13: 52-58.

Veraverbeke, S., Somers, B., Gitas, I., Katagis, T., Polychronaki, A. and Goossens, R., 2012a. Spectral mixture analysis to assess post-fire vegetation regeneration using Landsat Thematic mapper imagery: accounting for soil brightness variation. International Journal of Applied Earth Observation and Geoinformation 14: 1-11.

Veraverbeke, S., Gitas, I., Katagis, T., Polychronaki, A., Somers, B. and Goossens, R., 2012b. Assessing post-fire vegetation recovery using red-near infrared vegetation indices: accounting for background and vegetation variability. ISPRS Journal of Photogrammetry and Remote Sensing in press.

Veraverbeke, S., Verstraeten, W. W., Lhermitte, S., Van De Kerchove, R. and Goossens, R., 2012c. Spaceborne assessment of post-fire changes in vegetation, land surface temperature and surface albedo. International Journal of Wildland Fire in press.

Vicente-Serrano, S., Perez-Cabello, F. and Lasanta, T., 2008. Assessment of radiometric correction techniques in analyzing vegetation variability and change using time series of Landsat images. Remote Sensing of Environment 112: 3916-3934.

Viedma, O., Melia, J., Segarra, D. and Garcia-Haro, J., 1997. Modeling rates of ecosystem recovery after fires by using Landsat TM data. Remote Sensing of Environment 61: 383-398.

Vila, G. and Barbosa, P., 2010. Post-fire vegetation regrowth detection in the Deiva Marina region (Liguria-Italy) using Landsat TM and ETM+ data. Ecological Modelling 221: 75-84.

Wahren, C., Papst, W. and Williams, R., 2001. Early post-fire regeneration in subalpine heathland and grassland in the Victorian Alpine National Park, south-eastern Australia. Austral Ecology 26: 670-679.

White, J., Ryan, K., Key, C. and Running, S., 1996. Remote sensing of forest fire severity and vegetation recovery. International Journal of Wildland Fire 6: 125-136.

Wicks, T., Smith, G. and Curran, P., 2002. Polygon-based aggregation of remotely sensed data for regional ecological analysis. International Journal of Applied Earth Observation and Geoinformation 4: 161-173.

Wittenberg, L., Malkinson, D., Beeri, O., Halotzy, A. and Tesler, N., 2007. Spatial and temporal patterns of vegetation recovery following sequences of forest fires in a Mediterranan landscape, Mt. Carmel Israel. Catena 71: 76-83.

Wulder, M. A., White, J. C., Alvarez, F., Han, T., Rogan, J. and Hawkes, B., 2009. Characterizing boreal forest wildfire with multi-temporal Landsat and LIDAR data. Remote Sensing of Environment 113: 1540-1555.

8

The Science and Application of Satellite Based Fire Radiative Energy

Evan Ellicott and Eric Vermote
University of Maryland
USA

1. Introduction

Biomass burning has been a topic of research interest for many years due to the implications for climatic change as a result of landscape alteration and atmospheric loading of aerosols and trace gases from pyrogenic emissions (Crutzen & Andreae, 1990). Crutzen et al. (1979) first highlighted the variety of trace gas emissions from tropical forest fires and the potential these constituents could have in altering atmospheric chemistry and biogeochemical cycles. Subsequent research has demonstrated additional impacts on the biosphere, atmosphere, and directly upon humans. For example, ozone (O_3) is produced photochemically in the troposphere from hydrocarbon and nitrogen oxides released during vegetation burning and results in regional health hazards such as damage to human respiratory systems (Andreae, 2004; Levine, 2003). Cicerone (1994) emphasized that some byproducts of biomass burning, such as methyl chloride (CH_3Cl) and methyl bromide (MeBr), can escape to the stratosphere where they are responsible for ozone destruction; resulting in health risks at a much larger scale.

Fire is an integral part of many ecosystems (Kuhry, 1994; Cary and Banks, 2000), but the nature of this relationship may change according to some climate models which show fire frequency and intensity increasing with global warming trends (Intergovernmental Panel on Climate Change [IPCC], 2007). For example, boreal forests, one of Earth's larger biomes, are a key component in global carbon cycling. In particular, peatland in boreal and sub-arctic regions of Earth are estimated to contain 455Gt of carbon, translating to roughly a third of the world's soil carbon pool (Brady & Weil, 1999; Gorham, 1991; Moore, 2002; Pastor et al., 2003), and act as a sink for atmospheric carbon; accounting for an uptake of roughly 12% of the global anthropogenic emissions (Moore, 2002). Carbon sequestered through the process of photosynthesis by living vegetation does not exit the boreal system at the same rate since respiration via decomposition is retarded. However combustion of organic matter contributes to atmospheric loading of "greenhouse" gases (Kaufman et al., 1990; Page et al., 2002) and can affect carbon sequestration regimes (Kasischke et al., 1995). In addition, the influence of anthropogenic ignited fires, which accounts for 90% of all biomass burning (Levine, 2000), may increase with population growth and the added pressure for land and resources. A result of these driving forces will be greater biomass burning emissions, decreased sequestration of carbon, and the potential creation of feedback loops (Kasischke et al., 1995a; Chapman and Thurlow, 1998; Moore, 2002).

In order to understand the spatial and temporal global distribution of biomass burning, and ultimately the potential impacts to the biosphere and atmosphere, regular, broad scale monitoring is necessary. Satellite sensors provide daily, synoptic observations to detect and analyze fires (Justice et al., 2002) and therefore a great deal of research to characterize fires from remote sensing systems has been performed over the past several decades (e.g. Dozier, 1981; Kaufman et al., 1998; Giglio et al., 2003; Wooster et al., 2005; Ichoku & Kaufman, 2005).

2. Remote sensing of fire

Information derived from satellites has many advantages to traditional in situ data collection. A global perspective can be achieved to observe various Earth system processes allowing monitoring of spatially distinct, inaccessible, or remote locations. Regular monitoring is possible from polar orbiting satellite platforms such as National Aeronautics and Space Administration's (NASA) Aqua and Terra satellites and National Oceanic and Atmospheric Administration's (NOAA) Advanced Very High Resolution Radiometer (AVHRR). In the case of the former, the combination of Aqua and Terra provide nominally 4 daily "looks" of most locations on Earth. In addition, geostationary satellites such as NOAA's Geostationary Operational Environmental Satellites (GOES) and the European Organization for the Exploitation of Meteorological Satellites' (EUMETSAT) Meteosat provide high temporal resolution (15-30 minute), continental wide observations.

Many of the channels available from a particular satellite sensor are useful for fire monitoring, for example aerosols can be monitored using the the visible and near-infrared bands or burn scars can be monitored with the visible, near, and middle infrared bands. Burned area mapping, a commonly used metric, is important for estimating total biomass consumed and thus emission estimates. Advanced new algorithms for accurate estimation of burned area now address the effect of bi-directional reflectance (Roy et al., 2005); an effect which is a function of the sun-target-view geometry influencing the directional dependence in reflectance and a potential source of error when using time-series data.

The retrieval of fire hot-spots provides additional monitoring and measurement capabilities. The foundation for fire detection is the enhanced middle infrared radiance emitted during flaming or smoldering combustion as described by the Planck function. In the case of an observed fire, radiance values generally peak around 3.7 – 4.0µm, whereas the peak for background terrestrial surface temperatures is near 10.0 – 11.0µm; thus temperature anomalies can be flagged as potential fire "hot-spots". As a result of this rather simple relationship, remotely sensed data has had a significant contribution in fire science and monitoring. Heritage systems such as AVHRR and GOES, though not necessarily intended to include fire detection or monitoring missions, have proven valuable for this task nonetheless (Boles & Verbyla, 2000). Fire characterization from satellites, such as subpixel temperatures and flaming area, was obtained from a method developed by Dozier (1981), who introduced a theoretical procedure that exploits the different responses of two channels (3 and 4) aboard AVHRR (3.75µm and 10.8µm, respectively) for sub-pixel hot spot detection; an approach that set the framework for future sensor fire detection and characteristic methodologies (Giglio & Kendall, 2001; Justice et al., 2002; Giglio et al., 2003; Wooster et al., 2003, 2005).

The application of hot-spot detections has been employed in numerous studies and for a variety of uses. Legg & Laumonier (1999) demonstrated the effectiveness of hot spot detections using AVHRR and ATSR (Along Track Scanning Radiometer), as well as burned area estimates using SPOT (Satellite Pour Observation de la Terra) imagery, during the 1997 Indonesian fire season. Legg & Laumonier (1999) also employed the United States Defense Meteorological Satellite Program (US DMSP) to help eliminate spurious daytime hotspots by detecting highly reflective pixels at night, presumably from fire, while excluding known human related bright spots. Kaufman et al. (1990) used AVHRR fire counts to assess trace gas and aerosol emissions in the tropics. Although some of the assumptions about burned area and fire detections would later be shown to be erroneous, their research set in motion the development of new approaches for using remotely sensed fire information for emission estimates (Kaufman, 1998). Aragão et al. (2007) examined drought and fire spatial distribution in Amazonia using NOAA-12's AVHRR and MODIS hot spot detections and later Aragão et al. (2008) examined specific interactions between precipitation, deforestation, and fires related to the 2005 Brazilian Amazon drought. More recently, Aragão & Shimabukuro (2010), again using MODIS and NOAA-12 hot spots, showed the co-varying nature of deforestation and fire activity trends over the past several years in the Brazilian Amazon. Their research has implications for fire and emission policies such as the United Nation's REDD+ (Reducing Emissions from Deforestation and Degradation). Giglio et al. (2010) used active fire detections to expand their burned area product to pre-MODIS data using the ATSR and Visible and Infrared Scanner (VIRS) aboard the Tropical Rainfall Measuring Mission (TRMM). The active fire burned area was developed using relationships based on regression between MODIS-Terra 500 meter burned area reference maps and Terra active fire counts. Morton et al. (2008) showed a clear correlation between fire hot spot frequency and land use patterns in Amazonia. Their research concluded that trends in land use intensity and fire frequency were linked. Such work offers promise for developing monitoring schemes to characterize land use transitions to inform policy makers.

In addition to the above research, a variety near-real time applications of active fire detections exist. The U.S. Forest Service Active Fire Mapping Program (http://activefiremaps.fs.fed.us/) is an operational system providing invaluable, near-real time information about location and timing of fire activity in the United States and Canada allowing fire managers to efficiently monitor fires and allocate resources. The Fire Information for Resource Management System (FIRMS, http://maps.geog.umd.edu/firms/) delivers timely fire detections, made by MODIS and processed through the Rapid Response System (http://rapidfire.sci.gsfc.nasa.gov/), to fire managers around the world.

Development of "new tools" such as fire radiative energy (FRE) can aid in estimating the biomass combusted and rates of atmospheric loading trace gases and aerosols. Calculated by determining the amount of energy emitted during fire, FRE may offer an accurate measurement of the fire intensity and vegetation consumed per unit time, as will be discussed in more detail below.

3. Combustion, fire energy, and emissions

3.1 Burned area

In theory, remotely sensed data should offer the capability to directly quantify atmospheric emissions from fire events, but in practice this requires determining the source of emissions which involves complex, computationally demanding inversion and geochemical transport modeling. Therefore most current approaches, referred to as the "bottom up" method in this paper, involve multiplying the fuel consumed by an emission factor for the atmospheric species of interest. Emission estimates for natural and anthropogenic ignited vegetation fires are generally calculated using spatially explicit measures of pre-fire fuel loads, fuel consumption, and the areal extent of fire impact. The model presented by Seiler & Crutzen (1980) has been used extensively to quantify the mass of fuel consumed:

$$M = A \bullet B \bullet \beta \qquad (1)$$

M is the total dry biomass consumed (kg); A is the burned area (km²); B is the biomass or fuel load (kg km⁻²); and β is combustion efficiency (fraction of available fuel burned). Adapting this algorithm to calculate the emission of a particular species requires *a priori* information about the emission factor for a given species for the type of vegetation being burned, expressed as grams of species x per kilogram of dry fuel burned (Andreae & Merlet, 2001). The equation to estimate emission is then rather straightforward given the fuel consumed, as calculated in equation (1):

$$E_x = EF_x \bullet M \qquad (2)$$

E_x is the emission load of species x (g); EF_x is the emission factor for species x for the specific vegetation type or biome (g kg⁻¹); and M is the biomass burned in equation [1].

Traditionally, estimates have been made using statistical information such as FAO data on population and land use practices (Robinson, 1989; Seiler & Crutzen, 1980). Statistical information reported at national scales often requires extrapolation due to incomplete information, sporadic reporting, and highly variable estimates, especially within developing countries undergoing rapid land use change (Andreae, 1991). Advances in satellite technology offer the opportunity to make relatively accurate estimates for several of the parameters in equation [2] at synoptic scales (Justice et al., 2002). For example, Michalek et al. (2000) effectively combined Landsat TM data and field measurements to estimate carbon release from Alaskan spruce forest fires because of the high spatial resolution (<30m) and spectral ability to separate between pre-burn biomass and burn severity of Landsat. Page et al. (2002) estimated 0.19-0.23 Gt of carbon were released to the atmosphere from peat combustion during 1997 Indonesian fires. Their estimates were based on peat thickness, pre-fire land cover, and burnt area data collected from ground measurements and Landsat TM/ETM imagery. Satellite imagery proved useful for classifying land cover and determining burn scars, but Page et al. (2000) discovered that due to residual haze after fires and frequent cloud cover, the use of synthetic aperture radar (SAR) was necessary to determine the extent of burnt areas.

Limitations to the "bottom up" approach include fuel loads and burning efficiency, which cannot be directly estimated from satellite observations. In addition, there is a lack of agreement on the proper algorithm to characterize burned area from satellite data (Roy et al., 2005). Korontzi et al. (2004) showed that significant differences between burned area algorithm estimates can lead to differences as large as a factor of two in estimates of biomass consumed. Differences in spatial and temporal estimates of burned area products was demonstrated by Boschetti et al. (2004) who showed that the GLOBSCAR product had a burned area nearly twice as large as the GBA2000. They concluded that such discrepancies have serious implications for accurately quantifying emissions from fires (Boschetti et al., 2004). The difficulty in accurately measuring these variables leads to an uncertainty in emission estimates of at least 50%, and possibly much greater (Robinson, 1989; Andreae and

Merlet, 2001; van der Werf et al., 2003; French et al., 2004; Korontzi et al., 2004). Although datasets used for this application are always improving (Roy et al., 2005; van der Werf et al., 2006), due to the uncertainty in current estimates it is worthwhile to explore other approaches.

3.2 FRP

Vegetation fires can be thought of as the obverse of photosynthesis in which energy stored in biomass is released as heat (equation 3).

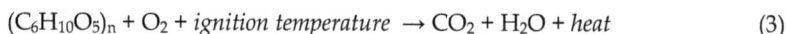

$$(C_6H_{10}O_5)_n + O_2 + \textit{ignition temperature} \rightarrow CO_2 + H_2O + \textit{heat} \tag{3}$$

The cascade of chain of reactions starts with the pre-heating of fuels ahead of the fire front and partial pyrolytic decomposition. Ignition signifies the transfer from pre-heating to combustion in which exothermic reactions start and the next phase, encompassing a combination of flaming and smoldering combustion, begins. Flaming combustion occurs when flammable hydrocarbon gases released during pyrolysis are ignited with wildfire flaming combustion temperatures in the range of 800 – 1400 K (Lobert & Warnatz, 1993). Pyrolytic action involves the thermal decomposition of fuel resulting in the release of water, CO_2, and other combustible gases (e.g. CH_4) and particulate matter. The heat produced, often measured as heat yield (MJ/kg), is thermal energy transferred via conduction, convection, vaporization, and radiation and provides a metric of the total potential energy released if complete combustion of the fuel occurs. Although other factors, including slope, fuel arrangement, and wind speed influence the actual heat yield in a fire event, the theoretical value varies very little between fuel types (Stott, 2000; Whelan, 1995). As described by Stefan-Boltzmann's Law, the radiant component is emitted as electromagnetic waves traveling at the speed of light in all directions and is proportional to the absolute temperature of the fire (assumed to be a black body) raised to the fourth power. The relationship between fire temperature and spectral radiance was shown to closely match the Stefan-Boltzmann law (Radiance = σT^4) and thus a simple equation incorporating the sample size, emissivity of the fire (with some assumptions needed), and Stefan's constant could provide the rate of fire radiative energy, or fire radiative power (FRP), emitted as shown in equation (4).

$$FRP = A_{sample} \, \varepsilon \sigma \sum A_n \, T_n^4 \tag{4}$$

where A_{sample} is the total area of the satellite pixel (m²), ε is the fire emissivity, σ is the Stefan Boltzmann constant ($5.67 \times 10^{-8} J^{-1}m^{-2}K^{-4}$), A_n is the fractional area of the ith thermal component, and T_n^4 is the temperature of the ith thermal component (K).

The foundation for using measurements of FRP is based on the fact that the rate of biomass consumed is proportional to the rate of FRE. Kaufman et al. (1996, 1998) suggested that estimates of fuel load combustion and emission rates could be made from satellite observations of the radiative energy liberated during fire events. The hypothesis is that the rate of emitted energy (i.e. FRP), and rate of fuel combustion are proportional to the fire size and fuel load (A and B, respectively) from equation [1]. It follows then that the rate of energy released is directly related to the rate of particulate matter and trace gas emissions. Integrating FRP over the lifespan of the fire event provides the total fire radiative energy (FRE) released, which in turn is directly proportional to the total fire emissions. It is the radiative component that is estimated from Earth observing satellite sensors, offering an alternative method to quantify the biomass consumed, and assuming an emission factor is known, it also offers the atmospheric emission load.

Unfortunately, sensors are unable to separate the spatially distinct components of the fire, potentially as small as millimeters, and the equation cannot distinguish between fractional areas of the entire fire which often are much smaller than the pixel itself. Thus, different methods have been tested and employed to overcome these limitations. The bi-spectral method, using two distinct channels (usually 4 and 11μm), can provide details about the fractional size and temperature of sub-pixel fire components (Dozier, 1981; Giglio & Kendall, 2001, Wooster et al., 2005), but is plagued by potential errors associated with channel misregistration and point spread function (PSF) differences between channels (Giglio & Kendall, 2001). Wooster et al. (2005) suggested that the bi-spectral method is effective, but primarily for high resolution sensors (<1km). The current method used aboard MODIS employs a single channel approach with fire and background components retrieved solely from the mid-infrared (4μm) channel (Justice et al, 2002). Kaufman et al. (1996, 1998) tested this single channel approach using the MODIS Airborne Simulator (MAS), model simulations of fire mixed-temperature pixels (to realistically mimic the non-homogeneous behavior of biomass burning temperatures), and *in situ* measurements. Based on the simulated fires, Kaufman et al. (1998) revealed that an empirical relationship exists between instantaneous FRE (i.e. FRP) and pixel brightness temperature measured in the Moderate Resolution Imaging Spectroradiometer (MODIS) middle infrared channel (4 μm). The result was a semi-empirical relationship which forms the basis for the current FRP algorithm (equation 5) used aboard MODIS. The authors also demonstrated the correlation between rates of smoke emission and the observed rate of energy released from airborne observations with the MAS (Kaufman et al., 1996, 1998).

$$\text{FRP [MW km}^{-2}\text{]} = 4.34 \times 10^{-19} (T^8_{MIR} - T^8_{bg, MIR}) \tag{5}$$

where FRP is the rate of radiative energy emitted per pixel (the MODIS 4μm channel has IFOV of 1km), 4.34×10^{-19} [MW km^{-2} Kelvin^{-8}] is the constant derived from the Kaufman et al. (1998) simulations, T_{MIR} [Kelvin] is the radiative brightness temperature of the fire component, $T_{bg, MIR}$ [Kelvin] is the neighboring nonfire background component, and MIR refers to middle infrared wavelength, typically 4μm.

Wooster et al. (2003) showed that FRP could also be derived using satellite-based middle infrared radiances and a simple power law to approximate Plank's law. The 'MIR radiance' method is applicable for temperatures covering the range of typical vegetation fires (600 – 1500 K). As with the 'MODIS' method, the MIR method relies on the difference between the fire pixel and background, but uses spectral radiance differences rather than brightness temperature. According to Wooster et al. (2005) the radiance methods allows perturbations, such as atmospheric effects and pixel area variation across the scan angles, to be accounted for after FRP has been derived.

4. The application of FRE

Kaufman et al. (1996, 1998) first showed the potential application of FRP and FRE for estimating fuel combustion rates and aerosol loading while examining prescribed fires during the Smoke, Clouds, and Radiation (SCAR) experiments. Wooster (2002) investigated the relationship between FRP/FRE and fuel consumption using small-scale experimental fires in which spectroradiometers recorded the radiative emission for the entire burning process at 5 to 10 second intervals. Wooster et al. (2003, 2005) expanded on their previous work, providing additional evidence of the effectiveness of using instantaneous and total FRE measurements to estimate biomass consumed from fire. Wooster & Zhang (2004) demonstrated the application of MODIS FRP observations by verifying the often proposed hypothesis that North American boreal fires are generally more intense than Russian boreal fires, while Ichoku & Kaufman (2005) used the MODIS FRP and aerosol products to derive near real time rates of aerosol emissions at regional scales. Research by Roberts et al. (2005) has shown the effectiveness of using geostationary satellite estimates of FRP from The Spinning Enhanced Visible and Infrared Imager (SEVIRI) to quantify rates of fuel consumption and characterize the fire intensity daily cycle. A laboratory investigation of FRE and biomass fuel consumption by Freeborn et al. (2008) supported the accuracy of Wooster et al.'s (2005) findings and lends credence to the application of satellite based measurements of FRE. Ichoku et al. (2008a), in a coordinated effort with research conducted by Freeborn et al. (2008), used laboratory investigations to examine rates and total fire radiative energy emitted and associated aerosol emissions. In both the case of Freeborn et al. (2008) and Ichoku et al. (2008a), the relationship between energy emitted, fuels consumed, and trace gas and aerosol emission demonstrated the efficacy of using FRE. Ichoku et al. (2008b) offered another example of using FRP, but at continental scales while investigating the global distribution of MODIS-based FRP estimates and revealed the regional distributions of fire intensity. Their research also showed significant differences in diurnal cycles and categorized intensities of FRP between regions which could not be explained by ecosystem type alone, suggesting perhaps that land use is a factor. Roberts & Wooster (2008) built upon their previous research (Roberts et al., 2005), showcasing the application of high temporal satellite based FRP measurements from the SEVIRI geostationary sensor to calculate FRE and estimate biomass combusted. Boschetti & Roy (2009) demonstrated a novel fusion approach to derive FRE based on temporal interpolation of MODIS FRP across independently derived burned area estimates. Their work was limited to Australia and the MODIS sensor, but as the authors suggest, the methodology could be expanded to other sensors and "is a fruitful avenue for future research and validation" (Boschetti & Roy, 2009). Freeborn et al. (2009) used frequency density distributions developed from MODIS and SEVIRI fire radiative power to synthesize the two sensors as a means for cross-calibration of their respective estimates. However, until Ellicott et al. (2009) and Vermote et al. (2009) no study had derived FRE at a global scale, in part due to limitations in temporal or spatial resolution of satellite sensors.

A current limitation of fire energy retrieval from satellites is that observations are of instantaneous energy (power) over some discrete length of time and space. To address this Ellicott et al. (2009) developed a unique approach to parameterize the temporal trajectory of FRP and calculate the integral (i.e. FRE) using MODIS. The parameterization was based on the long term ratio between Terra and Aqua MODIS FRP and diurnal measurements of FRP and fire detections made by satellites with greater temporal resolution. This included the geostationary sensor SEVIRI and the VIRS aboard TRMM. VIRS's low-inclination orbit (35°) provides observation times which precesses through 24 hours of local time every 23-46 days, depending on latitude, thus capturing the general diurnal trend of fire activity. In addition, high latitude (and thus high overpass frequency) daily observations by MODIS were included. The result was a global FRE product from MODIS at 0.5° spatial and monthly temporal resolution which currently spans from 2001 – 2010 (Figure 1).

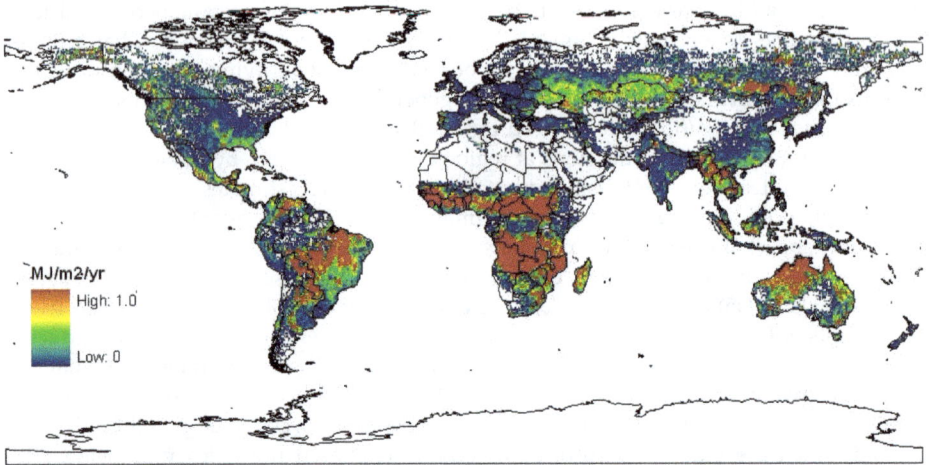

Fig. 1. Estimated annual mean FRE (MJ/m²) from Aqua (2003-2010) and Terra (2001-2002) MODIS. Integrated energy was calculated from FRP (MW) values derived from a Gaussian function using modeled parameters.

Based on their FRE estimates, Ellicott et al. (2009) estimated biomass consumption totals for Africa using FRE-based combustion factors derived by Wooster et al. (2005). Since Wooster et al.'s combustion factor is based on fuels typical for Africa, the estimates were limited to this continent. The results were compared with Roberts & Wooster (2008) who derived estimates of fuels consumed from the SEVIRI sensor. The results showed good agreement for a 12 month comparison of FRE-based estimates of dry matter consumed from SEVIRI (858 Tg DM) and MODIS (700 Tg DM). The GFEDv2 (van der Werf et al., 2006) though, showed nearly a factor of 3 greater fuel consumption for the same time period and area, suggesting that more work needs to be done to characterize the sources and magnitude of errors in these estimates.

Vermote et al. (2009) applied the FRE-approach described above to estimate organic and black carbon (OCBC) aerosols emitted from biomass burning in 2003. The relationship between the estimated FRE and a new MODIS-derived inversion product of daily integrated, biomass burning aerosol emissions was the foundation of their research. The inversion product (Dubovik et al., 2008) was generated from the MODIS fine mode aerosol optical thickness and inverse modeling transport processes adopted from the Goddard Chemistry Aerosol Radiation and Transport (GOCART) model. The process generated fine mode aerosol sources (locations and intensities) resulting from biomass burning which were then used to derive OCBC estimates. The relationship between FRE and OCBC was analyzed globally within 3 distinct vegetation zones (Figure 2).

The estimated FRE-based OCBC emission in 2003 was 20 Tg and the spatial pattern clearly shows areas of high fire activity and thus OCBC loading (Figure 3). Though lower than the 29.6 Tg and 26.1 Tg estimates made by Generoso (2007) and van der Werf et al. (2006), respectively, the estimate is still within the error bars of both datasets. Nevertheless, the underestimation raises questions about the sources of uncertainty and error in the components used to derive OCBC quantities.

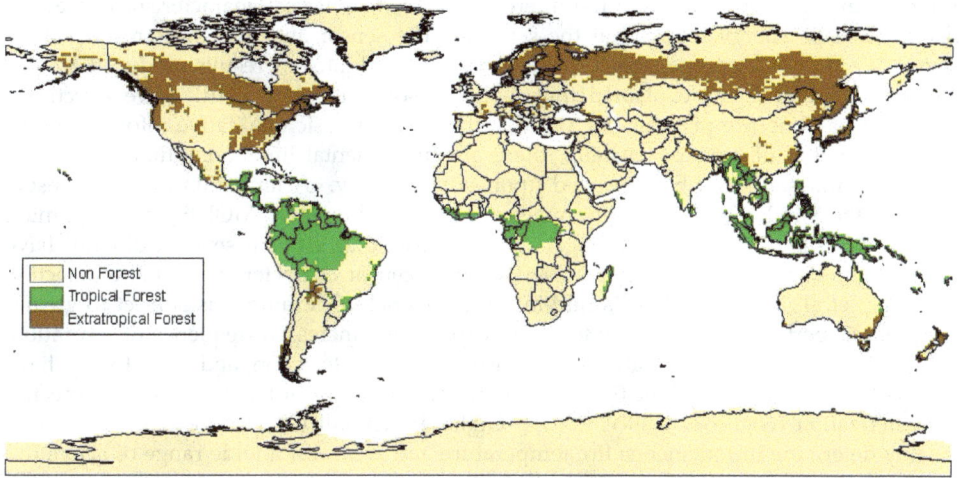

Fig. 2. Biome regions adopted from the Global Fire Emissions Database (GFED, van der Werf et al., 2006) used for analyzing the relationship between FRE and OCBC aerosol emissions (Vermote et al., 2009).

FRE-based Estimated OCBC : 2003

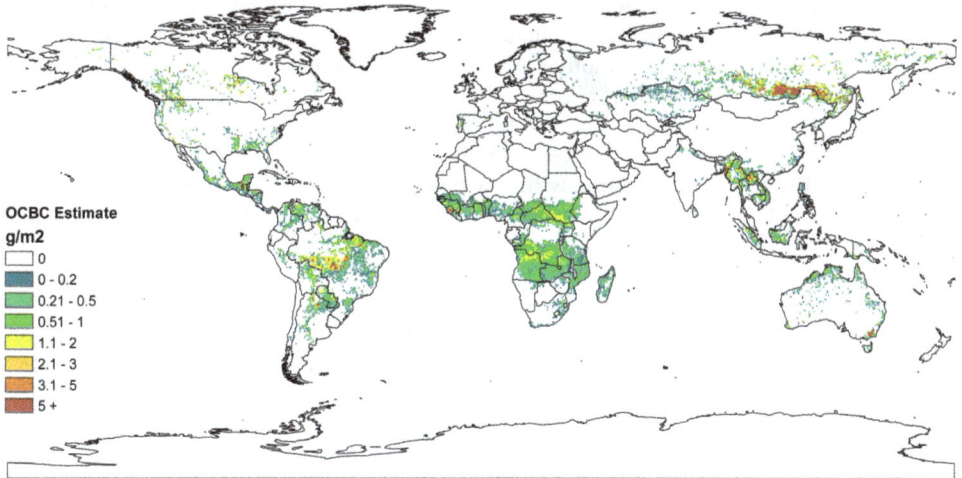

Fig. 3. Total OCBC (g/m²) emissions estimated from biomass burning for 2003. High source regions include east-central Brazil, central and southern Africa, Southeast Asia, Central America, and southeast Russia.

5. Uncertainty

Remote sensing science involves the inference of *in situ* physical characteristics based on electromagnetic energy received at the sensors. The sensor and inferences made have a degree of error, which propagates through processing and into any results produced.

Giglio & Kendall (2001) explained that while sensors such as AVHRR have effectively generated baseline fire products for fire distribution and basic qualitative information for parameterization of biomass burning, there are fundamental limitations that need to be addressed and could not be improved upon until recently. A limitation prior to sensors such as the Moderate Resolution Imaging Spectroradiometer (MODIS) and Spinning Enhanced Visible and Infrared Imager (SEVIRI) was that satellite sensors did not have dedicated fire channels or did not possess optimal sensor characteristics for fire detection (Kaufman et al., 1998). AVHRR's mid-infrared channel, for example, is subject to greater atmospheric perturbation and, due to sensor capabilities, increased frequency of saturation. Advancements of new generations of sensors for fire detection and monitoring have included refinements to specific wavelength selection in order to optimize spectral characterization (Giglio & Justice, 2003). Giglio & Kendall (2001) state that in order to reliably determine instantaneous fire temperature and area over a wide range of active fire sizes using Dozier's method, sensors with higher spatial resolution (~100m) and very high (~1000K) middle infrared band saturation would be necessary. However, replacement of the mid-IR channel aboard AVHRR (3.7μm) with a wavelength less sensitive to solar radiation (i.e. 3.9μm) would reduce the pixel saturation by half (Giglio & Justice, 2003; Kaufman et al., 1998). Defining higher pixel saturation temperatures and including wavelengths that can be used for false alarm detection on MODIS were improvements based on experiences with older systems (Justice et al., 2002).

Earlier work by Schroeder et al. (2008) showed that cloud obscuration in the Brazilian Amazon could lead to fire detection omission errors of roughly 11%. However, commission errors may occur as result of cloud shadows and semi-transparent clouds influencing surface thermal characteristics. Another source of commission errors may also result from the very efforts to correct for cloud obscuration by leading to an overestimate in the number of (assumed) detections. Schroeder et al. (2010) recently provided a thorough analysis of FRP, temperature estimates, and fire area estimates from moderate resolution sensors. Their results showed that location of fires within a pixel can be biased because of the sensor's point spread function leading to as much as a 75% underestimation in FRP. On the other hand, improper characterization of the ambient background surrounding fire pixel(s) can result in an overestimation of FRP up to 80%. This particular situation is mostly prevalent in areas of tropical deforestation.

The accuracy of the empirical formula for computing FRP was taken from the evaluation performed by Kaufman et al. (1998), who showed a potential error of 16% using 150 simulated mixed-energy fire pixels. Wooster et al. (2003) found a theoretical accuracy (RMSD) of 65×10^6 J over a range of 0 to 2000×10^6 J (or 6.5% for the average) using their MIR FRE approach.

MODIS omission rates of small active fires were also observed by Hawbaker et al. (2008). In their study, 73% of Aqua and 66% of Terra active fire detections were missed, primarily because of fast moving fire fronts, cloud cover, or spatial resolution. Hawbaker et al. (2008) was clear though that these small fires likely may have little impact in terms of total emissions as had been stated previously by Kaufman et al. (1998).

Freeborn et al. (2011) highlighted an issue with the MODIS Collection 5 FRP product. In the C5 FRP the calculation of the instantaneous energy (MW) derived from the brightness temperature includes a multiplication by the pixel area. Although this is fundamentally correct, since energy is measured per unit time and space, the adjustment leads to an overestimate with increasing scan angle because the pixel area grows as the scan moves off nadir. Interestingly, the opposite effect occurs when examining fire pixel counts (i.e. greater number of detections near nadir and decreasing detections with scan angle).

With regards to the application of FRE-based biomass consumption estimates published by Ellicott et al. (2009) and Roberts & Wooster (2008), there is some degree of uncertainty. Although the assumption that a single combustion factor is applicable for all fuel types and conditions (i.e. moisture content) will incur some bias, in general, heat yield does not vary much between fuels (Stott, 2000) and therefore until more research demonstrates otherwise, the two cited FRE-based combustion factors (Freeborn et al., 2008; Wooster, 2005) seem realistic.

Atmospheric attenuation is another component generally unaccounted for. In simulations conducted by Ellicott (unpublished), the MODIS FRP may be underestimated by as much as 20% (Figure 4). Similarly, Roberts & Wooster (2008) applied a constant correction factor (0.89) to SEVIRI FRP to account for atmospheric transmission loss.

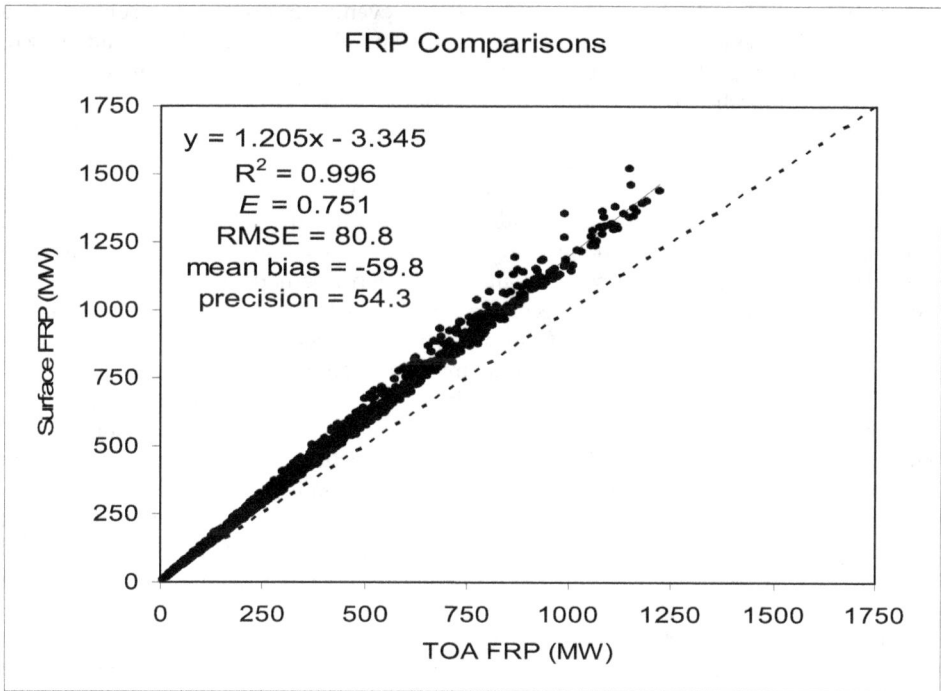

Fig. 4. Comparison of simulated surface and TOA FRP. Radiances were simulated from randomly generated fire pixel temperature and fractional area components (fire, smoldering, and background). MODIS Aqua profiles were used to provide realistic atmospheric parameters used in the radiative transfer modeling. The 1:1 (dashed) line is plotted for reference.

Finally, an error budget provided by Vermote et al. (2009) suggested that the FRE-parameterization approach developed by Ellicott et al. (2009) approaches 20% based on comparisons with the SEVIRI sensor.

6. Discussion

The atmosphere plays a fundamental role in regulating life on Earth. Changes in atmospheric composition can and do affect surface temperatures, hydrology, radiation budgets, weather, and even climate. Therefore, understanding the complex exchanges occurring between the atmosphere and surface requires accurate measurements of the variables characterizing both; for example atmospheric constituents, surface temperatures, and albedo. Quantifying these variables provides the necessary inputs for modeling the dynamic interactions and potential outcomes that result from changes in the relative proportions of atmospheric constituents. In light of the growing evidence for anthropogenic induced climate change, accurate characterization of the impact humans are having, both directly and indirectly, on altering Earth's systems is critical to guiding mitigation policy.

To that end, fire radiative energy may provide a efficient and accurate tool to monitor and measure biomass consumed and emissions from fire events. In order to be truly effective some of the idiosyncratic issues that plague sensors and algorithms need to be addressed, but perhaps most importantly, at least a global scale, is dealing with missed fire detections. For this, more comprehensive and distributed validation must occur.

In order to truly validate FRE estimates, greater spatial and temporal resolution data are needed. The evaluation of the FRE estimates with SEVIRI data offered a comparison with FRP retrievals made at higher temporal resolution, but incurred the downside of coarser spatial resolution. Future endeavors would include a scaling approach to test the temporal trajectory of instantaneous fire energy and total fire radiative energy released from a fire event. This would include the use of *in situ* observations, perhaps with a combination of field and laboratory experiments to reconcile differences between these two approaches. The next tier of retrievals would be from airborne observations, perhaps including both tower platforms (for small scale fires) and unmanned aircraft. The Ikhana unmanned airborne vehicle (UAV) used by the fire research at NASA AMES offers some opportunities in this regard. Recent field work demonstrated that while monitoring FRP from a helicopter seems ideal, many factors can limit the success of this tactic and that greater flexibility in choice of fires to observe and timing allowed for observation is needed. Moderate to high spatial resolution satellite observations would be employed in the next scaling layer and allow for greater spatial coverage while being constrained by higher spatial and temporal observations. To that end, geostationary satellite observations would cap the scaling approach, providing high temporal (15 – 30 minute) retrievals to aid in characterizing the diurnal cycle of fire radiative power as has been shown in this research. Incorporating sensors such as the Geostationary Operational Environmental Satellites (GOES) would offer greater spatial coverage beyond the SEVIRI sensor. Careful consideration of the limitations of comparison between sensors at multiple scales would obviously be needed (*Schroeder et al.*, 2005).

Other considerations worth pursuing to improve FRP retrievals from the MODIS sensor include parameterization of the sub-surface organic layer burning. According to French et al. (2004) surface organic layer burning is largest source of uncertainty in boreal biomass burning emission estimates. Page et al. (2002) estimated 0.19-0.23 Gt of carbon released to the atmosphere from peat combustion during 1997 Indonesian fires. Their estimates were based on peat thickness, pre-fire land cover, and burnt area data collected from ground measurements and Landsat TM/ETM imagery. Cloud cover has already been revealed by Schroeder et al. (2008) to limit fire detection capabilities for Brazilian fires. The spatial resolution of MODIS is another limitation to detecting fires in peatlands (and thus FRP estimation) as shown by Siegert et al. (2004). Developing a connection between field estimates of surface and sub-surface organic burning, burned area, and FRP would allow for parameterization of this component of fire radiative energy.

7. Acknowledgements

I thank Dr. Louis Giglio for his insight to the concepts, science, and application of fire radiative energy and technical assistance and in developing the FRE parametreization method, generously offering data, and providing critical assessment of our ideas and approaches. I thank Dr. Gareth Roberts for helping with SEVIRI data and giving feedback on my research and writing. I thank Dr. Wilfrid Schroeder for his helpful technical insight and fruitful discussions of the limitations and uncertainty in active fire detections and FRP. Finally, I thank Dr. Guido van der Werf for his support through offering insight on the GFED processes and data, providing advice on our research, and feedback on my analysis.
Dr. Ellicott would also like to thank NASA's Earth and Space Science Fellowship Program for recognizing the potential benefits of his research and providing financial support during his doctoral degree endeavor.

8. References

Aragao, L. E. O. C. & Shimabukuro, Y. E. (2010). The Incidence of Fire in Amazonian Forests with Implications for REDD, *Science, 328*(5983), 1275-1278.

Boles, S.H. & Verbyla, D.L. (2000). Comparison of three AVHRR-based fire detection algorithms for interior Alaska. *Remote Sensing Environment*, v72, p1-16.

Boschetti, L., Eva, H. D., Brivio, P. A., & Gregoire, J. M. (2004). Lessons to be learned from the comparison of three satellite-derived biomass burning products, *Geophysical Research Letters, 31*(21), L21501, doi:10.1029/2004gl021229.

Boschetti, L. & Roy, D. P. (2009). Strategies for the fusion of satellite fire radiative power with burned area data for fire radiative energy derivation, *J Geophys Res-Atmos, 114*.

Brady, N.C. & Weil, R. R. (1999). In: *The Nature and Properties of Soils*: 12th Edition." (N.C. Brady and R.R. Weil, Eds.) Prentice-Hall, London.

Cary, G. J., & Banks, J. C. G. (2000). Fire regime sensitivity to global climate change: An Australian perspective, in *Biomass burning and its inter-relationships with the climage system*, edited by J. L. Innes, M. Beniston, and M. M. Verstraete, pp. 233-245, Dordrecht, Kluwer Academic Publishers.

Chapman, S.J. & Thurlow, M. (1998). Peat respiration at low temperatures, *Soil Biology & Biochemistry, 30*, 1013-1021.

Cicerone, R. J. (1994). Fires, atmospheric chemistry, and the ozone layer, *Science, 263*, 1243-1244.

Crutzen, P. J., L. E. Heidt, J. P. Krasnec, W. H. Pollock, and W. Seiler (1979). Biomass burning as a source of atmospheric gases CO, H_2, N_2O, NO, CH_3Cl, and COS. *Nature, 282*, 253-256.

Crutzen, P. J. & Andreae, M. O. (1990). Biomass Burning in the Tropics: Impact on Atmospheric Chemistry and Biogeochemical Cycles, *Science, 250* (4988), 1669-1678.

Dozier, J. (1981). A method for satellite identification of surface temperature fields of subpixel resolution. *Remote Sensing of Environment.* v11, p221-229.

Dubovik, O., Lapyonok, T., Kaufman, Y. J., Chin, M., Ginoux, P., Kahn, R. A. & Sinyuk, A. (2008). Retrieving global aerosol sources from satellites using inverse modeling, *Atmospheric Chemistry and Physics, 8*(2), 209-250

Ellicott, E., Vermote, E., Giglio, L. & Roberts, G. (2009). Estimating biomass consumed from fire using MODIS FRE, *Geophys. Res. Lett., 36*, L13401, doi:10.1029/2009GL038581.

Freeborn, P. H., Wooster, M. J., Hao, W. M., Ryan, C. A., Nordgren, B. L., Baker, S. P. & Ichoku, C. (2008). Relationships between energy release, fuel mass loss, and trace gas and aerosol emissions during laboratory biomass fires, *Journal of Geophysical Research-Atmospheres, 113*(D1), D01301,doi:10.1029/2007jd008679.

Freeborn, P. H., Wooster, M. J., Roberts, G., Malamud, B. D. & Xu, W. D. (2009). Development of a virtual active fire product for Africa through a synthesis of geostationary and polar orbiting satellite data, *Remote Sensing of Environment, 113*(8), 1700-1711.

Freeborn, P. H., Wooster, M. J. & Roberts, G. (2011). Addressing the spatiotemporal sampling design of MODIS to provide estimates of the fire radiative energy emitted from Africa, *Remote Sensing of Environment, 115*(2), 475-489.

French, N. H. F., Goovaerts, P. & Kasischke, E. S. (2004). Uncertainty in estimating carbon emissions from boreal forest fires, *Journal of Geophysical Research, 109*(D14).

Generoso, S., Bey, I., Attie, J. L. &. Breon, F. M (2007). A satellite- and model-based assessment of the 2003 Russian fires: Impact on the Arctic region, *Journal of Geophysical Research-Atmospheres, 112*(D15), D15302, doi:10.1029/2006jd008344.

Giglio, L. & Kendall, J. D. (2001). Application of the Dozier retrieval to wildfire characterization. A sensitivity analysis. Remote Sensing of Environment, 77, 34-49.

Giglio, L. & Justice, C. O. (2003). Effect of wavelength selection on characterization of fire size and temperature, *Int. J. Remote Sens.*, 24, 3515-3520.

Giglio, L., Descloitres, J., Justice, C. O. & Kaufman, Y. J. (2003). An enhanced contextual fire detection algorithm for MODIS. *Remote Sensing of Environment*, 87, 273-282.

Giglio, L., Randerson, J. T., van der Werf, G. R., Kasibhatla, P. S., Collatz, G. J., Morton, D. C. & DeFries, R. S. (2010). Assessing variability and long-term trends in burned area by merging multiple satellite fire products, *Biogeosciences, 7*(3), 1171-1186.

Gorham, E. (1991). Northern peatlands: Role in the carbon cycle and probable responses to climatic warming. *Ecological Applications.* v1(2), p182-195.

Hawbaker, T. J., Radeloff, V. C., Syphard, A. D., Zhu, Z. & Stewart, S. I. (2008). Detection rates of the MODIS active fire product in the United States, *Remote Sens. Environ.*, 112(5), doi :10.1016/j.rse.2007.12.008.

Ichoku, C. & Kaufman, Y. J. (2005). A method to derive smoke emission rates from MODIS fire radiative energy measurements. *IEEE Transactions on Geoscience and Remote Sensing*, 43, 2636-2649.

Ichoku, C., Martins, J. V., Kaufman, Y. J., Wooster, M. J., Freeborn, P. H., Hao, W. M., Baker, S., Ryan, C. A. & Nordgren, B. L. (2008a). Laboratory investigation of fire radiative energy and smoke aerosol emissions, *J Geophys Res-Atmos*, 113(D14), -.

Ichoku, C., Giglio, L., Wooster, M. J., & Remer, L. A. (2008b). Global characterization of biomass-burning patterns using satellite measurements of fire radiative energy, *Remote Sensing of Environment, 112*(6), 2950-2962.

IPCC, 2007: Climate Change 2007: The Physical Science Basis. Contribution of Working Group I to the Fourth Assessment Report of the Intergovernmental Panel on Climate Change [Solomon, S., D. Qin, M. Manning, Z. Chen, M. Marquis, K.B.

Averyt, M. Tignor and H.L. Miller (eds.)]. Cambridge University Press, Cambridge, United Kingdom and New York, NY, USA, 996pp.

Justice, C. O., Giglio, L., Korontzi, S., Owens, J., Morisette, J. T., Roy, D., Descloitres, J., Alleaume, S., Petitcolin, F. & Kaufman, Y. J. (2002). The MODIS fire products. *Remote Sensing of Environment*, 83, 244-262.

Kasischke, E. S., Christensen Jr., N. L. & Stocks, B. J. (1995). Fire, Global Warming, and the Carbon Balance of Boreal Forests, *Ecological Applications*, 5(2), 437-451

Kaufman, Y., Tucker, C. & Fung, I. (1990). Remote sensing of biomass burning in the tropics, *Journal of Geophysical Research*, 95(D7), 9927-9939.

Kaufman, Y. J., Remer, L., Ottmar, R., Ward, D., Li, R. R., Kleidman, R., Fraser, R. S., Flynn, L., McDougal, D. & Shelton, G. (1996). Relationship between remotely sensed fire intensity and rate of emission of smoke: SCAR-C experiment, *Global Biomass Burning*, 685–696.

Kaufman, Y. J., Justice, C. O., Flynn, L. P., Kendall, J. D., Prins, E. M., Giglio, L., Ward, D. E., Menzel, W. P. & Setzer, A. W. (1998). Potential global fire monitoring from EOS-MODIS. *Journal of Geophysical Research*, 103, 32,215-32,238

Korontzi, S., Roy, D. P., Justice, C. O. & Ward, D. E. (2004). Modeling and sensitivity analysis of fire emissions in southern Africa during SAFARI 2000, *Remote Sensing of Environment*, 92(3), 376-396

Kuhry, P. (1994). The role of fire in the development of *Sphagnum*-dominated peatlands in western boreal Canada, *J. of Ecology*, 82, 899-910.

Levine, J. S. (2000). Global biomass burning: A case study of the gaseous and particulate matter emissions released to the atmosphere during the 1997 fires in Kalimantan and Sumatra, Indonesia, in *Biomass burning and its inter-relationships with the climate system*, edited by J. L. Innes, M. Beniston, and M. M. Verstraete, pp. 1-15, Dordrecht, Kluwer Academic Publishers.

Legg, C.A. & Laumonier, Y. (1999). Fires in Indonesia, 1997: A remote sensing perspective. *Ambio*. v28(6), 479-485.

Levine, J. S. (2003). Burning domestic issues, *Nature*, 423, 28-29.

Lobert, J. M., & Warnatz, J. (1993). Emissions from the combustion process in vegetation, in *Fire in the Environment: The Ecological, Atmospheric, and Climatic Importance of Vegetation Fires*, edited by P. J. Crutzen and J. G. Goldammer, pp. 15-38, John Wiley, New York.

Michalek, J. L., French, N. H., Kasischke, E. S., Johnson, R. D. & Colwell, J. E. (2000). Using Landsat TM data to estimate carbon release from burned biomass in an Alaskan spruce forest complex. *International Journal of Remote Sensing*. v21(2), p323-338.

Moore, P. (2002). The future of cool temperate bogs. *Environmental Conservation*. v29(1), p3-20.

Morton, D. C., Defries, R. S., Randerson, J. T., Giglio, L., Schroeder, W. & van Der Werf, G. R. (2008). Agricultural intensification increases deforestation fire activity in Amazonia, *Global Change Biol*, 14(10), 2262-2275.

Page, S.E., Siegert, F., Rieley, J. O., Boehm, H. V., Jaya, A. & Limin, S. (2002). The amount of carbon released from peat and forest fires in Indonesia during 1997. *Nature*. v420, p61-65.

Pastor, J., Solin, J., Bridgham, S. D., Updegraff, K., Harth, C., Weishampel, P. & Dewey, B. (2003). Global warming and the export of dissolved organic carbon from boreal peatlands. *Oikos*. v100, p380-386.

Roberts, G. J., & Wooster, M. J. (2008). Fire detection and fire characterization over Africa using Meteosat SEVIRI, *Ieee Transactions on Geoscience and Remote Sensing*, 46(4), 1200-1218Doi 10.1109/Tgrs.2008.915751.

Roberts, G. J., Wooster, M., Perry, G. L. W., Drake, N., Rebelo, L. M. & Dipotso, F. (2005). Retrieval of biomass combustion rates and totals from fire radiative power observations: Application to southern Africa using geostationary SEVIRI imagery, *Journal of Geophysical Research-Atmospheres,110*(D21),D21111, doi: 10.1029/2005jd006018.

Robinson, J. M. (1989). On uncertainty in the computation of global emissions from biomass burning, *Climatic Change, 14*(3), 243-261

Roy, D. P., Lewis, P. E. & Justice, C. O. (2002). Burned area mapping using multi-temporal moderate spatial resolution data – a bi-directional reflectance model-based expectation approach. *Remote Sensing of Environment, 83*, 263-286.

Schroeder, W., Morisette, J. T., Csiszar, I., Giglio, L., Morton, D. & Justice, C. O. (2005). Characterizing Vegetation Fire Dynamics in Brazil through Multisatellite Data: Common Trends and Practical Issues, *Earth Interactions, 9*(1), 26p.

Schroeder, W., Csiszar, I. & Morisette, J. (2008). Quantifying the impact of cloud obscuration on remote sensing of active fires in the Brazilian Amazon, *Remote Sensing of Environment, 112*(2), 456-470.

Schroeder, W., Csiszar, I., Giglio & Schmidt, C. C. (2010). On the use of fire radiative power, area, and temperature estimates to characterize biomass burning via moderate to coarse spatial resolution remote sensing data in the Brazilian Amazon, *J Geophys Res-Atmos, 115*, D21121, doi:10.1029/2009JD013769.

Seiler, W. & Crutzen, P. J. (1980). Estimates of gross and net fluxes of carbon between the biosphere and the atmosphere form biomass burning, *Clim. Change, 2*, 207-247.

Siegert, F., Zhukov, B., Oertel, D., Limin, S., Page, S. E. & Rieley, J. O. (2004). Peat fires detected by the bird satellite, *Int. J. Remote Sens., 25*(16), 3221-3230.

Stott, P. (2000). Combustion in tropical biomass fires: a critical review, *Progress in Phys. Geog., 24*, 355-377.

van der Werf, G. R., Randerson, J. T., Giglio, L., Collatz, G. J., Kasibhatla, P. S. & Arellano, A. F. (2006). Interannual variability in global biomass burning emissions from 1997 to 2004, *Atmospheric Chemistry and Physics, 6*, 3423-3441

van der Werf, G. R., Randerson, J. T., Collatz, G. J. & Giglio, L. (2003). Carbon emissions from fires in tropical and subtropical ecosystems, *Global Change Biology, 9*(4), 547-562.

Vermote, E., Ellicott, E., Dubovik, O., Lapyonok, T., Chin, M., Giglio, L. & Roberts, G. J. (2009). An approach to estimate global biomass burning emissions of organic and black carbon from MODIS fire radiative power, *J Geophys Res-Atmos, 114*, D18205, doi:10.1029/2008JD011188.

Whelan, R. J. (1995). The Ecology of Fire, Cambridge Univ. Press, New York.

Wooster, M. J. (2002). Small-scale experimental testing of fire radiative energy for quantifying mass combusted in natural vegetation fires. *Geophysical Research Letters, 29*, 2027-2030.

Wooster, M. J., B. Zhukov and D. Oertel. 2003 Fire radiative energy for quantitative study of biomass burning: derivation from BIRD experimental satellite and comparison to MODIS fire products. *Remote Sensing Environment, 86*, 83-107.

Wooster, M. J. & Zhang, Y. H. (2004). Boreal forest fires burn less intensely in Russian than North America. *Geophysical Research Letters, 31*, L20505, 1-3.

Wooster, M. J., Roberts, G. & Perry, G. L. W. (2005). Retrieval of biomass combustion rates and totals from fire radiative power observations: FRP derivation and calibration relationships between biomass consumption and fire radiative energy release. *Journal of Geophysical Research, 110*, D24311, 1-24.

Part 4

Models

9

Resilience and Stability Associated with Conversion of Boreal Forest

Jacquelyn Kremper Shuman and Herman Henry Shugart
University of Virginia, Department of Environmental Sciences
USA

1. Introduction

A clear understanding of boreal forest dynamics is critical to developing an accurate representation of the Earth's response to climate change. The Russian boreal forest is the largest continuous forest region on Earth and a tremendous repository of terrestrial organic carbon. The boreal forest has experienced significant warming over the past several decades and is expected to be impacted by global climate change (Chapin et al., 2000; McGuire et al., 2002; Soja et al., 2007). Siberian summers in the past century were warmer than any century in the past millennium, and future climate scenarios indicate that the region will continue warming, by some accounts between 2° and 10°C by 2100 (IPCC 2007; Soja et al., 2007). Warming climate will likely exert influence on species distributions and land cover types in the boreal forest regions (Ustin and Xiao 2001; Tchebakova et al., 2005; Tchebakova et al., 2009). In particular, these temperature increases have led to the shift of treelines northward or upslope of previous climate limits, and a reduction in cone and seed yield for *Larix sibirica* and *Pinus slyvestris* which changes forest composition and structure (Kharuk et al., 2009; Soja et al., 2007). These changes are important indicators of how Eurasian boreal forests may respond to, and ultimately amplify, increases in average global temperature.

These land cover changes can force alterations in regional climate through modifications in surface albedo and land/atmosphere energy fluxes (Bonan et al., 1992; Chapin et al., 2000; Baldocchi 2000; Amiro 2001; Beringer et al., 2005; Soja et al., 2007), as well as in global climate through changes in carbon sequestration and release patterns (Bonan 2008; Snyder et al., 2004). Global climate model (GCM) results have shown that clearing boreal forest alters surface albedo, and substantially cools the Earth, not only in the boreal region but across the Northern Hemisphere (Bonan et al., 1992), and has the greatest effect on global mean temperature when compared to the removal of other biomes (Snyder et al., 2004). Betts (2000) found surface albedo changes associated with the growth of coniferous evergreen trees led to significant increases in average global temperature large enough to overshadow the effect of carbon storage by growing evergreen forest in that region. Bioclimatic modeling predicts that by 2090 vegetation change across Siberia will create an albedo shift and increase overall net radiation, thereby producing enhanced warming above that already predicted for the high latitudes (Vygodskaya et al., 2007). Larch (*Larix* spp.) forest, dominated by both *L. sibirica* and *L. gmelinii,* covers extensive regions in Siberia. Field observations have documented shifts from larch to evergreen conifer forests, dominated by trees such as spruce (*Picea* spp.) or fir (*Abies* spp.) that are tolerant of higher temperatures

(Kharuk et al., 2007). Because larch is a deciduous conifer, this shift to evergreen dominance would lead to an albedo decrease, particularly in winter, when evergreen trees tend to mask laying snow relative to deciduous species (Betts and Ball 1997). The difference in summer albedo is smaller but also significant, with larch albedo measured at approximately 0.13 and evergreen species around 0.09 (Hollinger et al., 2010) This reduction of albedo associated with the shift in forest type indicates that increased temperatures may lead to a positive feedback response: a warmer climate accelerates the natural succession from larch to evergreen conifer forest and the resultant albedo promotes additional warming. Areas of southern Siberia identified as vulnerable to premature replacement of larch by evergreen conifers would undergo a local significant albedo shift of approximately 5.1 W m^{-2} following conversion from dominant larch to evergreen conifer stands (Shuman et al., 2011).

Identification of areas prone to vegetation change is crucial in efforts to mitigate the effects of potential forest type conversion. Remote sensing technology has advanced to a point which allows for estimation of biomass and detailed evaluation of land cover and land use change. Estimation of Russian forest biomass directly from Moderate Resolution Imaging Spectroradiometer (MODIS) data provided estimates of a distribution of biomass classes that correlated well to ground measured forest biomass with signatures from a minimum of training sites (Houghton et al., 2007). Detailed characterization of vegetation by remote sensing technology provides land cover maps for areas within Russia that are not easily accessible, and at a more frequent temporal resolution than is possible to obtain using field-based mapping techniques. A NOAA/AVHRR-derived vegetation map for a remote section of northern Siberia provides detailed information regarding latitudinal transition zones, vegetation differences inside each zone and variability along vertical transects for mountainous areas (Kharuk et al., 2003). This type of high quality map for a remote area provides land cover information essential to evaluating vegetation changes in response to climate change. Remotely sensed data can be used to identify areas undergoing a change in land cover type and assess the direction and magnitude of any albedo shift associated with such a change. Vegetation models can be used to provide information regarding the specific type of vegetation change and the location where this change is most likely to occur, and can thus inform vegetation monitoring efforts based on remotely sensed data.

In the past 20 years, individual-based models (IBMs) have been used to provide increasingly accurate predictions and simulations of forests (Mladenoff 2004; Scheller and Mlandenoff 2007). The model used in this study, FAREAST, is in a class of IBMs called "gap models" (Shugart and West 1980) that simulate individual trees, specifically their growth, mortality, and decomposition into litter in a relatively small area, typically the size of a forest gap. Forest gap models established according to the approach of Botkin et al., (1972) and Shugart and West (1977) are based upon the concept of "gap phase" replacement (Watt 1947). Gap models account for competition among individuals of multiple tree species for light and other resources, with the outcome determining the composition and structure of the forest through aggregation of homogenous mosaic patches through time (Shugart 1984). Testing of gap models is divided into verification and validation (Mankin et al., 1977; Cale et al., 1983; Rykiel 1996; Sargent 1984), and involves evaluating the ability of the model to predict species successional dynamics and biomass accumulation for the region of model development. Gap models have been verified and validated for a variety of forests world-wide (Botkin et al., 1972; Shugart and West 1977; Shugart, 1984 and its reprinting 1998, 2003; Kienast 1987; Leemans and Prentice 1989; Kienast and Krauchi 1991; Bugmann 2001). The initial tests of the FAREAST model included a simulation of forest composition and basal area at different elevations on Changbai Mountain in China, with statistical comparison to inventory data and then qualitative comparisons to observed forest type at 31 sites in the Russian Far East and Siberia (Yan and Shugart 2005). Further validation of the model using linear regression of model generated and independent forest inventory data indicated that FAREAST successfully captures the natural biomass dynamics of mixed-species forests across the vast geographic area and varied climatic conditions of Russia (Shuman and Shugart 2009; Shuman 2010).

The use of gap models allows for the evaluation of novel conditions or the addition of a new species for the purpose of evaluating the impact on existing vegetation. The impact that changing climate has on forests at local and regional scales has been explored with several different forest gap models (Shugart 1984; Solomon 1986; Pastor and Post 1988; Dale and Franklin 1989; Urban et al., 1993; Lasch and Lindner 1995; Bugmann 1996; Yan and Zhao 1996; Bugmann and Solomon 2000; Zhang et al., 2009). IBMs can be used to develop a vegetation "signature" for the response of ecosystems to change, especially climate change. Using a boreal forest gap model to assess climate change effects, Bonan (1989a,b) investigated the responses to several climate change predictions from global climate models along the north-facing and south-facing slopes of boreal forests near Fairbanks, Alaska. The black spruce forests growing on cold north-facing slopes were largely unaffected by the climatic warming, but white spruce forests on the relatively warmer south-facing slopes were strongly affected by the change in climate. Conditions predicted in the climate change scenarios for south-facing slopes were outside the ecological conditions under which the common tree species near Fairbanks are known to be able to persist. For white spruce, the limiting condition identified by the model results appeared to be moisture stress and not the direct effects of temperature change. A decade later, Barber et al., (2000) investigated tree ring data to determine the effect that several decades of warmer than usual temperatures in the Fairbanks area had had on white spruce stands and confirmed Bonan's model predictions with evidence for moisture-stress effects in the tree ring dataset.

In this study, the FAREAST model is used to simulate forest composition and biomass at 372 sites across Siberia and the Russian Far East for the purpose of evaluating forest response to climate change. Climate sensitivity analysis is performed in order to assess the resilience and stability of forest structure and composition to altered climate at multiple spatial scales. The model was used to simulate the impact of changes in temperature and precipitation on both total and genus-specific biomass at sites across Siberia and the Russian Far East, and for six different regions representing areas of high and low diversity. Comparisons of regions within areas of high and low diversity provide a tool to evaluate the relationship between diversity and the response of the system to changing climate. Model runs with and without European Larch (*Larix decidua*) are compared in order to assess the potential for the introduction of this species to mitigate the effects of climate change, especially the positive feedback among temperature, forest type and surface albedo.

2. Methods

2.1 Model simulation across Siberia and Russian Far East

FAREAST was run at a total of 372 sites across Siberia and the Russian Far East (RFE) from the eastern coast to the western border of the range limits of *L. sibirica*. FAREAST uses monthly climate parameters derived from historical station data to compute daily temperature and update soil water content. In particular, at each site, the model's climate inputs are drawn from a statistical distribution of monthly values for minimum and maximum mean temperature and precipitation that is derived from 60 years of data recorded at local weather stations (NCDC 2005a, 2005b). The model also uses values for soil field capacity, and soil carbon and nitrogen from Stolbovoi and McCallum (2002) for each site.

The birth, growth, and eventual death of individual trees are determined in response to competition for light and local site parameters such as soil moisture and nutrient availability, which are updated annually with bio-environmental conditions and available nutrients. Complete model processing details are available in Yan and Shugart (2005). Fifty-eight different tree species are included in FAREAST simulations, and can be grouped into ten genera (*Abies* spp., *Betula* spp, *Larix* spp., *Picea* spp., *Pinus* spp. *Populus* spp., *Tilia* spp., *Quercus* spp., *Fraxinus* spp., and *Ulmus* spp.) and two collections of less common species (other deciduous and other coniferous). These species represent the genera that dominate Northern Eurasian forests, and include species that were added when the original geographic area of interest for the model in Yan and Shugart (2005) was expanded to cover all of Russia (Shuman and Shugart 2009). Twenty-five parameters describe each species and determine which species has an advantage in terms of competition for light or nutrients, or tolerance to lack of water. Tree growth and regeneration is limited by functions describing local light, temperature, nutrients and drought dynamics determined through interaction and annual update of soil water and available carbon and nitrogen. Tree mortality each year is a consequence of a Monte Carlo realization of individual species probability of mortality plus added probability of mortality on that individual from stress or disturbances. Successional dynamics are therefore a result of competition between tree species for light and nutrients, as well as limitations to growth imposed by local environmental conditions. Each site uses a unique species list drawn from species range information created in ESRI ArcGIS (2008) using range information adapted from Nikolov and Helmisaari (1992) and Hytteborn et al., (2005). At each site, 200 independent twelfth-hectare plots were simulated for 500 years and the modeled biomass values were averaged for each species in each year of the model run. This average produces a landscape-level approximation of succession, which includes the natural disturbance associated with the death of individual trees.

2.2 Climate sensitivity analysis

The authors evaluated biomass for the total forest, *Larix* spp., and evergreen conifers at 372 sites across Siberia in response to changes in temperature and precipitation, and to the inclusion of European Larch (*Larix decidua*) (Table 1).

Treatments used in climate sensitivity analysis	
Change in temperature	Base + 4 degrees
Change in precipitation	Base ± 10%
Introduction of *Larix decidua*	Base + *Larix decidua*

Table 1. A total of 12 treatments including the base condition, combinations of temperature and precipitation change and *Larix decidua* addition to the available species pool were used.

L. decidua has a higher tolerance for an increased number of warm days than do species of larch native to Siberia and the RFE. This gives *L. decidua* an advantage over other species of larch for establishing in areas with warmer temperatures. The climate scenarios used in this analysis are based on the moderate predictions of temperature and precipitation increase that are made by global climate models for portions of Eurasia (IPCC 2007). For the base scenario, no changes are made to the distributions of monthly temperature and precipitation values derived from historical records. The remaining climate scenarios employ a linear increase in temperature or precipitation or both from the start of simulation, year zero, to year 200 of the simulation. This is followed by an additional 300 years of simulation during which the climate stabilizes around the conditions attained in year 200. For each of the 12 treatments, biomass (tC ha^{-1}) values were summed across species to obtain values for the total forest, *Larix* spp., and evergreen conifers at each site.

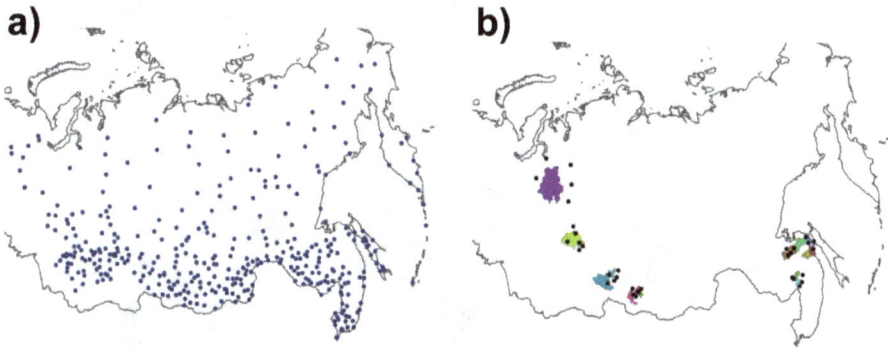

Fig. 1. Multi-scale analysis included data for 372 sites at the continental scale (a) and sub-sets from six regions **(b)** within northwest Siberia, the central border of Siberia, two sets from southern Siberia, and two eastern sets from high diversity areas in the Amur region of the Russian Far East

A non-parametric factorial ANOVA was performed at 10 year intervals and used to assess differences in total forest, *Larix* spp., and evergreen conifer biomass (tC ha^{-1}) between model runs that employed one of the 11 different climate and *L. decidua* treatments and the base climate scenario (SAS v. 9.1, SAS Institute Inc. 2002). This analysis was completed at the continental scale for a total of 372 sites (Figure 1a), and for six regional subsets (Figure 1b) including northwest Siberia (NW Siberia), the central border of Siberia (Central Siberia), two sets from southern Siberia (E Irkutsk, and W Irkutsk), and two sets from the Amur region of the Russian Far East (N RFE, and SW RFE). These regions represent areas with a broad range of climatic conditions and offer a representative sample of different forest types. Within the six regional subsets, local scale results were evaluated for changes in successional dynamics resulting from the climate or *L. decidua* treatments.

3. Results

3.1 Model simulation across Siberia and RFE
Overall biomass dynamics across Siberia and the RFE for the baseline climate scenario shows the highest values across the Amur region of the RFE, moderate biomass within

southern Siberia and low biomass across the northernmost sites. The successional dynamics across western Siberia in response to baseline climate feature a larch-dominated system persisting over time (Figure 2a). In the warmer southern portions of the region, larch forest undergoes a transition to mixed evergreen conifer and deciduous broad-leaved species beginning around year 230 (Figure 2b). This transition continues and by year 500, the southern portion of Siberia becomes a mixed larch and evergreen conifer forest (Figure 2c).

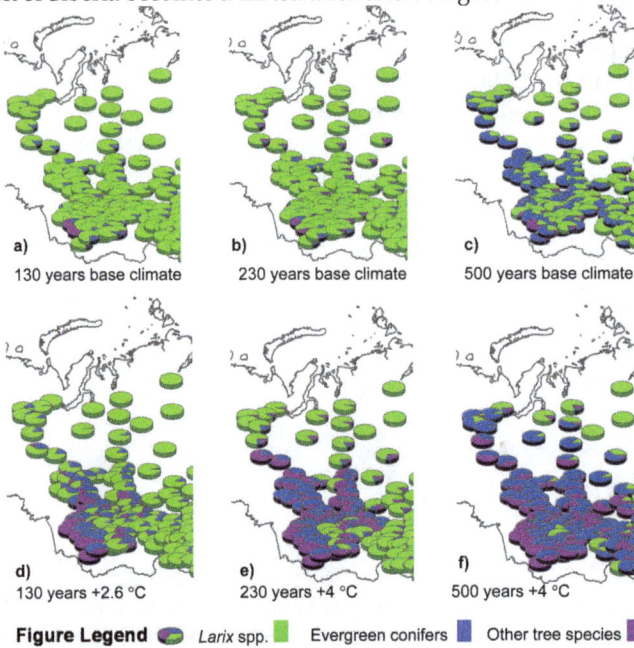

Fig. 2. Species distribution in western Siberia, a subset of the total area and dataset, for the baseline (**a, b, c**) and temperature increase (**d, e, f**) climates.

3.2 Climate sensitivity analysis
3.2.1 Continental scale results
Biomass response to the temperature treatment is significant (p<0.001), continues to the end of simulation for total forest and *Larix* spp. biomass (Figure 3), and is reflected in the shift in species distribution over time (Figure 2). By year 130, the effects of the response to warming can already be seen when compared to the base climate (Figure 2a) in the shift of species dominance from larch to evergreen conifer and other species at sites in southwestern Siberia (Figure 2d). The presence of evergreen conifers and other species, in what was larch-dominated forest under the historical climate (Figure 2b), expands across more of southern Siberia by year 230 (Figure 2e), and by year 500 Siberia is no longer a larch-dominated forest under increased temperature conditions (Figure 2f).

At the continental scale non-parametric factorial ANOVA results for biomass (tC ha^{-1}) under the temperature, precipitation, and European Larch (*Larix decidua*) treatments indicated that all classes of biomass were affected (p<0.001) (Figure 3). The temperature and *L. decidua* treatments have the strongest and most persistent effect on biomass throughout the simulation. *L.decidua* appears to be well-adapted to establishing across the region and affects biomass in all groups (total forest, *Larix* spp., and evergreen conifer) at the continental scale for the entire simulation. There is an effect of precipitation change on biomass, but this effect occurs only later in succession, at year 140, and alters total forest and evergreen conifer biomass, not *Larix spp.* biomass (Figure 3).

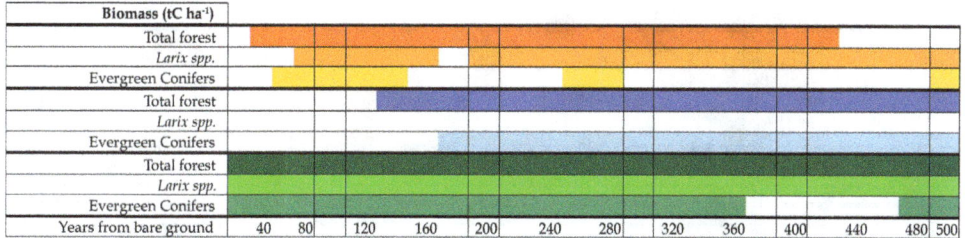

Biomass (tC ha⁻¹)													
Total forest													
Larix spp.													
Evergreen Conifers													
Total forest													
Larix spp.													
Evergreen Conifers													
Total forest													
Larix spp.													
Evergreen Conifers													
Years from bare ground	40	80	120	160	200	240	280	320	360	400	440	480	500

Figure Legend for treatments:

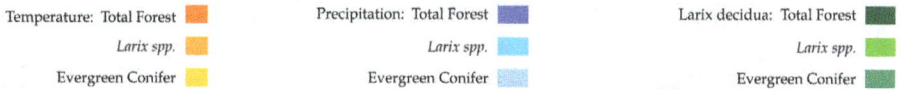

Temperature: Total Forest   Precipitation: Total Forest   Larix decidua: Total Forest

Larix spp.   Larix spp.   Larix spp.

Evergreen Conifer   Evergreen Conifer   Evergreen Conifer

Fig. 3. Non-parametric factorial ANOVA results for climate sensitivity analyses for 372 sites across Siberia and Russian Far East. Shown in colors corresponding to figure legend are comparisons to baseline biomass values that were significant to $p<0.001$ for treatment effects of temperature, precipitation, and *Larix deciuda* on total forest, *Larix* spp., and evergreen conifer biomass.

3.2.2 High diversity regional and local scale results

Sites in the Amur region of the Russian Far East (RFE) have an average of 38 individual tree species, and are classified for this analysis as high diversity. Within the high diversity regions, the non-parametric factorial ANOVA results showed a sporadic response ($p<0.001$) to the temperature and European Larch (*L. decidua*) treatments for the biomass classes measured (Figure 4). In contrast to the continental effect, the high diversity regions showed minimal response to the treatment effects of temperature increase and *L. decidua* addition, and no response to the effect of precipitation change.

Local results for the high diversity regions are variable depending on local climate conditions. The southwestern RFE (SW RFE) region under the base climate has mixed deciduous forests in the early successional stages which mature into mixed deciduous and evergreen conifer forests. Larch is present in SW RFE region, but is not a dominant species at any point during succession. Local scale analysis in SW RFE shows that the increased temperature alters the late successional dynamics by drastically reducing or replacing evergreen conifers with mixed deciduous species as early as year 150; this is reflected in the response ($p<0.001$) of evergreen conifer biomass from year 120 until the end of simulation (Figure 4). The *L. decidua* treatment does not significantly affect biomass in SW RFE region. With base climate conditions at sites in the northern RFE (N RFE) there is an initial pioneering stage dominated by *Larix* spp. which then transitions to evergreen conifer (*Picea* spp.) dominant forest (Figure 5a). The effect of temperature increase of 4°C across 200 years accelerates and alters the transition to a mixed-species forest dominated by *Pinus* spp. rather than *Picea* spp. (Figure 5b). This is reflected in the effect of the temperature treatment ($p<0.001$) on total forest biomass and *Larix spp.* biomass in late succession after year 200 for sites in N RFE (Figure 4). Even with the increased biomass of the larch canopy created with the introduction of *L. decidua* the transition from larch to *Pinus* spp. in late succession still occurs under increased temperature conditions (Figure 5c), with strong similarity to the species shift seen without inclusion of *L. decidua* (Figure 5b).

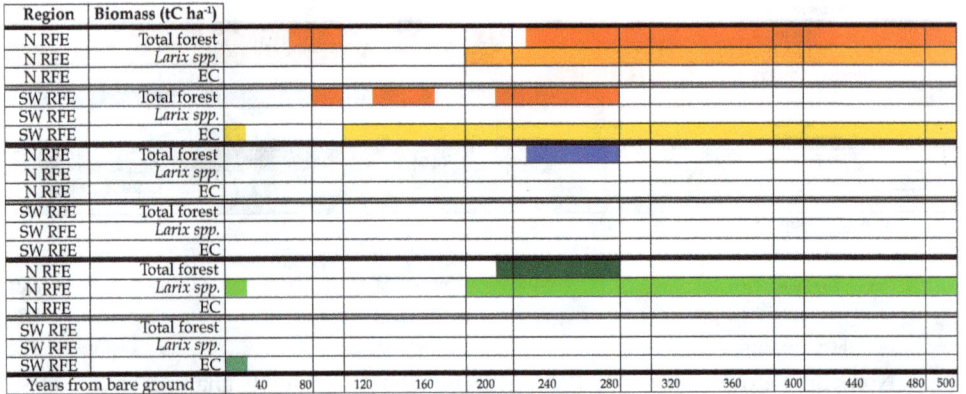

Region	Biomass (tC ha⁻¹)
N RFE	Total forest
N RFE	*Larix spp.*
N RFE	EC
SW RFE	Total forest
SW RFE	*Larix spp.*
SW RFE	EC
N RFE	Total forest
N RFE	*Larix spp.*
N RFE	EC
SW RFE	Total forest
SW RFE	*Larix spp.*
SW RFE	EC
N RFE	Total forest
N RFE	*Larix spp.*
N RFE	EC
SW RFE	Total forest
SW RFE	*Larix spp.*
SW RFE	EC
Years from bare ground	40 80 120 160 200 240 280 320 360 400 440 480 500

Figure Legend for treatments:

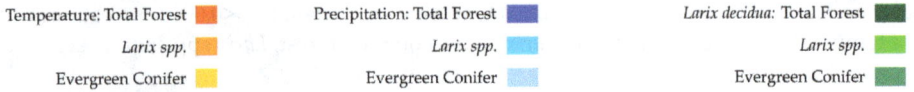

Temperature: Total Forest	Precipitation: Total Forest	*Larix decidua*: Total Forest
Larix spp.	*Larix spp.*	*Larix spp.*
Evergreen Conifer	Evergreen Conifer	Evergreen Conifer

Fig. 4. Non-parametric factorial ANOVA results for climate sensitivity analyses in two high diversity sites of the Amur region in the Russian Far East: northern Russian Far East (N RFE) and southwestern Russian Far East (SW RFE). Shown in colors corresponding to figure legend are comparisons to baseline biomass values that were significant to $p<0.001$ for treatment effects of temperature, precipitation, and *Larix deciuda* on total forest, *Larix* spp., and evergreen conifer (EC) biomass.

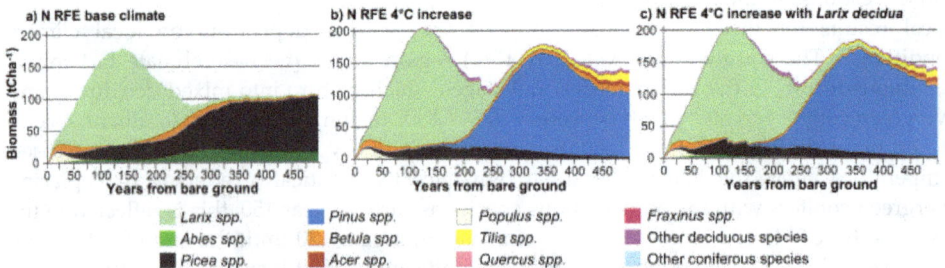

Fig. 5. Simulated species biomass dynamics (tC ha⁻¹) for high diversity Burukan site in northern Amur region of the Russian Far East (N RFE). Species composition by dominant genera is shown over 500 simulated years starting from bare ground for the base historical climate (a), temperature increase (b), and temperature increase with *Larix decidua* (c).

3.2.3 Low diversity regional and local scale results

The 279 sites across Siberia and the remainder of the RFE have an average of 9 individual tree species, and are classified as low diversity for this analysis. Similar to the continental response, within the low diversity regions the non-parametric factorial ANOVA results showed a consistent response ($p<0.001$) to temperature and *L. decidua* treatments for biomass of the total forest, *Larix* spp., and evergreen conifers when compared to baseline biomass values (Figure 6). Specifically the temperature increase in low diversity regions affects total forest and *Larix* spp. biomass early in succession and prior to year 200 (Figure 6). These regions also display a synchrony or lag response with total forest and *Larix* spp. biomass being closely connected in terms of the timing of the significant departure from baseline biomass. In all low diversity regions analyzed, except central Siberia, the response of evergreen conifer biomass to warming occurs after the response of total forest and *Larix* spp. biomass. The effect of the precipitation treatment was significant ($p<0.001$) in only one low diversity region analyzed, central Siberia.

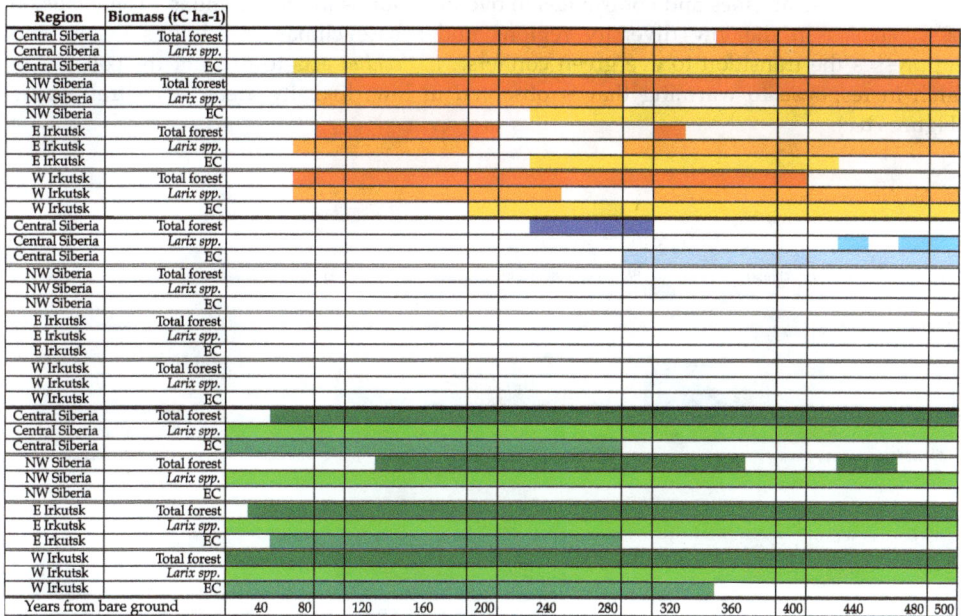

Region	Biomass (tC ha-1)													
Central Siberia	Total forest													
Central Siberia	Larix spp.													
Central Siberia	EC													
NW Siberia	Total forest													
NW Siberia	Larix spp.													
NW Siberia	EC													
E Irkutsk	Total forest													
E Irkutsk	Larix spp.													
E Irkutsk	EC													
W Irkutsk	Total forest													
W Irkutsk	Larix spp.													
W Irkutsk	EC													
Central Siberia	Total forest													
Central Siberia	Larix spp.													
Central Siberia	EC													
NW Siberia	Total forest													
NW Siberia	Larix spp.													
NW Siberia	EC													
E Irkutsk	Total forest													
E Irkutsk	Larix spp.													
E Irkutsk	EC													
W Irkutsk	Total forest													
W Irkutsk	Larix spp.													
W Irkutsk	EC													
Central Siberia	Total forest													
Central Siberia	Larix spp.													
Central Siberia	EC													
NW Siberia	Total forest													
NW Siberia	Larix spp.													
NW Siberia	EC													
E Irkutsk	Total forest													
E Irkutsk	Larix spp.													
E Irkutsk	EC													
W Irkutsk	Total forest													
W Irkutsk	Larix spp.													
W Irkutsk	EC													
Years from bare ground		40	80	120	160	200	240	280	320	360	400	440	480	500

Figure Legend for treatments:

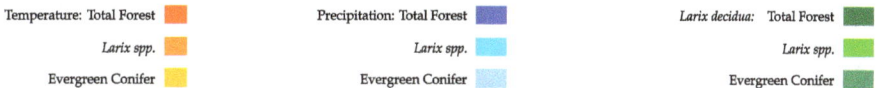

Temperature: Total Forest	Precipitation: Total Forest	Larix decidua: Total Forest
Larix spp.	Larix spp.	Larix spp.
Evergreen Conifer	Evergreen Conifer	Evergreen Conifer

Fig. 6. Non-parametric factorial ANOVA results for climate sensitivity analyses in low diversity sites of Siberia. Shown in colors corresponding to figure legend are comparisons to baseline biomass values for treatment effects of temperature, precipitation, and *Larix decidua* that were significant to $p<0.001$ for total forest, *Larix* spp., and evergreen conifer (EC) biomass. NW Siberia is the northwestern Siberia region. E and W Irkutsk regions are in southern Siberia.

The low diversity regions under historical site conditions have a successional pattern of increasing *Larix* spp. biomass to year 200, followed by the slow establishment of evergreen conifers with *Larix* spp. maintaining a significant presence to the end of simulation (Figure 7a). The temperature treatment accelerates the establishment of evergreen conifers and at some sites causes a complete collapse of larch biomass in many of the low diversity sites around year 200 when the temperature has increased by 4°C (Figure 7b). Successional dynamics in northwestern Siberia represent an exception to this general pattern. The colder regional temperatures in northwestern Siberia do not promote transition from *Larix* spp. to evergreen conifer, rather there is persistent *Larix* spp. dominance (Figure 7d). Northwestern Siberia does not experience the collapse of larch that is seen at sites further south (Figure 7b), but does transition to forests dominated by evergreen conifers in late successional stages in response to warming (Figure 7e). This late successional transition is similar to the natural succession dynamics of central and southern Siberia under base climate conditions (Figure 7a). The effect of the *L. decidua* treatment on *Larix* spp. biomass is immediate and continues to the end of simulation in all low diversity regions (Figure 6), indicating that *L. decidua* easily establishes and contributes to overall biomass in these regions. The inclusion of *L. decidua* in the low diversity regions under base climate conditions delays and suppresses the transition to evergreen conifers. *L. decidua* acts to prevent the collapse of larch in response to warming that is observed in low diversity areas in central Siberia (Figure 7b,c).

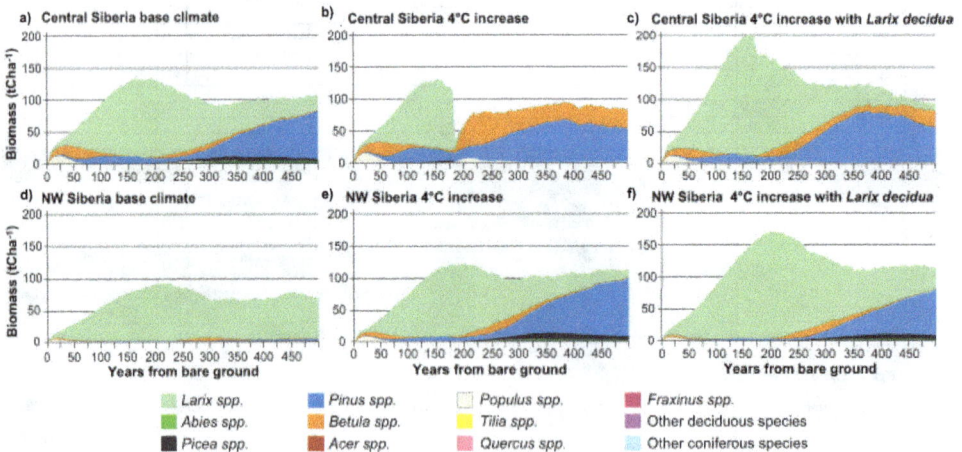

Fig. 7. Simulated mixed species biomass dynamics (tC ha-1) for low diversity sites in Siberia. Species composition by the dominant genera over 500 simulated years starting from bare ground for the base historical climate **(a,d)**, temperature increase **(b,e)**, and temperature increase with *Larix decidua* **(c,f)**.

4. Discussion

4.1 Model simulation across Siberia and RFE

Biomass patterns from simulation of mature forest under historical climate conditions reflect the idea that areas with increased plant diversity have increased productivity (Tilman and Downing 1994; Chapin et al., 1997, Bengtsson et al., 2000), with areas of higher biomass located in the areas of increased diversity in the Amur region of the RFE. The 93 high diversity sites are all located in the Amur region of the RFE and have an average of 38 individual tree species. The remaining 279 sites have an average of 9 individual tree species. The Amur region of the RFE also has higher average temperatures and precipitation values than across Siberia and the remainder of the RFE which allows a more diverse group of species to actively compete and achieve optimal biomass without climate limitations. Similar biomass results from past simulations which allow 44 individual tree species to grow at all sites without range limitation across Siberia and the RFE suggest it is the severe climate, and not a decreased species diversity, which limits the amount of total biomass across the interior of Russia (Shuman and Shugart 2009).

Successional dynamics across the study area under base climate reflect fundamental competition dynamics among species. Larch (*Larix* spp.) is highly tolerant of cold temperatures, but is one of the most shade-intolerant genera in the region (Nikolov and Helmisaari 1992). As the forest matures, competition for light becomes a key factor in determining which species becomes dominant. In northwestern Siberia, the cold temperatures prevent many species from competing with the cold-tolerant larch. Central and southern Siberia do not experience the severely cold temperatures of northwestern Siberia, and evergreen conifers actively compete with larch. Due to the shade-intolerance of larch, these forests transition to evergreen dominance as seen in the base climate simulation (Figure 2a,b,c). The transition from larch to evergreen conifer is also a product of the lack of insect or wildfire disturbance in these simulations. At each site the results are a landscape-level approximation of succession, which includes the natural disturbance caused by the death of individual trees. Warming climate is expected to cause increases in total area burned, fire-season length, and the severity of fire (Overpeck et al., 1990; Kasischke et al., 1995; Stocks et al., 1998; Soja et al., 2004; Soja et al., 2007). Similarly, the incidence of insect disturbance is also expected to become more prevalent with warming conditions (Holling 1992, Volney and Fleming 2000; Logan et al., 2003). Understanding the intrinsic successional dynamics isolates the direct response of the system to changing climate. Establishing the response of the system without the added changes of disturbance provides a strong basis for deconstructing the complexities of the system response to climate change.

4.2 Climate sensitivity analysis
4.2.1 Continental scale discussion

Larch is shade-intolerant and, in all but the coldest regions, evergreen conifers naturally replace larch over time, especially when no disturbance occurs that can rejuvenate the larch by providing open gaps of sunlight (Nikolov and Helmisaari 1992). The shift from deciduous larch to evergreen conifer forest is accelerated across Siberia under warming conditions (Figure 2), and implies a significant change in albedo. Following 200 years of forest development, larch-dominated forests are replaced with evergreen conifer-dominated forests in areas across Siberia. In southern Siberia, where forests are vulnerable to early replacement of larch by evergreen conifer, there would be a local significant albedo shift of approximately 5.1 W m^{-2} if the larch stands are replaced by evergreen conifers (Shuman et al., 2011). This represents a local increase in average annual absorbed surface radiation of between 2 and 7%. Albedo difference was the driver of the results of the modeling experiments completed by Bonan et al., (1992), Betts (2000), and Snyder et al., (2004) all of whom predicted the effects of such an albedo shift would extend beyond the boundaries of the boreal region. Chapin et al., (2000) documented similar albedo differences between forest and shrub tundra in Alaska with an increase in absorbed radiation leading to a warming trend that extends beyond the tundra. Similar to the modeling results for the boreal region, and the finding from the Alaskan tundra, the albedo shift implied by the successional dynamics shown in our forest simulations have the potential to increase temperatures across the region and create a positive feedback of regional warming. In particular, our results establish that there will be local shifts from larch to evergreen conifer. The resultant increase in the amount of absorbed incoming radiation has the potential to impact surrounding regions and set off a cascade of species shifts towards evergreen conifers, which in turn promote more warming. This positive feedback is of great concern across Siberia.

The response at the continental scale is most similar to the results for the low diversity regions for all treatments. This is not surprising given that 75% of the sites included in the continental-scale analysis are classified as low diversity, with an average of 9 tree species. Historically, this system maintains existing vegetation through cycles of predictable disturbance and succession, which support the regeneration of larch following fire (Chapin et al., 2004). The repetition of this successional sequence across the broad range of climatic conditions found in the boreal region creates a resilient vegetation composition with stable cycles of vegetation states (Chapin et al., 2004). The climate change scenarios considered here modify the vegetation composition in a new way, which in turn alters the successional history and reduces the resilience of vegetation, thereby forcing a new vegetation state to emerge. The larch-dominated forests appear sensitive to an increase in temperature very early in succession when the overall stand age is also young. The evergreen conifer dominated forests seem to be sensitive to changes in precipitation at mid-succession when there is a mix of stand ages from young to mature trees. Mid-succession is also the natural transition point, under base conditions, between larch and evergreen conifer, so the response to precipitation is likely connected with the emergence of evergreen conifers as the dominant species. The connection between precipitation and evergreen conifers is explored in more detail with the regional scale analysis.

At the continental scale, total forest and *Larix* spp. biomass are highly responsive to the *L. decidua* treatment and show a pattern similar to that of the forests in low diversity regions. These results suggest that low species diversity makes the system vulnerable to establishment by a new species, but highlight the potential of the introduction of *L. decidua* to be used in the mitigation or management of the albedo shift caused by transition to evergreen conifer dominance. *L. decidua* has the same characteristics as existing Siberian larch species, and thus forests dominated by this species have higher albedo, and decreased absorbed incoming radiation, when compared to stands of evergreen conifers in the same region. Maintaining larch-dominated stands across the region would delay the positive feedback triggered by the albedo shift that is associated with conversion from deciduous larch to evergreen forest.

4.2.2 High diversity regional and local scale discussion

The response to the temperature and *L .decidua* treatments in the high diversity regions highlights the importance of analysis of local climate and successional dynamics. The climate and *L. decidua* treatments do not have the strong effect on biomass in these high diversity regions of the RFE that is seen at the continental scale and in low diversity regions. Sites in the northern RFE (N RFE) region fall within larch's optimal growth ranges for both temperature and precipitation. Larch establishes strongly and competitively in N RFE in early succession followed by a transition to *Picea* spp. dominance under base climate conditions. The high diversity in the area indicates there is strong competition, and *L. decidua* does not have unique characteristics which allow it to establish in this region.

Under warmer conditions in the N RFE in late succession there is a transition, not to *Picea* spp., but to *Pinus* spp. dominance. This highlights the importance of genus-level analysis. These genera are typically combined into a single evergreen conifer group, both for the purposes of global climate models and for the regional scale analysis presented here. Such a grouping prevents the detection of this shift between the two evergreen conifer species. This shift has the potential for further exploration in association with altered albedo values. Measured in the boreal forest of Canada, the summer albedo difference between *Pinus* spp. (0.086) and *Picea* spp. (0.081) is negligible, but the winter albedo for *Pinus* spp. (0.150) and *Picea* spp. (0.108) is more significant (Betts and Ball 1997). The albedo values in the RFE are likely similar to these Canadian species, and there is potential, with the winter albedo difference, to alter total annual absorption of surface radiation in response to the shift from *Picea* spp. to *Pinus* spp. with warming. Unlike the low diversity regions, the successional dynamics resulting from including *L. decidua* as a species in simulation are similar to those observed in response to warming, with late successional dominance by *Pinus* spp. This result suggests that introduction of *L. decidua* would not be a useful strategy for mitigation of vegetation shifts in this region.

The response to the climate and *L. decidua* treatments in southwestern (SW) RFE is also a product of local climate conditions. The SW RFE region has a climate with temperatures which place it in the upper limit of tolerance for larch, creating a climate in which the native larch species cannot compete and establish. Under base climate conditions, it is difficult for larch to compete, so it is not surprising that larch continues to be absent under warmer conditions. Even with increased tolerance for warmer conditions, *L. decidua* cannot effectively compete with other species in the SW RFE, and does not have a significant impact on biomass. Under base climate conditions, the SW RFE has mixed evergreen and deciduous species in the mature forest late in succession. With warming temperatures, there is an increase in biomass of mixed deciduous trees, and a decrease in evergreen conifer biomass. The decrease of evergreen conifers is balanced by the increase of mixed deciduous trees, so the effect on total biomass is not consistent and lasts only 60 years.

The high diversity areas in the RFE have high ecological resilience and stability, which allow them to maintain basic ecosystem function following climate change and avoid irreversible shifts to another vegetation state. Stability is the ability of a system to return to equilibrium following temporary disturbance, and resilience is the persistence of a system and its ability to absorb change and disturbance without changing state (Holling 1973). In other terms, stability is a persistence of the system state and a consequence of interactions within the system where the next state of the system is predictable from within the system (Margalef 1968; Lewontin 1969; Child and Shugart 1972). Ecological resilience is therefore related to the magnitude of disturbance or change that can be absorbed before the system transitions to another stability domain (Folke et al., 2004; Gunderson 2000, Peterson et al., 1998). In the N RFE, the successional cycle from larch-dominance to evergreen-dominance is the expected cycle between vegetation states, because this is the response for base climate conditions. Both N RFE and SW RFE respond to an increase in temperature with a shift in the dominant species during late succession, but these late successional groups are functionally similar to the assemblage of species that exist under the base climate. This is an indication that this group of high diversity regions in the Amur region of the RFE does not experience a change in vegetation state with in response to climate change, and is resilient to the perturbations associated with this amount of climate change.

The concept of response diversity adds to the conclusion that the system in the Amur region of the RFE is resilient by defining the range of reactions to environmental change among species contributing to the functioning of a given ecosystem (Elmqvist et al., 2003; Folke et al., 2004). The RFE region has high response diversity, which means that it has functionally similar species sets which respond differently to environmental change and provide a buffer that protects the system against failure and increases tolerance to disturbance or climate change (Elmqvist et al., 2003; Folke et al., 2004). In other words, there are species in the system capable of maintaining the original state of ecosystem function under the new conditions following change; a concept also known as the insurance hypothesis (Folke et al., 1996, Naeem and Li 1997). High response diversity within an ecosystem increases the chances of reorganization or restart of the system into the desired state after disturbance (Chapin et al., 1997; Bengtsson et al., 2000; Elmqvist et al., 2003). The altered climate disturbed the ecosystem, but because of the high diversity of the SW and N RFE, the species can reorganize and maintain the same vegetation state and ecosystem function that is observed throughout succession when the system is not disturbed by climate change. Therefore the diversity of species in this region allows for replacement of one species with another functionally similar one under new climate conditions. It is also the adaptability of the system under altered climate which prevents a substantial contribution to biomass from the introduction of *L. decidua*. Neither a change in climate, nor the addition of a single species (*L. decidua*) leads to a change in ecosystem function in this high diversity system.

It is important to note that the RFE regions analysed display ecological resilience and high response diversity for the both temperature increase and precipitation change treatments evaluated. There is a response to temperature in both regions, but not at the same magnitude as that of the low diversity areas. These results suggest that increased amounts of climate change may have a stronger impact on the system. Further analysis is necessary to determine if the RFE system is equally resilient when the temperature is increased by more than 4°C, or if it can restart under this altered climate condition following disturbance, such as fire or insect outbreak, that take the system back to bare ground. The results seen in this study indicate only a slight sensitivity of the mid- to late-successional stages, at and beyond year 200, which correspond to the time in the simulation when the temperature had increased to 4°C. The early successional stages are comprised of a different set of species which may not show the same high resilience or response diversity displayed by the mid- to late-successional stages in the RFE.

4.2.3 Low diversity regional and local scale discussion

The response of the low diversity regions to the altered climate treatment is similar to that observed at the continental scale. These low diversity regional scale responses, in conjunction with the low average diversity across the 372 sites considered at the continental scale, further emphasize the difference of the response in the high and low diversity areas. Similar to analysis at the continental scale, the response of the low diversity regions to treatments is connected to .e transition from larch to evergreen dominance, which occurs across the southern and central portion of Siberia, and is accelerated by warming. Northwestern Siberia has the coldest temperatures compared to the other regions, and these cold temperatures naturally suppress evergreen conifer establishment during late succession under base climate. The patterns of successional dynamics in central Siberia under base conditions and in northwestern Siberia in response to warming climate suggest that the model is predicting consistent transitions. Under warming conditions, northwestern Siberia is exposed to temperatures more similar to those in the base condition in central Siberia. Thus, the forests in northwestern Siberia have similar dynamics under warming conditions to those seen in the central Siberia region for the base climate (Figure 7a,e). The temperature treatment results suggest that, with 4°C of warming, the larch-dominated system across southern and central Siberia will be prematurely replaced with evergreen conifer and other deciduous trees, and the forests of northwestern Siberia will warm enough that evergreen conifers will be able to effectively establish.

The lack of response in northwestern and southern Siberia, and the late response in central Siberia, to the effect of precipitation change suggests a connection between evergreen conifer presence and seasonal precipitation. The precipitation treatment is significant in only one of the low diversity regions analyzed (i.e., central Siberia). Within the central Siberia region there is a short-lived response of total forest biomass to precipitation change at year 240, and a longer period of response of evergreen conifer biomass from year 300 to 500. Of the six regions analyzed, central Siberia has one of the lowest average seasonal precipitation curves; it is closest to that of northwestern Siberia. Forests within central Siberia naturally transition from a larch-dominated to an evergreen-conifer dominated system, with larch as a secondary species (Figure 7a). This successional transition is similar to local dynamics at sites in both East and West Irkutsk, regions in southern Siberia that both have higher precipitation than central Siberia. Northwestern Siberia, which has similar precipitation to central Siberia, does not experience a natural shift from larch- to evergreen conifer-dominance. The colder regional temperatures in northwestern Siberia suppress evergreen conifer growth, thereby helping larch to maintain dominance (Figure 7d). These observations suggest that the precipitation treatment response in central Siberia is a result of both the lower seasonal precipitation, and the presence of evergreen conifers late in succession in this region. In other words, the transition from larch- to evergreen conifer-dominant in central Siberia, combined with the low annual precipitation, creates a region which is responsive to precipitation change (Figure 6). The variability across the low diversity regions suggests that differences in response to climate treatments are a result of local conditions and species composition over time. The continental scale analysis shows a response to precipitation for both total forest and evergreen conifer biomass. This suggests that many of the sites considered in this continental scale analysis have low seasonal precipitation, and are dominated by moisture-sensitive evergreen conifers later in succession.

There are marked differences between the results for the high and the low diversity regions, which demonstrate differences in the stability and resilience of these regions. The high diversity areas of the RFE have high ecological resilience and maintain basic ecosystem function as a result of similarly functioning species replacing one another following climate change. The low diversity regions, however, have low resilience and cannot maintain basic ecosystem function following climate change, specifically temperature increase. This is shown by the collapse of dominant larch species following the 4°C increase in sites in central and southern Siberia, and the fact that the northwestern Siberia sites shift to an entirely new stable state, which is not seen under base conditions in this region, in response to warming climate conditions. The collapse of larch in southern regions in response to a temperature increase suggests that larch is particularly vulnerable, and that the systems' response threshold may be exceeded with the 4°C increase. With fewer species present, it is more likely that extinctions will alter ecosystem processes (Chapin et al., 1997). Furthermore, the diversity of the area is so low that there are no species capable of fulfilling the original ecosystem function under the new conditions following change. Additionally, differences in sensitivity among functionally different species, in this case larch and evergreen conifers, make the ecosystem vulnerable to change (Chapin et al., 1997).

Holling (1992) hypothesized that the vegetation of the boreal forest would buffer initial climate changes, but that there would be a limit to the buffering and an abrupt vegetation change would follow. These results follow Holling's hypothesis and suggest an abrupt shift in vegetation in response to temperature increase. This abrupt vegetation change is congruent with the identification of this system as having low resilience and stability when compared with the high diversity areas of the RFE. Chapin et al., (2004) suggested that vegetation within central portions of the boreal forest would remain stable for long periods followed by abrupt changes to a new state, which is what we see in the results for the low diversity areas in central Siberia. Low diversity areas do not have the appropriate pool of species to continue the cycle of succession and reorganization following change, thus the system in these areas is flipped into a different state (Bengtsson et al., 2000). The results presented here are consistent with field measurements documenting the shift of treelines northward or upslope of previous climate limits, and a reduction in cone and seed yield for *L. sibirica* (Kharuk et al., 2009; Soja et al., 2007). They are also consistent with bioclimatic model results predicting a replacement of taiga with forest-steppe or steppe environments across southern Siberia (Tchebakova et al., 2005; Vygodskaya et al., 2007; Tchebakova et al., 2009). These results also suggest that warming temperatures will lead to a shift in the ability of larch to establish and may signal a collapse of the species in this genus.

The introduction of *L. decidua* to the low diversity sites may help to buffer the perturbations associated with a warming climate. Unlike the high diversity regions in the RFE, the low diversity areas showed a strong response of biomass to the inclusion of *L. decidua*. Local scale analysis with warming conditions shows that the inclusion of *L. decidua* prevents the collapse of larch in central Siberia and delays transition to evergreen conifer dominance in northwestern Siberia. *L. decidua* is competitive in this low diversity area, and fills an important functional niche when temperatures are increased. Existing larch species cannot tolerate the warmer conditions, and their collapse opens functional space for the warm adapted *L. decidua*. These results, though theoretical, provide evidence that it is possible to address the issue of species replacement, and associated albedo shift, with techniques involving species management or introduction.

5. Conclusion

The FAREAST model was used to simulate forest successional dynamics across a region with broad geographic and climatic variability, and examine the behavior of forests at different scales in response to altered climate. The model simulated forest growth at high diversity sites in the Amur region of the RFE and low diversity sites across Siberia and the remainder of the RFE. The model successfully captures the natural successional dynamics of forests for base climate conditions across the area.

Results of the climate sensitivity analysis indicate that a 4°C increase impacts the biomass of the total forest, *Larix* spp., and evergreen conifers at the continental scale, and in low diversity regions, early in succession. The effect of temperature is highly significant throughout most of the simulation in low diversity areas, whereas there is little effect of temperature in high diversity regions. Results at the continental scale suggest that the forests across much of Siberia and the RFE behave as a low diversity system. The early effect of temperature, across areas where larch is naturally dominant in early succession, suggests that larch is particularly vulnerable to temperature increase. In areas outside the cold northern portion of Siberia, larch is shown to abruptly collapse in response to warming. The effect of altered precipitation was significant at the continental scale, and in one low diversity region in central Siberia in mid- to late-succession following evergreen conifer establishment. Central Siberia, in addition to experiencing a late successional shift from larch- to evergreen conifer-dominance, has low seasonal precipitation. This suggests that sites that respond to the precipitation treatment have low annual precipitation, and are dominated by moisture-sensitive conifers. The precipitation effect was not significant in the high diversity regions, which have different dominant evergreen species in late succession.

Concepts of ecological stability and resilience are used to explain the variable response of the high and low diversity areas to altered climate. The high diversity regions showed high stability and resilience for they maintained overall species and biomass dynamics in response to changing climate, and had replacement of one species by a functionally similar species under new climate conditions. Unlike the high diversity regions, the low diversity regions across Siberia show low ecological stability and resilience. The low diversity areas displayed a strong response of biomass to the climate treatments, and locally showed the collapse of the dominant larch species under increased temperatures. It is this the lack of diversity in the response of functionally similar species to environmental change, and thus an inability to maintain ecosystem function with altered conditions, which creates the low ecological stability across the low diversity areas of Siberia.

L. decidua, the warmer adapted European larch, was added to the species list to gauge the potential of this species to prevent a premature shift to evergreen vegetation, and the associated albedo shift, in response to climate change. *L. decidua* established strongly in the low diversity system, but not in the high diversity areas of the RFE. Due to the increased diversity in the Amur region of the RFE, there are native species which can fill the same functional space under new climate conditions as the species which were dominant under base climate conditions, thereby creating high resilience and stability. It is this pool of locally available species which prevents *L. decidua* from significantly impacting biomass in the high diversity regions of the RFE. Within the low diversity regions, however, local scale results show that *L. decidua* becomes established and acts to prevent the collapse of larch in response to warming and delay the shift to an evergreen conifer-dominated forest. It is the low diversity which contributes to the lack of resilience and stability under altered climate, but also allows for strong establishment of *L. decidua*. Therefore *L. decidua* is uniquely adapted to establish in this low diversity system as well as to prevent the positive feedback associated with a premature shift to evergreen conifer-dominance.

This study establishes that larch-dominated forests across Siberia will transition to a different vegetation state, and have an altered species composition, in response to climate change, especially increased temperature. *L. decidua* has been identified as a species capable of strong establishment across Siberia, and with the capacity to prevent the positive feedback associated with vegetation shift, from larch to evergreen conifer, in the region. These results deal with the response of the current system to controlled climate change. Future studies need to address the ability of the system to restart following disturbance, under altered climate conditions. Early successional species, such as larch may not be able to establish dominance in conditions, which have already warmed. In this case, the expected successional dynamics will be completely altered.

These results highlight potential for the use of remote sensing data in areas identified as vulnerable to vegetation change. Modeling studies offer the opportunity to identify a signature of climate change in vegetation dynamics in advance of those changes occurring on the ground. Remote sensing technology can be used to track land cover changes in areas identified by model results as vulnerable to vegetation shift. Furthermore, the results of this study indentify a positive feedback cycle where warming creates vegetation shift, which then creates further warming. The detailed vegetation maps derived from remote sensing data offer a capability to evaluate locations where vegetation shift has occurred in an effort to track the progress of this positive feedback cycle and assess the direction and magnitude of any albedo shift associated with such a change. Vegetation monitoring informed by modeling efforts provide a robust tool in responding to and identifying vegetation changes due to climate change.

6. Acknowledgments

This work was supported by the following NASA grants to H.H. Shugart: NNG-05-GN69G, NNX-07-A063G, NNX-07-AF10G, NAG-11084. We greatly appreciate the support and encouragement of Dr. Pavel Groisman and the Northern Eurasian Earth Science Partnership Initiative (NEESPI) in regard to this ongoing research. We extend our thanks to Dr. Paolo D'Odorico and Dr. Virginia Seamster for suggestions and feedback on earlier versions of this manuscript.

7. References

Amiro, B. D. 2001. Paired-tower measurements of carbon and energy fluxes following disturbance in the boreal forest. *Global Change Biology* Vol 7,pp.253-268.

Baldocchi, D., Kelliher, F. M., Black, T. A. and Jarvis, P. 2000. Climate and vegetation controls on boreal zone energy exchange. *Global Change Biology* Vol 6,pp.69-83.

Barber, V.A., G.P. Juday and B.P. Finney. 2000. Reduced Growth of Alaskan White Spruce in the Twentieth Century from Temperature-Induced Drought Stress. *Nature* Vol 405,pp. 668-673.

Bengtsson J, S G Nilsson, A Franc, P Menozzi. 2000. Biodiversity, disturbances, ecosystem function and management of European forests. *Forest Ecology and Management* Vol 132,pp. 39-50.

Beringer, J., Chapin III, F. S., Thompson, C. C., and McGuire, A. D. 2005. Surface energy exchanges along a tundra-forest transition and feedbacks to climate. *Agricultural and Forest Meteorology* Vol 131, pp.143-161.

Betts, R A. 2000. Offset of the potential carbon sink from boreal forestation by decreases in surface albedo. *Nature* Vol 408, pp.187-190.

Betts A K and J H Ball. 1997. Albedo over the boreal forest. *Journal of Geophysical Research* Vol 102, No (D24), pp. 28901-28909.

Bonan, G.B. 1989a. Environmental factors and ecological processes controlling vegetation patterns in boreal forests. *Landscape Ecology* Vol 3, pp.111-130.

Bonan, G.B. 1989b. A computer model of solar radiation, soil moisture and soil thermal regime. *Ecological Modeling* Vol 45,pp.275-306.

Bonan, G. 2008. Forests and climate change: Forcings, feedbacks, and the climate benefits of forests. *Science* 320:1444-1449.

Bonan, G B, D Pollard and S L Thompson. 1992. Effects of boreal forest vegetation on global climate. *Nature* Vol 359, pp. 716-718.

Botkin, D B, J F Janak, and J R Wallis. 1972. Some ecological consequences of a computer model of forest growth. *Journal of Ecology* Vol 60, pp.849-872.

Bugmann, HK. 1996. A simplified forest model to study species composition along climate gradients. *Ecology* Vol 77, pp. 2055-2074.

Bugmann, HK. 2001. A review of forest gap models. *Climatic Change* Vol 51, pp.259-305.

Bugmann, HK and A M Solomon. 2000. Explaining Forest Composition and Biomass across Multiple Biogeographical Regions. *Ecological Applications* Vol 10, No 1,pp. 95-114.

Cale, W G, R V O'Neil, and H H Shugart. 1983. Development and Application of Desirable Ecological Models. *Ecological Modeling* Vol 18,pp. 171-186.

Chapin, F S; B H Walker, R J Hobbs, D U Hooper, J H Lawton, O E Sala, D Tilman. 1997. Biotic Control over the Functioning of Ecosystems. *Science* Vol 277,pp. 500-504.

Chapin F S, T Callaghan, Y Bergeron, M Fukuda, J F Johnstone, G Juday and S A Zimov. 2004. Global Change and the Boreal Forest: Thresholds, Shifting States or Gradual Change? *Ambio* 3 Vol 3, No 6, pp. 361-365.

Chapin F S, W Eugster, J P McFadden, A H Lynch and D A Walker. 2000. Summer Differences among Arctic Ecosystems in Regional Climate Forcing. *Journal of Climate* Vol 13, No 12, pp. 2002-2010.

Chapin, F. S., McGuire, A. D., Randerson, J., Pielke, R., Baldocchi, D., Hobbie, S. E., Roulette, N., Eugster, W., Kasischke, E., and Rasteter, E. B. 2000. Arctic and boreal ecosystems of western North America as components of the climate system. *Global Change Biology* Vol 6, No 1, pp. 211-223.

Child, G I and H H Shugart. 1972. Frequency Response Analysis of Magnesium Cycling in a Tropical Forest Ecosystem. in Patten B C (editor) *Systems Analysis and Simulation in Ecology Vol II* Academic Press, New York, USA

Dale, V H, and J F Franklin. 1989. Potential Effects of Climate Change on Stand Development in the Pacific Northwest. *Can J Forest Res* Vol 19, pp. 1581-1590.

Elmqvist T, C Folke, M Nyström, G Peterson, J Bengtsson, B Walker, and J Norberg. 2003. Response diversity, ecosystem change, and resilience. *Ecol Environ* Vol 1, No 9, pp. 488-494.

ESRI. 2008. ESRI ArcGIS version 9.3 [computer program] ESRI, Redlands, CA, USA

Folke C, C S Holling and C Perrings. 1996. Biological Diversity, Ecosystems, and the Human Scale. *Ecological Applications* Vol 6, No 4, pp. 1018-1024.

Folke C, S Carpenter, B Walker, M Scheffer, T Elmqvist, L Gunderson and C S Holling. 2004. Regime Shifts, Resilience, and Biodiversity in Ecosystem Management. *Annu Rev Ecol Evol Syst* Vol 35, pp. 557-581.

Gunderson, L H. 2000. Ecological Resilience – In Theory and Application. *Annu Rev Ecol Evol Syst* Vol 31, pp. 425-439.

Holling C S. 1973. Resilience and Stability of Ecological Systems. *Annu Rev Ecol Evol Syst* Vol 4, pp. 1-23.

Holling C S. 1992. The role of forest insects in structuring the boreal landscape. pp 170-191. In: Shugart H H, R Leemans, G B Bonan (eds) *A systems analysis of the global boreal forest* Cambridge University Press Cambridge UK

Hollinger, D.Y., S.V. Ollinger, A.D. Richardson, T.P. Meyers, D.B. Dail, M.E. Martin, N.A. Scott, T.J. Arkebauer, D.D. Baldocchi, K. Clark, P.S. Curtis, K. Davis, A. Desai, D. Dragoni, M.L. Goulden, L. Gu, G.G. Katul, S. Pallardy, K.T. Paw U, H. Schmid, A.E. Suyker, and S.B. Verma. 2010. Albedo estimates for land surface models and support for a new paradigm based on foliage nitrogen concentration. *Global Change Biology*, Vol 16, No 2 pp. 696-710.

Houghton, R A, D Butman, A Bunn, O N Krankina, P Schlesinger and T A Stone. 2007. Mapping Russian Forest Biomass with Data from Satellites and Forest Inventories. *Environ. Res. Lett.* 2 (045032): 7 pp URL http://www.iop.org/EJ/abstract/1748-9326/2/4/045032/

Hytteborn, H, A A Maslov D I Nazimova and L P Rysin. 2005. Boreal Forests of Eurasia pp23-99 In: F Andersson (editor) *Ecosystems of the World 6 Coniferous Forests* Elsevier B V, Amsterdam, Netherlands

IPCC. 2007. Climate change 2007: the physical scientific basis. Contribution of Working Group I to the Fourth Assessment Report of the Intergovernmental Panel on Climate Change Cambridge University Press, New York, USA

Kasischke, E S, N L Christensen and B J Stocks. 1995. Fire, global warming, and carbon balance of the boreal forests *Ecological Applications* Vol 5, No 2,pp. 437-451.

Kharuk, V, K Ranson and M Dvinskaya. 2007. Evidence of Evergreen Conifer Invasion into Larch Dominated Forests During Recent Decades in Central Siberia. *Eurasian Journal of Forest Research* Vol 10, No 2, pp. 163-171.

Kharuk, V I, K J Ranson, T A Burenina and E V Fedotova. 2003. Mapping of Siberian forest landscapes along the Yenisey transect with AVHRR. *Int J of Remote Sensing* Vol 24, No 1, pp. 23-37.

Kharuk, V I, K J Ranson, T I Sergey and M L Dvinskaya. 2009. Response of Pinus sibirica and Larix sibirica to climate change in southern Siberian alpine forest-tundra ecotone. *Scandinavian Journal of Forest Research* Vol 24, No 2, pp. 130-139.

Kienast, F. 1987. FORECE- a forest succession model for southern central Europe. ORNL/TM-10575, Oak Ridge National Laboratory, Oak Ridge, Tennessee, USA

Kienast, F and N Krauchi. 1991. Simulated successional characteristics of managed and unmanaged low-elevation forests in central Europe. *Forest Ecology and Management* Vol 42, pp. 49-61.

Lasch, P and M Lindner. 1995. Application of two forest succession models at sites in north east Germany. *Journal of Biogeography* Vol 22, pp. 485-492.

Leemans, R and I C Prentice. 1989. FORSKA, a general forest succession model Institute of Ecological Botany, Uppsala, Sweden

Lewontin, R C. 1969. The Meaning of Stability. *Diversity and Stability of Ecological Systems Brookhaven Symposia in Biology* Vol 22, pp. 13-24.

Logan, J A, J Régnière and J A Powell. 2003. Assessing the impacts of global warming on forest pest dynamics. *Frontiers in Ecology and in the Environment* Vol 1, No 3, pp. 130-137.

Mankin, J B, R V O'Neil, H H Shugart and B W Rust. 1977. The importance of validation in ecosystem analysis. pp.63-71 In: G Innis (editor) *New Directions in the Analysis of Ecological Systems, Part I The Society for Computer Simulation*, La Jolla, CA

Margalef, R. 1968. *Perspectives in Ecological Theory* University of Chicago Press, Chicago, USA

McGuire, A.D., C. Wirth, M. Apps, J. Beringer, J. Clein, H. Epstein, D.W. Kicklighter, J. Bhatii, F.S. Chapin III, B. de Groot, D. Efremov, W. Eugster, M. Fukuda, T. Gower, L., Hinzman, B. Huntley, G.J. Jia, E. Kasischke, J.M. Melillo, V. Romanovsky, A. Shvidenko, E., Vaganov, and D. Walker. 2002. Environmental variation, vegetation distribution, carbon dynamics, and water/energy exchange in high latitudes. *Journal of Vegetation Science* Vol 13, pp. 301-314.

Mladenoff, D.J. 2004. LANDIS and forest landscape models. *Ecological Modelling* Vol 180, pp.7-19.

Naeem S and S Li. 1997. Biodiversity enhances ecosystem reliability. *Nature* Vol 390, pp. 507-509.

National Climate Data Center (NCDC). 2005a. TD-9290c-1 "Global Synoptic Climatology Network C The former USSR Version 1.0". Available from NOAA National Climatic Data Center, Asheville, NC, USA

National Climate Data Center (NCDC). 2005b. TD-9813 "Daily and Sub-daily Precipitation for the Former USSR Version 1.0" Available from NOAA National Climatic Data Center, Asheville, NC, USA

Nikolov N and H Helmisaari. 1992. Silvics of the circumpolar boreal forest tree species. pp 9-84 In: Shugart H H, R Leemans and GB Bonan (eds) *A systems analysis of the global boreal forest* Cambridge University Press Cambridge UK

Overpeck, J T, D Rind and R Goldberg. 1990. Climate induced changes in forest disturbance and vegetation. *Nature* Vol 343, pp. 51-53.

Pastor, J and W M Post. 1988. Response of Northern Forests to CO_2-Induced Climate Change. *Nature* Vol 334, pp. 55-58.

Peterson G, C R Allen, and C S Holling. 1998. Ecological Resilience, Biodiversity, and Scale. *Ecosystems* 1 Vol, pp. 6-18.

Rykiel, E J. 1996. Testing ecological models: the meaning of validation. *Ecological Modelling* Vol 90, pp. 229-244.

Sargent, R G. 1984. A tutorial on verification and validation of simulation models. pp. 115-122. In: S. Sheppard, U Pooch, and D Pegden (Editors) *Proceedings of the 1984 Winter Simulation Conference* IEEE, Piscataway, NJ.

SAS Institute Inc. 2002. SAS release 9.1 [computer program] SAS Institute Inc Cary, NC, USA

Scheller R M and D J Mladenoff. 2007. An ecological classification of forest landscape simulation models: tools and strategies for understanding broad-scale forested ecosystems. *Landscape Ecology* Vol 22, pp. 491-505.

Shugart, H H. 1984. *A Theory of forest dynamics* Springer Verlag, New York

Shugart, H H. 1998. *Terrestrial ecosystems in changing environments* Cambridge University Press, Cambridge

Shugart, H H. 2003. *A theory of forest dynamics: the ecological implications of forest succession models* Blackburn Press, Caldwell, NJ

Shugart, H H and D C West. 1977. Development of an Appalachian deciduous forest succession model and its application to assessment of the impact of the chestnut blight *Journal of Environmental Management* Vol 5, pp. 161-179.

Shuman, J K. 2010. Russian forest dynamics and response to changing climate: a simulation study. PhD Thesis, University of Virginia, Charlottesville, VA

Shuman, J K and H H Shugart. 2009. Evaluating the sensitivity of Eurasian forest biomass to climate change using a dynamic vegetation model. *Environ. Res. Lett.* 4 (045024): 7 URL http://iopscience.iop.org/1748-9326/4/4/045024/

Shuman, J K, H H Shugart, and T L O'Halloran. 2011. Sensitivity of Siberian Larch forests to climate change. *Global Change Biology* (Accepted Article) doi: 10.1111/j.1365-2486.2011.02417.x

Snyder, P. K., Delire, C., and Foley, J. A. 2004. Evaluating the influence of different vegetation biomes on the global climate. *Climate Dynamics*, Vol. 23 No 3/4, pp. 279-302.

Soja, A J, W R Cofer, H H Shugart, A I Sukhinin, P W Stackhouse, D J McRae, S G Conard. 2004. Estimating fire emissions and disparities in boreal Siberia (1998-2002). *Journal of Geophysical Research 109* (D14S06) doi:10.1029/2004JD004570

Soja, A J, N M Tchebakova, N H French, M D Flannigan, H H Shugart, B J Stocks, A I Sukhinin, E I Parfenova and F S Chapin III. 2007. Climate-induced boreal forest change: Predictions *versus* current observations. *Global and Planetary Change* Vol 56, pp. 274-296.

Solomon, A M. 1986. Transient Response of Forests to CO2 Induced Climate Change: Simulation Modeling Experiments in Eastern North America. *Oecologia* Vol 68, pp. 567-579.

Stocks, B J, M A Fosberg, T J Lynham, L Mearns, B M Wotton, Q Yang, J Z Jin, K Lawrence, G R Hartley, J A Mason and D W McKenney. 1998. Climate change and forest fire potential in Russian and Canadian boreal forests. *Clim Change* Vol 38, No 1, pp. 1-13.

Stolbovoi, V and I McCallum. (eds) 2002. *CD-ROM Land Resources of Russia* International Institute for Applied Systems Analysis and the Russian Academy of Science, Laxenburg, Austria

Tchebakova, N M, G E Rehfeldt and E I Perfenova. 2005. Impacts of climate change on the distribution of *Larix* spp. and *Pinus Sylvestris* and the climatypes in Siberia. *Mitigation Adaptation Strategies for Global Change* Vol 11, No 4, pp. 861-882.

Tchebakova N M, G E Rehfeldt and E I Parfenova. 2009. From Vegetation Zones to Climatypes: Effects of Climate Warming on Siberian Ecosystems .pp 427-447 in A Osawa O A Zyranova, Y Matsurura, T Kajimoto and R W Wein (eds) Permafrost Ecosystems Siberian Larch Forests Springer Verlag

Tilman D and J A Downing. 1994. Biodiversity and stability in grasslands. *Nature* Vol 367, pp. 363-365.

Urban, D L, M E Harmon, and C B Halpern. 1993. Potential Response of Pacific Northwestern Forests to Climatic Change, Effects of Stand Age and Initial Composition. *Climatic Change* Vol 23, pp. 247-266.

Ustin, S. L. and Xiao, Q. F. 2001. Mapping successional boreal forests in interior central Alaska. *International Journal of Remote Sensing*, Vol. 22, No 6, pp. 1779-1797.

Volney W J A and R A Fleming. 2000. Climate Change and impacts of boreal forest insects. *Agriculture, Ecosystems and Environment* Vol 82, pp. 283-294.

Vygodskaya N N, P Ya Groisman, N M Tchebakova, JA Kurbatova, O Panfyorov, E I Parfenova, A F Sogachev. 2007. Ecosystems and climate interactions in the boreal zone of northern Eurasia. *Environ. Res. Lett.* 2 (045033): 7 pp URL http://www.iop.org/EJ/abstract/1748-9326/2/4/045033/

Watt A S. 1947. Pattern and Process in the Plant Community. *The Journal of Ecology* Vol 35, pp.1-22.

Yan, X and H H Shugart. 2005. A forest gap model to simulate dynamics and patterns of Eastern Eurasian forests. *Journal of Biogeography* Vol 32, pp. 1641-1658.

Yan, X and S Zhao. 1996. Simulating the response of Changbai Mountain forests to potential climate change. *Journal of Environmental Sciences* Vol 8, pp. 354-366.

Zhang, N, H H Shugart and X Yan. 2009. Simulating the effects of climate changes on Eastern Eurasian forests. *Climatic Change* Vol 95, pp. 341-361.

10

Reconstructing LAI Series by Filtering Technique and a Dynamic Plant Model

Meng Zhen Kang[1,2], Thomas Corpetti[2,3], Jing Hua[1,2] and Philippe de Reffye[4]
[1]State Key Laboratory of CompSys, Institute of Automation, CAS,
[2]LIAMA, Institute of Automation, CAS,
[3]CNRS, National Center for Scientific Research,
[4]Cirad-Amis, Montpellier Cedex 5
[1,2]China
[3,4]France

1. Introduction

Remote sensing images provide very rich and useful information linked to relevant biophysical parameters such as the LAI (Leaf Area Index), fCOVER (fraction of vegetation cover) or fAPAR (fraction of Absorbed Photosynthetically Active Radiation). At the moment, several techniques for estimating such variables are available and widely used in many applications, such as estimation of the total biomass and monitoring the dynamics in canopy vegetation (Baret et al., 2007; Lecerf et al., 2008). For several years, a large number of Very High Spatial Resolution (VHSR) satellites, such as Quickbird, Geoeye and Ikonos, have been launched, and very important missions such as the Venus and the Sentinel-2 are expected in 2012 and 2013. This provides possibility of having more or less temporal consistency in VHSR observations of the land use on relevant agricultural sites. However, because of the heterogeneity of the available VHSR data, in particular due to their different wavelengths sensibility and of the intrinsic errors induced by the estimation processes, the resulting time series of biophysical parameters are more or less noisy. As a matter of fact, the estimated variables may only poorly fit their actual dynamics. The estimation of the complete sequence of such parameters is then of prime importance, in particular if one wants to analyze the evolution of the biomass.

In this chapter, we propose to explore the possibilities of using tools issued from tracking techniques, in particular particle smoothing, to recover time-consistent series of LAI (Doucet et al., 2001; Kitagawa, 1996) from noisy and incomplete observations. Such techniques, based on Monte-Carlo strategies, allow performing the estimation of an unknown state function, LAI in current case, according to a given dynamical model and to possibly corrupted measurements. The dynamical model on which we rely on is GreenLab model, a functional-structural plant model simulating plant development and growth (Yan et al., 2004). Given model parameters, GreenLab can compute the evolution of LAI, the biomass production and partitioning, the organ size and biomass. Inverse method can be applied to estimate hidden model parameters by fitting model output with measured data (Kang et al., 2008). We suppose that from remote sensing data observing agricultural parcels, the type of crops can be identified, on which some GreenLab hidden parameters can be initialized. Besides, we suppose that the noisy time series of LAI has been estimated from remote sensing images, which have occluded area due for instance to cloud coverage, aerosols, etc. The objective is to construct a continuous LAI series by re-estimating GreenLab model parameters. This finally enables to simulate the complete plant growth and to output 3D evolution of the observed crops, using empirical geometrical parameters for the given crop. The overall strategy is illustrated in Fig. 1. The different steps of the methodology are presented in following sections.

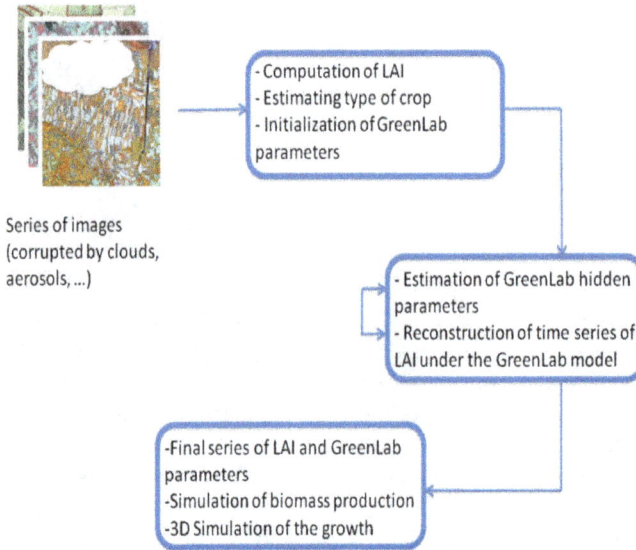

Fig. 1. Illustration of overall process of estimating biomass dynamics from remote sensing images

2. Overall framework

In section 2.1, we first introduce methods of estimating parameters from remote sensing images. In section 2.2, the concepts of GreenLab model are presented. Section 2.3 introduces the filtering technique used to recover time consistent series of LAI.

2.1 Estimating type of crop and LAI from remote sensing data
Estimating the type of crop of an observed field from remote sensing data is an old problem that has been widely studied by the computer vision and remote sensing community. This belongs to the classification issue where various families of approaches exist. One can roughly identify two methodologies: pixel-based and region-based classification techniques, see for instance Congalton (1992) and Yan et al. (2006). The first family aims at assigning to each pixel of the image a label corresponding to the nature of the culture, independently to its neighborhood. This performs efficiently when the spatial resolution of the data is low, yielding more or less homogeneity of the pixel reflectance for a given culture. On the other hand, the region-based techniques are useful when dealing with images of high or very high

resolution. In this case, the variability of the image luminance inside a given culture is high, and the pixel reflectance is not informative. Texture analysis strategies can be used to characterize and label the different crops, which use 1st or 2nd order statistical criteria or more advanced techniques like wavelets (Lefebvre et al., 2010). Many commercial software (e.g. Idrisi, ENVI or eCognition) allow this kind of classification.

Estimating the biophysical variables, such as LAI, fCOVER or fAPAR, from satellite observations can provide crucial information for numbers of applications, for instance, monitoring changes in canopy vegetation at global or regional scales, identifying bare soils, or detecting grassland areas. Among the different techniques available, there is a technique based on the inversion of the SAIL+PROPSPECT radiative transfer model (Verhoef, 1984; Jacquemoud and Baret, 1990) using training samples and neural networks, as introduced in (Baret et al., 2007). The SAIL model deals with light scattering by leaf layers with application to canopy reflectance model, and PROSPECT is about leaf optical properties spectra. It has been proved in (Baret et al., 2007; Lecerf et al., 2008) that this approach performs efficiently for low, medium, high and very high-resolution data and is therefore adapted to the variability of satellite images available.

2.2 GreenLab model

GreenLab model simulates the two basic processes of plant: development (organogenesis) and growth (organ expansion). In GreenLab, the organogenesis is simulated by an automaton, which gives the dynamics of number, age and type of organs in plant architecture (Yan et al., 2004). The organ expansion is simulated by a source-sink approach. At each time step t, the source function gives the biomass production of a plant, Q_t, as a function of plant leaf area S_t and environmental factor E_t, see Eqn. (1).

$$Q_t = E_t \cdot r \cdot S_P \left(1\text{-exp} \left(-\frac{S_t}{S_P} \right) \right)$$ (1)

In Eqn. (1), S_P represents the projection area of an individual plant, which is equivalent to the inverse of planting density d when crop canopy is closed, i.e, $S_P = 1/d$. r is a model parameter that can be estimated inversely (Kang et al., 2008; Guo et al., 2006; Dong et al., 2008). In case that E_t represents the intercepted light by crop, r means light use efficiency. The ratio S_t/S_P gives a LAI series.

The produced biomass is shared among all growing organs in proportion to their current sink strength, based on common pool hypothesis. For an organ of type O and age j, its increment is biomass in computed as in Eqn. (2).

$$\Delta q_{t,j}^O = P^O f^O{}_j \cdot Q_t / D_t$$ (2)

In GreenLab, each type of organs (e.g. blade, sheath, internode, female organ) has certain relative sink strength P^O, which may vary during the expansion of an individual organ, described by an empirical function $f^O{}_j$. Total plant demand D_t is sum of sink strength from all growing organs. The organ biomass, which is the accumulation of biomass increment during its life time, is dependent on the ratio between biomass production and demand (Q_t/D_t), called source-sink ratio. According to the appearance time of each individual organ given by the automaton, and its increment in biomass since appearance, the biomass of all individual organs in plant can be computed. As the ratio between blade biomass and specific leaf weight, at any time, leaf area S_t and consequently LAI, can be computed recursively in GreenLab model. For some simple case, analytical equation can be written explicitly. For example, for maize, LAI at a given time t can be written as in Eqn. (3).

$$LAI_t = \frac{S_t}{S_P} = \frac{\sum\limits_{i=1}^{t} N_{t,i}^B q_{t,i}^B}{\lambda_t S_P} = \frac{\sum\limits_{i=1}^{t} N_{t,i}^B \sum\limits_{j=1}^{\min(i,T_B)} P^B f^B_j \cdot Q_{t-i+j} / D_{t-i+j}}{\lambda_t S_P}$$

$$= \frac{P^B \cdot r \cdot \sum\limits_{i=1}^{t} N_{t,i}^B \sum\limits_{j=1}^{\min(i,T_B)} f^B_j E_{t-i+j} \left(1 - \exp\left(-LAI_{t-i+j}\right)\right) / D_{t-i+j}}{\lambda_t}$$

(3)

Where $q^B_{t,j}$ is biomass of a j-aged leaf blade, $N^B_{t,j}$ is the number of such leaf (being one or zero for maize of single stem structure), λ_t is specific leaf weight, T_B is expansion duration of a leaf.

GreenLab model is a generic model that has been applied to different crops, such as wheat (Kang et al., 2008), maize (Guo et al., 2006), and tomato (Dong et al., 2008; Kang et al., 2010), from which the dynamic growth and development process of crop are rebuilt. In model calibration, the hidden model parameters controlling the source and sink function, such as organ sink strength P^O, were estimated inversely from the measured plant data, such as total organ biomass and individual organ size, using weighted root square error as the criterion (Guo et al., 2006). LAI can be thus reconstructed by the calibrated model and compared with measured data, as in Fig. 2.

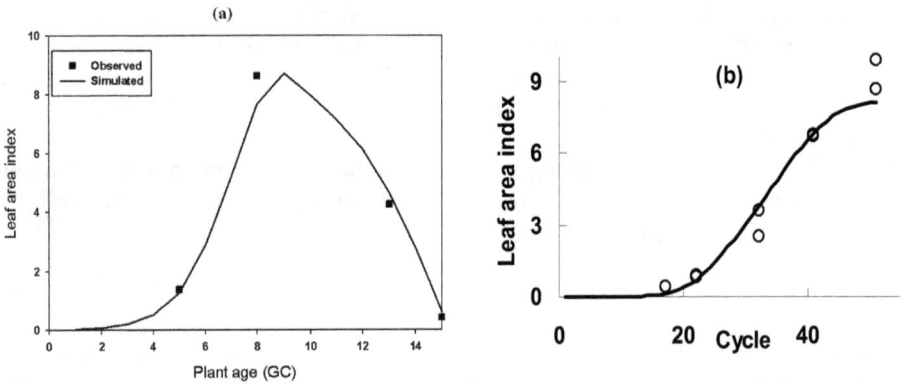

Fig. 2. LAI from measurement (dots) and GreenLab simulation (solid lines): (a) wheat (Kang et al., 2008), (b) chrysanthemum (Kang et al., 2006).

Initialisation of GreenLab parameters is dependent on the crop type, for which a common development and growth pattern exists. For example, a maize plant generally starts by vegetative growth with short internodes and finishes with a tassel, although the amount of leaves and position of cob in main stem may vary. Such development pattern is considered in initializing GreenLab organogenesis model. As to functional parameters, it is found that some are more or less stable for different seasons and population densities (Ma et al., 2008) for a given cultivar. Initialization of these parameters may be done according to previous modelling experience, and they will be re-estimated in following stage. In building 3D plant, an organ geometry library will be called and transformed according to their computed size and position to assemble the full plant structure. Geometrical parameters, such as organ insertion angle, are set empirically.

Compared to traditional processed-based models (PBM), the feature of GreenLab lies in the modelling of plant architecture and its effect of biomass production, thereby making simulation of plant plasticity possible. Besides, GreenLab simulates plant growth as a whole dynamic system, and different variables like leaf area, stem height, spike weight are linked to each other, instead of being modeled independently. Nevertheless, the two types of model can build interface by fitting LAI from GreenLab simulation with LAI from a PBM (Feng et al., 2010). On this basis, crop fields can be simulated and visualized dynamically, giving the expected LAI. The similar principle can be applied to fitting noisy LAI from remote sensing data.

2.3 Filtering technique

To cope with temporal inconsistencies likely to occur when one estimates LAI on each image independently, we suggest the imposition of a temporal constraint, here, the GreenLab plant growth model.

The dynamic coherence can be enforced by embedding the estimation problem within a filtering process. Roughly speaking, two main families can be used: variational or stochastic approaches. Variational techniques, also known as variational data assimilation, perform the estimation by minimizing a cost-function issued from a deterministic formalization in a Bayesian framework of the problem. It extracts the best compromise between observations, dynamic model and confidence measure. Due to a rewriting in a dual space (Lions, 1971), the gradient of the cost-function can efficiently be obtained using a forward-backward integration of the dynamical model, and such techniques are very adapted when one deals with large system states. On the drawback, non-linear dynamic models need to be managed in an incremental framework corresponding to a succession of linearized problems. On the other hand, when one deals with smaller system's state and non-linear dynamic models, stochastic techniques as the particle filter or the particle smoother, related to Monte Carlo approaches, are strongly adapted. The main idea consists in manipulating a set of particles more or less connected to the final state to estimate, the latter resulting from linear combination of such particles. In the next paragraph we introduce the main principles.

The system state to recover at time t, noted x_t, is submitted to a dynamical model f up to some uncertainties. This results in a stochastic process of the form in Eqn. (4)

$$x_t = f(x_{t-1}) + n_t \tag{4}$$

where n_t is a centered Gaussian noise of variance σ_n^2. In addition to this dynamic process, we are able to observe the system state as in Eqn. (5):

$$y_t = g(x_t) + v_t \tag{5}$$

where g is a (linear or not) observation operator and v_t a Gaussian noise of variance σ_v^2. Particle filtering, also known as Sequential Monte Carlo (Doucet et al., 2001), is an attractive way to recover the system state x_t for all $t \in [1,T]$. It can be shown that a sequential estimation of x_t can be obtained through the following system, starting from (a) an available sequential observation of sequence $y_{1:t} = y_1,\ldots, y_t$; (b) an initial distribution of the system's state $p(x_1)$; (c) transition model $p(x_t|x_{t-1})$ and observation model $p(y_t|x_t)$ respectively, related to the stochastic processes f and g presented above. See (Doucet et al., 2001) for details. Steps include:

1. Choose a set of N samples (or particles) x_1^i, $i=[1, \ldots, N]$ randomly taken from the initial distribution $p(x_1)$ and compute $p(y_1|x_1)$;
2. Prediction step: for all $t \in [2,T]$ and the set of samples $x_{t-1|t-1}^i$, generate the predicted samples $x_{t|t-1}^i = f(x_{t-1|t-1}^i) + n_t^i$ for each particle and get Eqn. (6):

$$p(x_t|y_{1:t-1}) = \int p(x_t|x_{t-1})p(x_{t-1}|y_{1:t-1})dx_{t-1} \tag{6}$$

3. Correction step: from the distribution $p(x_t|y_{1:t-1})$ and the new observation y_t, we have a result as in Eqn. (7):

$$
\begin{aligned}
p(x_t|y_{1:t}) &= \frac{p(x_t,y_{1:t})}{p(y_{1:t})} \\
&= \frac{p(y_t|x_t)p(x_t|y_{1:t-1})}{\int p(y_t|x_t)p(x_t|y_{1:t-1})dx_t} = w(t_{t|t-1})p(x_t|y_{1:t-1})
\end{aligned}
\tag{7}
$$

Eqn. (7) can be approximated from the set of particles using Eqn. (8):

$$p(x_t|y_{1:t}) \approx \sum_{i=1}^{N} w(x_{t|t-1}^i)\delta x_t^i \tag{8}$$

where δx_t^i is the dirac function in x_t^i and the weight function is as in Eqn. (9):

$$w(t_{t|t-1}) = \frac{p(y_t|x_{t-1}^i)}{\sum_{k=1}^{N} p(y_t|x_{t-1}^k)} \tag{9}$$

Therefore, once the system at time t-1 has been obtained, the process consists in generating a prediction of all particles at time t thanks to the transition $p(x_t|x_{t-1})$ and the available observations $y_{1:t-1} = y_1,\ldots, y_{t-1}$. These predictions are then corrected by taking into account the new observation y_t in a second step. The final estimated distribution is obtained from the set of particles and their associated weights $w(x_{t|t-1}^i)$. This is the main principle of the sequential particle filtering.

When the whole sequence is available, i.e. all observations $y_{1:T} = y_1,\ldots, y_T$ are available for all time t, a smoothing version of the previous technique can be applied by taking into account future observations. From the first estimation issued from the previous process, the idea consists in performing a backward exploration in order to correct the weights of the different particles. This is the so-called forward-backward smoother.

The idea consists in reweighting the particles recursively backward in time, starting from the end time T to the initial one. It can be shown that rewriting the distribution $p(x_t|y_{1:T})$ in terms of backward transitions $p(x_t \mid x_{t+1}, y_{1:T})$ yields, after several manipulations, to reweight all the particles with Eqn. (10):

$$w_{t|T}^i = w_t^i \left[\sum_{j=1}^{N} w_{t+1|T}^j \frac{p(x_{t+1}^j \mid x_t^i)}{\sum_{k=1}^{N} w_t^k p(x_{t+1}^j \mid x_t^k)} \right] \tag{10}$$

Therefore, the process consists in first performing a sequential filtering and then, to reweight the particles backward in time. More details can be found in (Doucet et al., 2001). This is the process we suggest to use in this application to recover consistent LAI values from noisy observed ones.

3. Computational experiment

In this computational experiment, we rely on the filtering technique to recover LAI sequence using the GreenLab model. We suppose the noisy LAI can be obtained from remote sensing data. Here we use synthetic data from GreenLab model so that the true values of y_t are known for evaluation.

3.1 Case study

We chose maize plant for a case study on application of the filtering method presented above. According to previous study on maize plant (Guo et al., 2006), we set model parameters in GreenScilab, an open source software for implementing GreenLab model. The LAI can is part of model output, as shown in Fig. 3 (a). By arranging the simulated 3D maize plant according to the given density, a virtual maize field can be simulated. Fig. 3 (b) shows such an image at a plant age. It is supposed that the aim is to recover this LAI sequence from noisy data of LAI obtained from remote sensing images at several different stages.

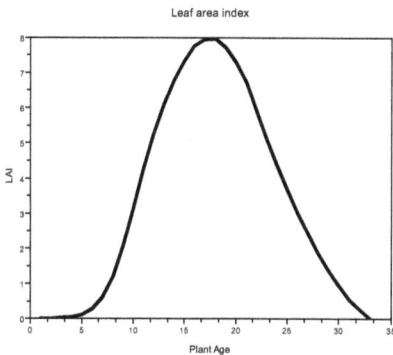

Fig. 3. Synthetic result from GreenLab model. (a) simulated LAI dynamics of virtual maize from GreenLab model; (b) top view of a virtual maize field;

3.2 Recovering consistent series of LAI

We tested our smoothing approach on the sequence of synthetic values of LAI generated from the GreenLab model, see Fig. 4 (blue line). From this sequence, we have randomly extracted some points on which we have added noise (black points). We assume that these points correspond to measurements obtained on the corresponding field from remote sensing images. They represent an incomplete time series of noisy values of LAI. From these inconsistent series, we have generated a smooth version from the forward-backward smoother presented in the previous section, using GreenLab as a dynamic model. This is depicted in Fig. 4 (green line). From this synthetic example, it is obvious to observe that the new series is consistent with the expected ground truth. This is confirmed when observing the quantitative values of the Root Mean Square Error between the ground truth, the noisy and reconstructed data shown in Table 1.

In order to highlight the benefit of the use of a dynamical model, we have blurred in a stronger way the series. We have assumed that during a long time period in which the variation of LAI is maximal, no observations are available (due, for instance, to the maintenance of the sensor, a too large cloud covering during the winter, etc.). In addition, to take into account the errors related to the acquisition process itself, we have also blurred the remaining data. This results in a strongly noisy sequence of LAI, as shown with the black points in Fig. 5.

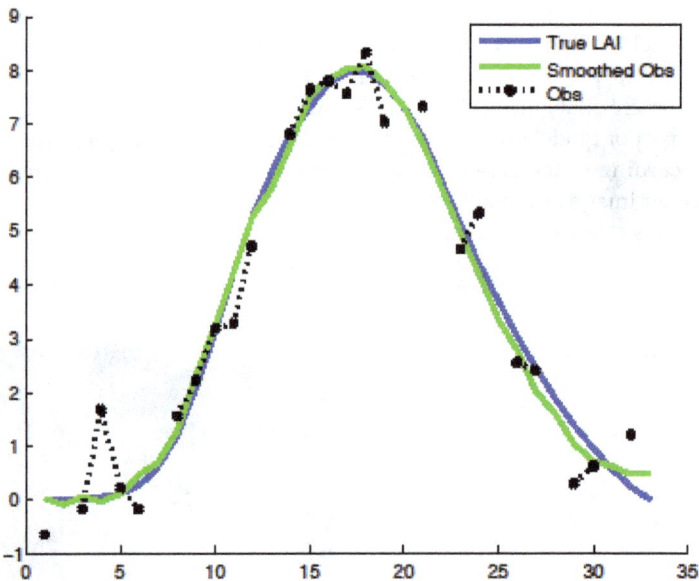

Fig. 4. Recovering LAI series from noisy data, with a synthetic LAI sequence (blue line), a noisy version (black points), and a recovered time series under the GreenLab model (green line)

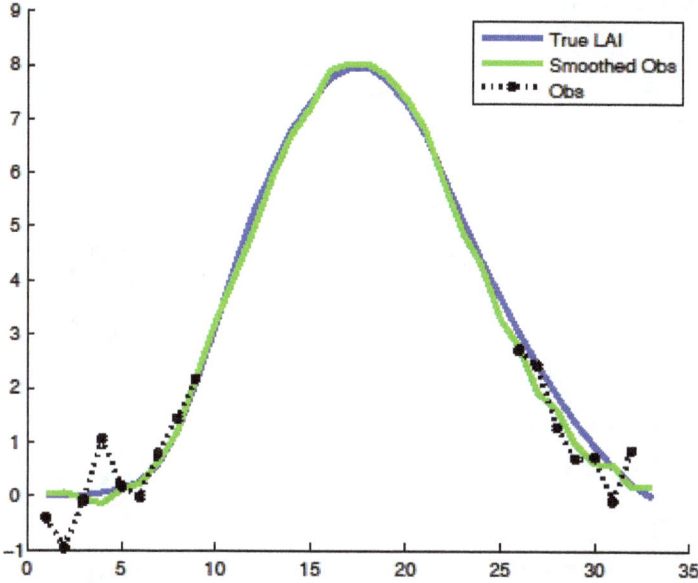

Fig. 5. Recovering LAI series from strongly noisy data, with a synthetic LAI sequence (blue line), a strongly noisy version (black points), and a recovered time series under the GreenLab model (green line)

From Fig. 5, one can immediately observe that during the period corresponding to the main evolution of LAI where no data are acquired, a simple interpolation technique without any prior model would result in incorrect values and dramatically underestimate the LAI. At the opposite, using the forward-backward smoother, the recovered series of Fig. 5 (green line) is very closed to the original one, despite this difficult testing situation. The RMS error is slightly higher than the previous situation (when observations are less noisy) but is still very competitive. These results demonstrate the great benefit of recovering the data under the constraint of a dynamic model.

	RMSE Original data	RMSE Reconstructed data
Noisy series (Fig. 4)	0.6145 for 25 observed values	0.09341 for 256 recovered time steps
Strongly noisy series (Fig. 5)	0.9951 for 18 observed values	0.10055 for 256 recovered time steps

Table 1. Root Mean Square Errors for the two noisy sequences tested

These experiments on noisy and strongly noisy synthetic data demonstrate the possibility of the framework presented in this chapter (schematized in Fig. 1) to recover time consistent series of LAI with fine time resolution from a small set of noisy image observations. The practical issue with real images is under development.

3.3 Plant growth reconstruction

Using the estimated hidden GreenLab parameters in recovering the series of LAI, one can simulate the dynamics of biomass production and partitioning, as shown in Fig. 6(a), and the 3D model of the observed plants, as shown in Fig. 6(b). As for a maize plant, Fig. 6(a) gives the total plant biomass (g) in black line, and the allocation to leaf blade (green), sheath (blue), internode (brown), cob (red) and male flower (purple). Recall that empirical geometrical parameters are used in building the 3D structure.

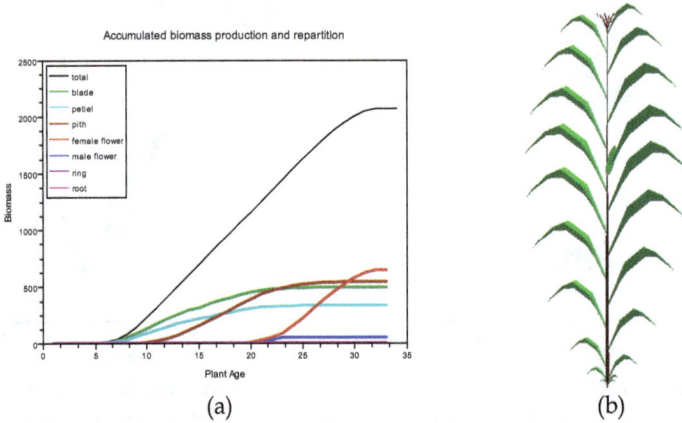

(a)   (b)

Fig. 6. Computed result from calibrated model. (a) biomass production and its allocation among different type of organs; (b) a 3D maize plant.

4. Conclusion

In this chapter, we have presented a framework of estimating LAI series from remote sensing images using filtering techniques and GreenLab model. Computational experiment was done on synthetic data to show the feasibility of full process. The result shows that by embedding a dynamic model, the LAI series can be recovered even if the source data from remote sensing images are very noisy and sparse. And in doing so, the GreenLab model can be calibrated partially to simulate biomass production and allocation. The advantage of embedding a crop model is that the knowledge on crop development and growth can be used in recovering LAI series, and the link between LAI and biomass production make it possible to estimate biomass production from remote sensing data, which is the ultimate aim of estimating LAI.

Yet this theoretical work needs to be further tested by real remote sensing sources. Challenges include the initialization of model parameters, such as the setting on topological parameters and initial source and sink parameter. Detection of crop type can help to solve this issue by providing empirical parameters. Yet their values are not necessary to be accurate, and other information from remote sensing, such as leaf chlorophyll content and leaf water content, may compensate. On the other hand, the combination of a functional-structural plant model as GreenLab brings many possibilities. For example, as the three-dimensional structures of crop are built, it is possible to run radiative transfer model in virtual canopy. Although the result will be dependent on the definition of geometrical structure and optical properties of individual organs, it provides a possibility of validating the reconstructed canopy dynamics by comparing the virtual canopy with the obtained high resolution source images. The development of remote sensing technique and advance in plant modelling are increasing the interdisciplinary research of these two areas.

5. References

Baret F., Hagolle O., Geiger B., Bicheron P., Miras B., Huc M., Berthelot B., Nino F., Weiss M., Samain O., Roujean J.L., and Leroy M.. "LAI, FAPAR, and FCover CYCLOPES global products derived from Vegetation. Part 1: principles of the algorithm", Remote Sensing of Environment, 110:305-316, 2007.

Congalton R. G., A review of assessing the accuracy of classifications of remotely sensed data, Remote Sensing of Environment, vol. 37, p. 35-46, 1991.

Dong Q.X., Louarn G., Wang Y.M., Barczi J.F., de Reffye P. 2008. Does the Structure-Function Model GREENLAB deal with crop phenotypic plasticity induced by plant spacing? A case study on tomato. Annals of Botany 101: 1195-1206.

Doucet A., Freitas N., and Gordon N. Sequential Monte Carlo Methods in Practice. SpringerVerlag, first edition, 2001.

Feng L., Mailhol J. C., Rey H., Griffon S., Auclair D. and de Reffye P. Combining a process based model with a functional structural plant model for production partitioning and visualization. 6th International workshop on Functioanl-Structural Plant Models (FSPM 10), Sept. 12-17, 2010. University of California, Davis, USA.

Guo Y., Ma Y.T., Zhan Z.G., Li B.G., Dingkuhn M., Luquet D., de Reffye P. 2006. Parameter optimization and field validation of the Functional-Structural Model GREENLAB for maize. Annals of Botany 97: 217-230.

Jacquemoud, S. and Baret F., 1990. PROSPECT: A model of leaf optical properties spectra. Remote Sensing of Environment, 34: 75-91.

Kang M.Z., Heuvelink E., and de Reffye P. 2006. Building virtual chrysanthemum based on sink-source relationships: Preliminary results. Acta Horticulturae 718: 129–136.

Kang M.Z., Evers J.-B., Vos J., de Reffye P. 2008. The derivation of sink functions of wheat organs using the GreenLab model. Annals of Botany 101: 1099-1108.

Kang M.Z., Yang L.L., Zhang B.G., de Reffye P. 2011. Correlation between dynamic tomato fruit set and source sink ratio: a common relationship for different plant densities and seasons? Annals of Botany 107: 805-815.

Kitagawa G. Monte carlo filter and smoother for non-gaussian nonlinear state space models.Journal of Computational and Graphical Statistics, 5(1):1–25, 1996

Lecerf R., Hubert-Moy L., Baret F., Abdel-Latif B., Corpetti T., Nicolas H.. Estimating biophysical variables at 250m with reconstructed EOS/MODIS time series to monitor fragment landscapes. In IEEE Int. Geoscience and Remote Sensing Symp, IGARSS '08, Volume 2, Pages 954-957, Boston, USA, July 2008.

Lefebvre A., Corpetti T., Hubert-Moy L. Segmentation of very high spatial resolution panchromatic images based on wavelets and evidence theory. In SPIE International Conference on Image and Signal Processing for Remote Sensing, 78300E Proc. SPIE 7830 (ed.), Toulouse, France, September 2010

Lions J.-L. 1971. Optimal control of systems governed by PDEs. Springer-Verlag.

Ma Y, Wen M, Guo Y, Li B, Cournède P-H, de Reffye P. 2008. Parameter optimization and field validation of the functional structural model GREENLAB for maize at different population densities. Annals of Botany 101: 1185–1194.

Yan G., Mas J.-F., Maathuis B.H.P., Xiangmin Z., Van Dijk P.M., Comparison of pixel- based and object-oriented image classification approaches - a case study in a coal fire area, Wuda, Inner Mongolia, China, International Journal of Remote Sensing, vol. 27, no. 18, p. 4039-4055, September, 2006.

Yan H.P., Kang M.Z., de Reffye P., Dingkuhn M. 2004. A dynamic, architectural plant model simulating resource- dependent growth. Annals of Botany 93: 591–602.

Verhoef W. 1984. Light scattering by leaf layers with application to canopy reflectance modeling: the SAIL model. Remote Sensing of Environment, 16: 123.

Part 5

Applications

11

Rice Crop Monitoring with Unmanned Helicopter Remote Sensing Images

Kishore C. Swain[1] and Qamar Uz Zaman[2]
[1]*Department of Agricultural Engineering,*
Triguna Sen School of Technology, Assam University,
[2]*Engineering Department, Nova Scotia Agricultural College, Nova Scotia,*
[1]*India*
[2]*Canada*

1. Introduction

Agricultural crop, one of the biological entities, is sensitive to its environmental condition including various soil and crop inputs. Alteration in environmental condition causes reduction in crop productivity (such as crop yield and total biomass etc.). Ultra-modern technology such as, precision agriculture (PA) is capable to prevent crop damage and maintain crop productivity. PA is the technology of applying correct amount of crop input at the exact place and time of requirement. Application of PA technology has become increasingly prevalent among the farmers from developed countries as well as developing countries due to its capability for optimizing crop yield by facilitating sound crop status monitoring (Zhang and Taylor, 2001). Mostly, satellite images have been used as the primary source of information for analyzing crop status in precision agriculture. However, obtaining up-to-date aerial photography is very expensive, the quality is variable, and data processing is also intensive and complicated. Innovative new technologies to acquire timely and accurate crop information are required for the success of PA technology.

Assessment of leaf radiation has the potential to detect nitrogen (N) deficiency and is a promising tool for N management and monitoring. Moreover, over-fertilization may result in surface runoff and pollute subsurface water (Wood et al., 1993; Auernhammer et al., 1999; Daughtry et al., 2000; Zaman et al., 2006). Chlorophyll is an indirect indicator of nitrogen status and is used in optical reflectance-based variable-rate nitrogen application technology (Lee and Searcy, 2000; Jones et al., 2004; Alchanatis et al., 2005; Kim and Reid, 2006; Min et al., 2008). Biermacher et al. (2006) used sensor-based systems to determine crop nitrogen requirements and estimated that the variable-rate system had the potential to achieve a net profit of about $22 to $31 per ha. The ability to accurately estimate plant chlorophyll concentration can provide growers with valuable information to estimate crop yield potential and to make decisions regarding N management (Kahabka et al., 2004; Reyniers and Vrindts, 2006; Zaman and Schumann, 2006).

Spectroradiometry has been useful in the research environment for determining principal wavebands and spectral patterns that relate to nutrient stress (Noh et al., 2004; Tumbo et al. 2001). High spectral resolution and the ability to account for temporal changes are distinct advantages. Okamoto et al. (2007) used a hyperspectral line-scanning camera for weed detection. This system produced hyperspectral images from a Specim ImSpector V9 imaging spectrograph mounted on a tractor that was set to move slowly through the field. Principal spectral components could be extracted and analyzed using various discrimination schemes. However, on-the-go hyperspectral sensing is slow and impractical, since enough area must be covered per sweep for timely data acquisition over large field areas.

Biomass is an important trait in functional ecology and growth analysis. The typical methods for measuring biomass are destructive, laborious and time consuming. Thus, they do not allow the development of individual plants to be followed and require many individuals to be cultivated for repeated measurements. Non-destructive method may be an option to overcome these limitations. Crop residue estimation has been accomplished using RADARSAT images (Jensen et al., 1990; McNairn et al., 1998), using LANDSAT images (Thoma et al., 2004), and using images captured by radio-controlled model aircraft (Hunt et al., 2005).

Prediction of yield using remote sensing images has been practiced by many researchers (Fablo and Felix, 2001; Alvaro et al., 2007). Rice crop area has been estimated from Landsat images (Tennakoon et al., 1992) for wide-scale yield prediction. Canopy reflectance was estimated at panicle initiation stage using a portable spectroradiometer (LI-1800, LICOR) with a remote cosine receptor attached to a 1.5 m extension arm for smaller-scale yield prediction (Chang et al., 2005). Yield prediction has also been accomplished for corn (Chang et al., 2003; Kahabka et al., 2004), cotton (Thomasson et al., 2000) wheat (Doraiswamy et al., 2003), citrus (Zaman et al., 2006) and wild blueberry (Zaman et al., 2010). Tea leaf yield was estimated using vegetation indices such as normalized difference vegetation index (NDVI) and triangular vegetation index (TVI) (Rama Rao et al., 2007).

Rice (*Oriza sativa* L.), which is the staple food of most Asian countries, accounts for more than 40% of caloric consumption worldwide (IRRI, 2006). Annual rice production was approximately 590 million tons and yield was 4.21 ton ha^{-1} in Asia for 2006 (FAOSTAT, 2007). The profit from cultivating a rice crop is derived from the crop grain yield and total biomass produced. Predicting rice yield at or around the panicle initiation stage would provide valuable information for future planning and yield expectations.

2. Low altitude remote sensing (LARS) system

Site-specific management of inputs characteristic of PA promotes conservation of agricultural resources while maintaining crop viability. However, the application of satellite-based images still cannot fulfill the specific requirements of PA technology. Stafford (2000) observed that satellite images for application of PA are handicapped in terms of spectral and temporal resolution and can be affected by variable weather conditions. Lamb and Brown (2001) indicated that the low-resolution satellite images only beneficial for large-scale studies, are not appropriate for the small-scale farms prevalent in many areas of Asia, for example. Additionally, satellites providing higher-resolution images, e.g., QuickBird (DigitalGlobe, Longmont, Colo.) and ASTER (National Aeronautics and Space Administration, Washington, D.C.), have long revisit times, making them of limited utility for any application that might require frequent images (nutrient stress monitoring, for example). In the past, researchers had used manned aerial vehicles (helicopters and aero planes) to acquire surface images (Table 1). Though a large area can be mapped within short time, cost involved in the aerial vehicles is very high and also requires sophisticated system, trained operators, and professionals. Therefore, an unmanned helicopter is used for aerial image acquisition.

Systems/ Facilities	Equipment			Applications (size and structure)			
	GPS /INS	Laser	Camera	Large Areas	Small areas (< 2-4 km²)	Route mapping	Complex buildings/ structures
Aircraft	Yes	Yes	Film based and digital	Yes	Yes/No	Yes	No
Helicopter	Yes	Yes	Digital	No	Yes	Yes	No
Terrestrial system (car, train)	Yes	Yes	Digital	No	Yes/No	Yes	No
RC-Helicopter	Yes	Yes	Digital	No	Yes	Yes	Yes

Table 1. Comparative benefits of remote image acquisition platforms

LARS is a relatively new concept of remote image acquisition currently discussed by agriculturists implementing precision agriculture technology. As the name suggests, it is a system of acquiring images of the earth surface from a lower altitude as compared to the commercial remote sensing satellites. In this system, the images are acquired mostly below cloud cover and very near field features of interest. Low-altitude remote sensing using unmanned aerial vehicles can be an inexpensive and practical substitute for sophisticated satellite and general aviation aircraft, and it is immediately accessible as a tool for the farmer.

Various unmanned LARS systems have been developed and used in the remote image acquisition for PA applications. Some LARS platforms, kites (Aber et al., 2002), balloons (Amoroso and Arrowsmith, 2000; Seang and Mund, 2006), high-clearance tractors (Bausch and Delgado, 2005), and unmanned airplanes and helicopters (Sugiura et al., 2002; Fukagawa et al., 2003; Eisenbiss, 2004; Herwitz et al., 2004; Sugiura et al., 2004; Hunt et al., 2005; MacArthur et al., 2005, 2006; Xiang and Tian, 2006, 2007a, 2007b; Huang et al., 2008) have been successfully using for PA applications in different cropping systems.

These platforms were mounted with image acquisition devices and location measuring receivers, which can fly over agriculture farms and targeted areas for capturing images. As indicated by Sugiura et al. (2002) the major drawbacks of unmanned helicopters are limited payload capacity and precise control over working speed of the system. Thus, mounted systems operation has to be programmed properly to neutralize the effect of wind speed. The low payload capacity of the system was adjusted by selecting light weight mounting equipment and tools. Stombaugh et al. (2003) suggested replacing heavy weight professional digital cameras with light weight, low cost, commercial digital cameras. As the individual images acquired by the LARS system covers small area, geo-referenced images can be mosaic for mapping entire farmland and targeted areas. Global positioning system (GPS) was used in aerial platforms for obtaining aircraft location information (Hayward et al., 1998), for geo-referenced videobased remote sensing images (Thomoson et al., 2002) and in VRT system guidance (Fadel, 2004). Buick (2002) proposed the guidelines to select proper GPS receivers for specific applications.

Thomson and Sullivan (2006) observed that both agricultural aircraft and unmanned aerial vehicles (UAVs) may potentially more easily scheduled and accessible remote sensing platforms than the remote sensing satellites and general aviation aircraft customarily used in the U.S. However, use of agricultural aircraft is limited to those areas where aerial crop spraying is prevalent. Hunt et al. (2005) used a radio-controlled helicopter-mounted image acquisition system to estimate biomass and nitrogen status for corn, alfalfa, and soybean crops. Digital photographs have been used for site-specific weed control for grassland swards (Gebhardt et al., 2006; Beerwinkle, 2001), for tomato (Zhang et al., 2005) and for wild blueberry (Chang et al., 2011). Chen et al. (2003) using an high-elevation tractor system, indicated that multi-spectral images at 555, 660, and 680 nm wavelength band centers demonstrated good prediction ability for determining the nitrogen content of rice plants.

This chapter is intended to focus on the effectiveness of low-altitude remote sensing (LARS) images obtained by a multispectral imaging platform mounted in a radio-controlled unmanned helicopter to estimate rice crop parameters as a function of varying nutrient availability. Non-destructive image analysis technique is used to estimate rice yield and total biomass. It also examines the effectiveness of near-real time estimation of protein content from nutrient availability with rice leaf. Consistent with the fact that most multispectral cameras small enough to be used in unmanned aerial vehicles utilize pre-defined wavebands for feature detection, applicability of the widely used NDVI incorporating these wavebands is evaluated.

2.1 System components

A radio-controlled model helicopter (X-Cell Fury 91, Miniature Aircraft, Orlando, USA) is equipped with a Tetracam agricultural digital camera (ADC) (Tetracam, Inc., Chatsworth, Cal.), (Table 2). It is also equipped with various sensors, such as: C-100 Magnetic compass (to obtain platform orientation angle from North), Inertial Measurement Unit (to obtain roll and pitch orientation angles), Barometric sensor (to measure pressure variation for altitude measurement), COM-1288 GPS receiver (to provide position information: latitude and longitude), digital camera (to acquire multispectral (G-R-NIR) images) etc., monitored by a PC-104 based CPU-1232 microprocessor. A PC-104 compatible Power Supply Unit (ACS-5150), being powered from an external 12Vdc battery, is used to supply the necessary power to all the sensors including microprocessor (Figures 1 and 2).

The camera is a wideband multispectral camera utilizing a CMOS CCD (charge-coupled device) with a Bayer filter mask for multispectral imaging (Table 2). The unmanned helicopter weighed about 6 kg with a payload capacity of 5 kg. The radio console is capable of controlling the unmanned helicopter within a 1 km radius. The system uses a battery-initiated glow fuel (250 mL) engine, supporting 15 min of flight at length. A spectroradiometer with wavelength range of 350 to 2350 nm (Spectra Co-op, Inc., Tokyo, Japan) can be used to estimate reflectance at ground level in the red (at 660 nm) and NIR bands (at 800 nm). Bandwidth at each center is 2.5 nm.

A control program, developed in "C" programming language, was used for the DOS operating system based microprocessor, to coordinate, the simultaneous clicking of digital camera and obtaining the readings from the sensors, and to store the information as a file in the storing device. The program enabled the system to acquire image and sensor reading at minimum time interval of 12 seconds. The images and corresponding sensor readings as digital number (0-255) were supplied to the image processing algorithm.

(a)   (b)

NIR Camera   Micro processor   GPS Receiver   IMU

(c)

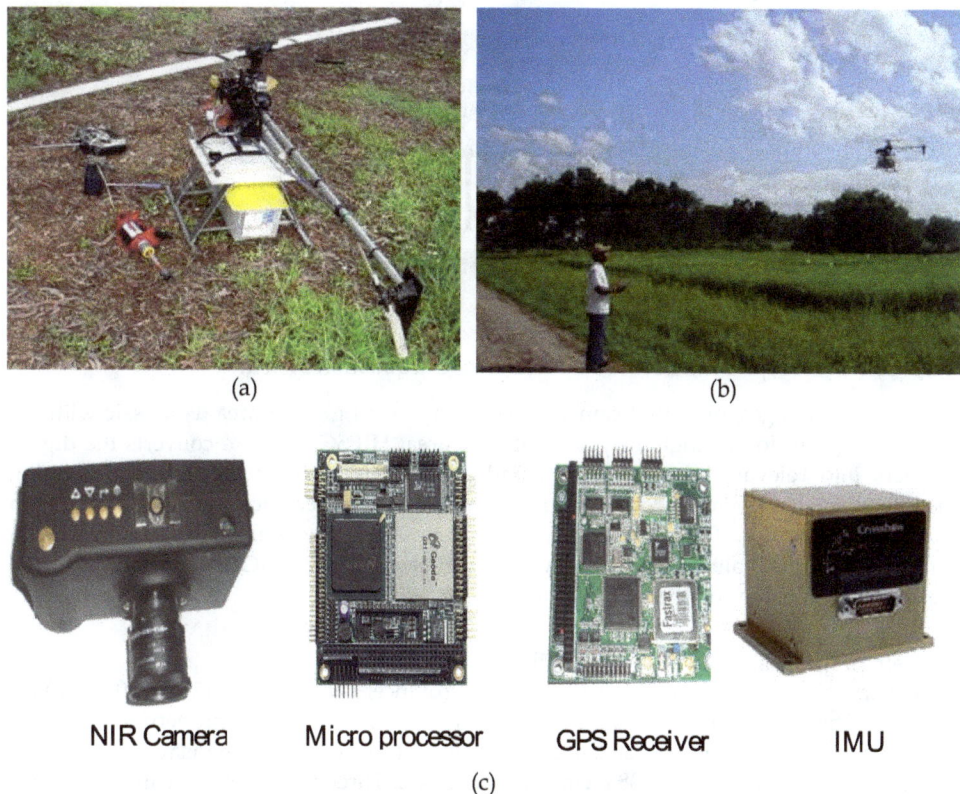

Fig. 1. LARS system operation: (a) R/C helicopter mounted with image acquisition system, (b) acquiring image in rice crop, and c) major components.

Characteristics	Values
Image size (resolution)	1280 × 1024 (1.3 Mpixel)
Pixel size	6.01 micron
Ground pixel resolution	0.000707 m/pixel (estimated)
Spectral bands	3 (green, red, and NIR); band centers and bandwidths are fundamentally equivalent to Landsat bands TM2, TM3, and TM4
Lens type	C-mounted
Lens	8.5 mm
Triggering	Manual/cable switch triggering

Table 2. Specification of the Tetracam ADC Green-Red-NIR sensors

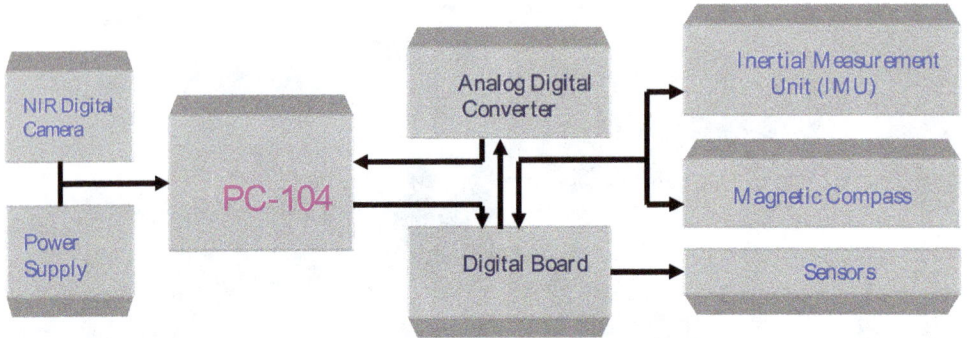

Fig. 2. Schematic representation of the LARS image acquisition system

As individual images, of digital camera covers very small ground area, is mosaic with the algorithm to develop a single map of the study area. HIPSC software converts the digital numbers into relevant sensor readings and used them to carry out image processing operations, such as: image rotation, image mosaic and reflectance index (NDVI, Green NDVI etc.) estimation. The software can develop site-specific zone maps based on variation in reflectance index values and also provide ground control points (GCPs) for mosaic image geo-registration using commercial software.

2.2 Field preparation and data acquisition

The experimental site is located in Pathumthani Province, Thailand (14° 12' N, 100° 37' E) for the study. The site may be located anywhere in the world, but, the soil properties have to be measured accurately. The soil of the experimental site belonged to the clay textural class with average bulk density of 1.38 g cm^{-3} and pH of 4.2. Three replicates were made, and the treatment plots, each of size 10 m × 10 m, randomly distributed within each replicate. To estimate the nitrogen application rate, the total nitrogen present in the soil was tested using standard methods (Kjeldahl apparatus). At the experimental site, the concentration of pre-existing nitrogen was classified as low (<0.18%) level for all the plots, as per the local Agricultural Extension Service guidelines.

For the underlined experiment, the plots were well-watered using flood irrigation and carefully maintained for pest control to ensure uniform yield potential. The rice seeds were broadcasted (on 14 Dec. 2006) in accordance with local practices under irrigated farming conditions. Nitrogen fertilizer was applied at five rates: 0%, 25%, 50%, 75%, and 100% of recommended values, representing 0, 33, 66, 99, and 132 kg ha^{-1}, respectively. Plots with different nitrogen rates were maintained to promote a wide range of rice yield so the effectiveness of LARS images could be evaluated for varying nutrient availability. This follows a similar technique by Chen et al. (2003), who used four N rates (0, 45, 90, and 135 kg ha^{-1}) in field experiments with a Tainung 67 rice crop for multispectral image analysis. An early rice variety, Supanburi-1 (95 day period), was used in the study, as this is one of the most popular variety in central Thailand. Urea (46-0-0) was applied as the source of nitrogen for the study. Different nitrogen rates along with recommended phosphorous fertilizer were applied 30 days after sowing rice. Images were obtained twice with the LARS system just before panicle initiation stage (45 and 65 days after planting, Figure 1b).

The altitude has to be selected considering the camera's field of view to acquire a single image for each treatment plot. Images with effective dimensions of 18 m × 14 m were collected from a 20 m flying height, covering a single plot. For field application the height can be varied as per the suitability of the researcher, to cover wider area in each image. Flight altitude was recorded with a height sensor (MPXAZ4115A barometric sensor, Freescale Semiconductor, Austin, Tex.) mounted on the LARS system. Images are obtained at five different heights, and the images obtained closest to the 20 m height were selected for analysis. Five ground-based reflectance readings were obtained for the rice canopy and BaSO$_4$ standard white reference board using the *Spectroradiometer* in each of the experimental plots. The ground-based readings were obtained immediately after the LARS system-based image acquisition. The plot-wise ground-based reflectance value is calculated as the mean of the five readings.

2.3 Image pre-processing

Multispectral images acquired by the Tetracam ADC camera (.dcm format) were converted into *.tiff* format for analysis. The *.tiff* format reduces the storage space and also effectively retains the image quality for image processing. Images were uploaded to Pixelwrench software (Tetracam, Inc., Chatsworth, Cal.), which contains programs for deriving one of several vegetation indices (.hdr format) from raw image data. An NDVI image was produced for each test plot, and the average NDVI index was estimated using the custom-developed program in the C programming language from images acquired by the LARS-mounted sensors (Figure 3). Ground-based reflectance data were collected to estimate mean NDVI of the experimental plots (NDVI$_{SPECTRO}$). NDVI$_{SPECTRO}$ was estimated using the software provided by the *Spectroradiometer* manufacturer. Linear regression models can be developed in SAS (ver. 9.1, SAS Institute, Inc., Cary, N.C.) or any standard software.

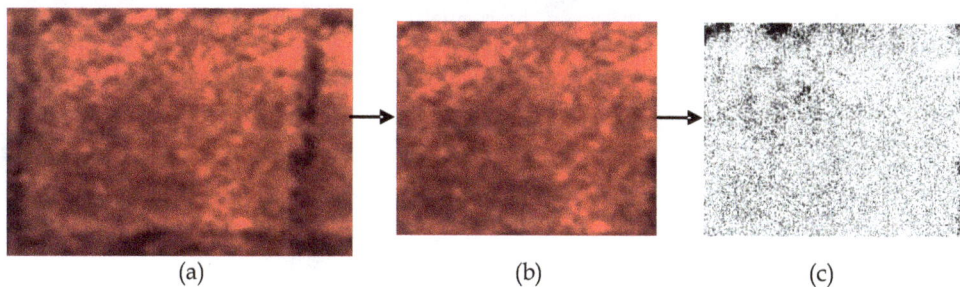

(a)   (b)   (c)

Fig. 3. Stages of image processing: (a) raw image with plot boundaries (as taken by the image acquisition system), (b) plot-scale image of the rice crop, and (c) NDVI image

3. Validation of LARS setup

The normalized difference vegetation index (NDVI) is the mostly adopted reflectance index for agricultural cropping and vegetation studies (Rouse et al., 1973) given as;

$$NDVI = \frac{NIR - R}{NIR + R}$$

(1)

Where, *NIR*: Radiance value for *Near-infrared* band; *R*: Radiance value for *Red* band.

The Green Normalized Differential Vegetation Index (GNDVI) to establish the suitability of reflectance index for rice cropping with variable nitrogen rates (Gitelson et al., 1996) was also used. GNDVI, based on the greenness level, represented by the chlorophyll content determining the radiance level of the leaf surface, was very significant for the rice crop monitoring. The GNDVI was estimated as follows,

$$GNDVI = \frac{NIR - G}{NIR + G} \tag{2}$$

Where, *NIR*: Radiance value for *Near-infrared* band; *G*: Radiance value for *Green* band.
The NDVI index was also calculated from ground level *Spectrophotometer* radiance values using the Eq[n]. 1 for establishing suitability of LARS system. Around five readings were taken from each plot in order to estimate the average NDVI for each treatment plots. The *SPAD 502 meter* readings of leaf greenness can be converted into Chlorophyll content by the following equation for rice cropping (Markwell et al., 1995).

3.1 Relationship between reflectance indices and variable N-treatments
The graph of $NDVI_{SPECTRO}$ and $NDVI_{LARS}$ plotted for the different N-treatments showed positive response with increased recommended nitrogen rates. The NDVI index, taken 45 days after sowing, showed weak relationship with nitrogen treatment rates, attaining coefficient of determination (r^2) of 0.60. As, the fertilizer application, just two weeks before date of testing, response time may not be enough to influence plant leaf radiance level to greater extent (Figure 4). However, the relationship was stronger ($r^2 \approx 0.85$) with higher NDVI values, ranging from 0.70 to 0.90, for second set of *Spectrophotometer* reading taken at booting stage (for 65 days old plants). $NDVI_{LARS}$, estimated from LARS images, were very low, after 45 days of sowing, ranging from 0.2 to 0.6, due to the lower radiance value of soil, exposed in gaps between the plants' leaves. The radiance level of the crop leaves, covering the whole plot area with least exposed area at booting stage, attained their original values (with NDVI between 0.85 to 1.0). $NDVI_{LARS}$ at booting stage showed strong relationship with r^2 of 0.73 for different N-treatment rates (Figure 4b). The greenness index (GNDVI) plotted against variable nitrogen rates showed, lower correlation with r^2 of 0.66 and 0.7, for the images taken at 45 days and 65 days respectively, with slightly strong relationship for the later. The lower range of $GNDVI_{LARS}$ index had values ranging from 0.5 to 0.6 at booting stage, maintained positive response with higher nitrogen rates (Figure 4c).

3.2 Suitability of reflectance indices determined from LARS images
Cross comparison analysis was carried out to testify the applicability of LARS images through indices such as $NDVI_{LARS}$ and $GNDVI_{LARS}$ with the *Spectrophotometer* reading index such as $NDVI_{SPECTRO}$ by plotting graphs between them (Figure 5). The $NDVI_{LARS}$ was proportional to that of $NDVI_{SPECTRO}$ with r^2 of 0.72 and 0.79 for 45 days and 65 days old rice crop, respectively. The $NDVI_{LARS}$ ranges from 0.85 to 1.0 showing sound crop coverage throughout the plot at booting stage of crop. The lower range $NDVI_{LARS}$ value (0.2~ 0.5) for 45 days crop made the reading unsuitable to represent the crop in crop modeling and predictions. The higher r^2 value (≈ 0.7) for indices estimated from LARS images ($NDVI_{LARS}$, $GNDVI_{LARS}$) with index from ground *Spectrophotometer* reading ($NDVI_{SPECTRO}$) showed the suitability of the proposed system for crop status studies.

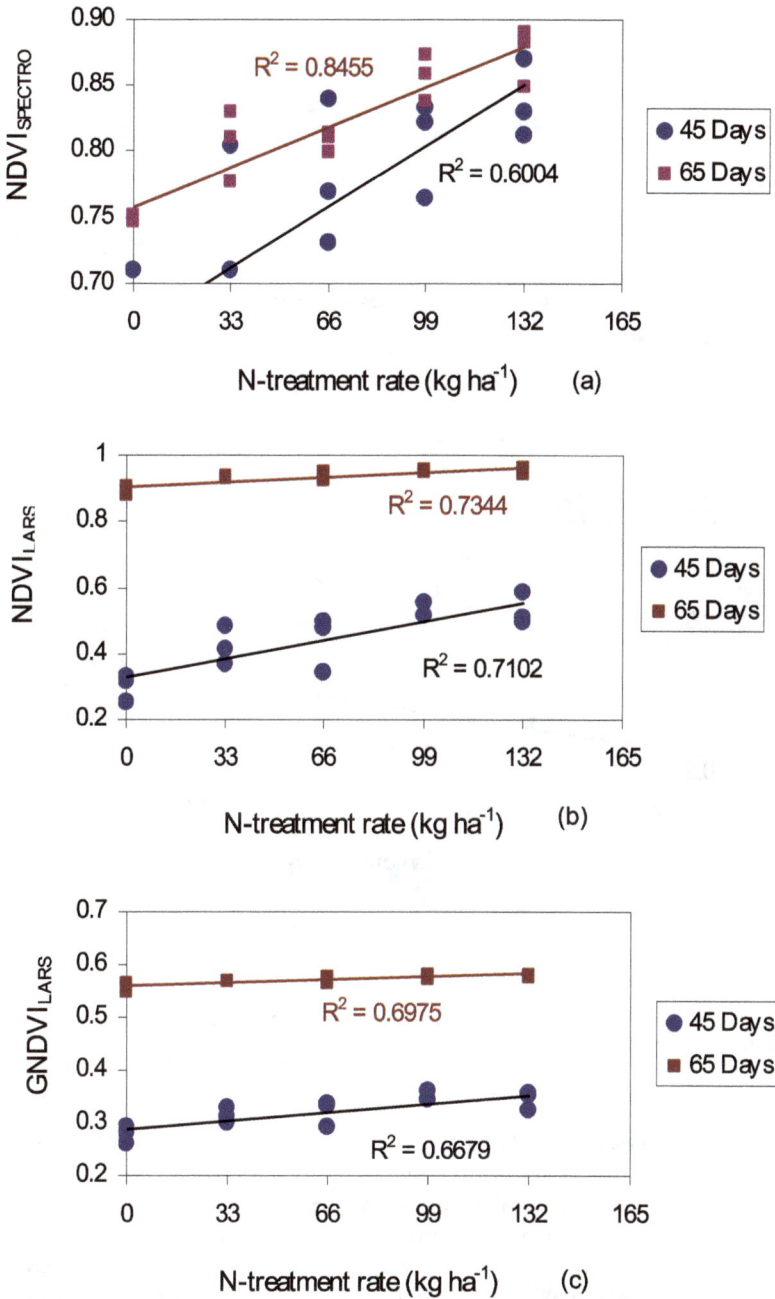

Fig. 4. Variation of vegetation index with N-treatment rates; a) NDVI$_{SPECTRO}$; b) NDVI$_{LARS}$; c) GNDVI$_{LARS}$

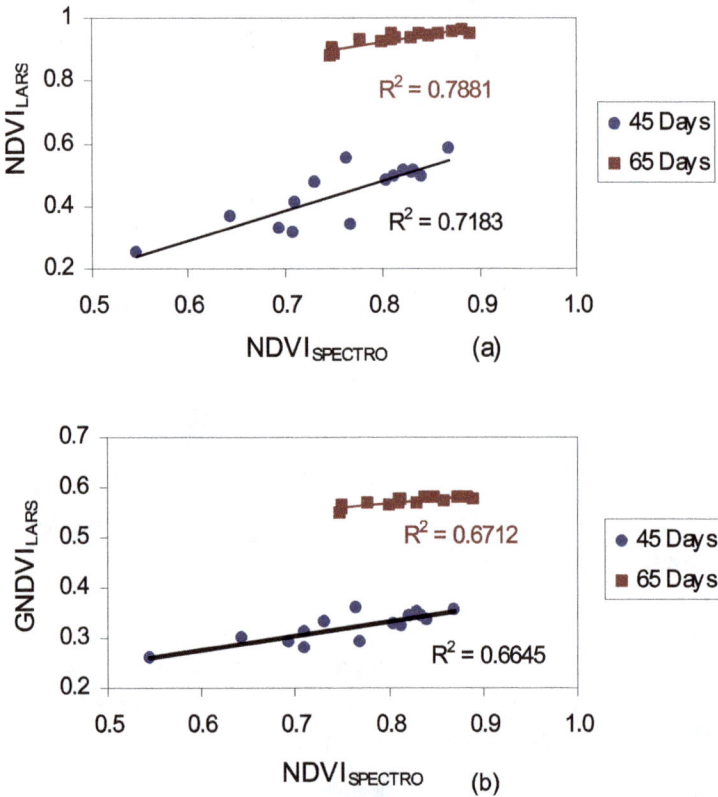

Fig. 5. Comparison of indices based on groundtruthing data and LARS images; $NDVI_{SPECTRO}$ with; a) $NDVI_{LARS}$; b) $GNDVI_{LARS}$

3.3 Discussions

For the experiment, the recommended amount of fertilizer was applied to 40 days old crop and the first set images and groundtruthing were taken at 45 days. The leaf coverage was low with a major share of exposed soil resulting in lower correlation of green indices (NDVI and GNDVI) values. The coefficient of determination (r^2) was improved visibly for 65 days old crop with denser crop leaving little exposed soil. As observed, 65 days old crop, LARS and ground measurements, was better suited, hence selected for crop status monitoring studies. Variation in green indices (NDVI and GNDVI) showed symmetry with the variation of nitrogen level for different treatments.

4. Estimation of crop parameters

4.1 Estimation of total biomass

Biomass is a plant attribute that is time consuming and difficult to measure or estimate, but easy to interpret. Biomass is regarded as an important indicator of ecological and management processes in the vegetation. Biomass estimation facilitates accurate management decisions regarding chemical and fertilizer applications, estimation of yield, and post harvest handling of stover (Pordesimo et al., 2004).Quantifying spatial variation in pasture and crop biomass can help to direct management practices and improve farm productivity, through accurate and informed movements of grazing rotations, crop and pasture nutrient management and also yield prediction (Trotter et al., 2008). Measurement of plant biomass by harvesting is destructive, expensive and time consuming (Reese et al., 1980). de Matthaeis et al. (1995) used AIRSAR data collected over the agricultural fields to monitor biomass variation. They found that the L-band is more effective for crops with low plant density, while C-band is better for high plant density crops.

4.1.1 Calculations

The rice biomass (threshed rice plant without the grain) of three sampled areas, 4 m² each, were collected and weighted. The moisture content (w.b.) of the threshed rice plant was estimated using standard method. The dry weight of the threshed rice plant was estimated and converted into the total biomass weight per ha i.e. (ton/ha).

$$\text{Total biomass (ton/ha)} = (100 - M.C.) \times BiomassWt. \times \frac{10000}{12 \times 1000} \qquad (3)$$

Where,

Total biomass: Weight of rice plant (without rice grain) in ton/ha
BiomassWt: Weight of threshed rice plant (without rice grain)
M.C.: Moisture content of weighed rice plant (w.b.)

Total oven-dried (Abdullah et al., 1992) biomass was ranged from 3.58 to 7.36 ton ha⁻¹ for the different treatments (Table 4). Total dry biomass weight between the treatments showed significant differences at the 0.10 level but no significant difference between replicates.

N Rate Treatment	Replicate			Average
	1	2	3	
0 kg ha⁻¹	3.58	4.25	6.30	4.710
33 kg ha⁻¹	5.51	5.84	5.64	5.660
66 kg ha⁻¹	5.57	5.97	5.77	5.771
99 kg ha⁻¹	6.50	7.36	5.97	6.611
132 kg ha⁻¹	5.57	6.63	7.30	6.501

Table 3. Total biomass (ton ha⁻¹) of the experimental plots

Linear calibrations curves were developed in SAS 9.1 to estimate the biomass from NDVI index values calculated from LARS images. From these results, $NDVI_{LARS}$ could explain 76% of the variation in biomass weight ($r^2 = 0.760$, RMSE = 0.598 ton ha⁻¹, Figure 6).

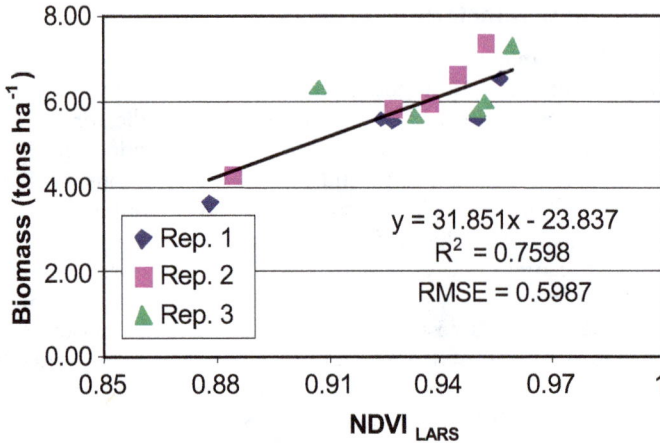

Fig. 6. Estimation of biomass with NDVI$_{LARS}$ values

4.2 Estimation of rice yield

The rice crop was harvested from three sample areas of 4 m² from each plot, 102 days after sowing for this experiment. The moisture content (% w.b.) at the time of weighing was estimated using a field moisture meter (Kett PM600, Ohta-Ku, Tokyo, Japan). The yield of each plot (100 m² area) was estimated as the average of three sampled areas and converted to a ton-per-hectare area using the following equation. Rice yield was estimated at 14% moisture content (MC) for each treatment (Field crop report, 1998).

$$Yield \ (ton \ ha^{-1}) \ = \ \frac{(100 - MC) \times RW \times 10000}{86 \times A \times 1000} \qquad (4)$$

Where,
　　MC = moisture content (% wet basis)
　　RW = weight of rice (kg)
　　A = harvested area (m²)
Rice yield, ranged from as low as 1.88 ton ha⁻¹ (0 kg ha⁻¹ N) to 3.68 ton ha⁻¹ (132 kg ha⁻¹ N) based on a 14% MC, illustrates the effectiveness of the fertilizer treatment rates on rice yield (Table 4). The crop yield variation was also tested for statistical significance (Johnson and Bhattacharyya, 2001). Yield data between the treatments showed significant differences at the 0.10 and 0.05 levels, whereas differences were not significant among the replicates

N Rate	Replicate			Average
Treatment	1	2	3	
0 kg ha⁻¹	1.88	1.97	1.64	1.83
33 kg ha⁻¹	2.13	2.87	3.28	2.76
66 kg ha⁻¹	2.78	2.70	3.44	2.97
99 kg ha⁻¹	2.37	3.85	3.52	3.25
132 kg ha⁻¹	3.52	3.36	3.68	3.52

Table 4. Rice yield (ton ha⁻¹) of the experimental plots

The regression model, developed for rice yield with NDVI index value in SAS 9.1, indicated a good fit (r^2 = 0.728, RMSE = 0.458 ton ha^{-1}, Figure 7). Variation among the replicates might be due to initial nutrient levels present in the soil from randomly selected plots.

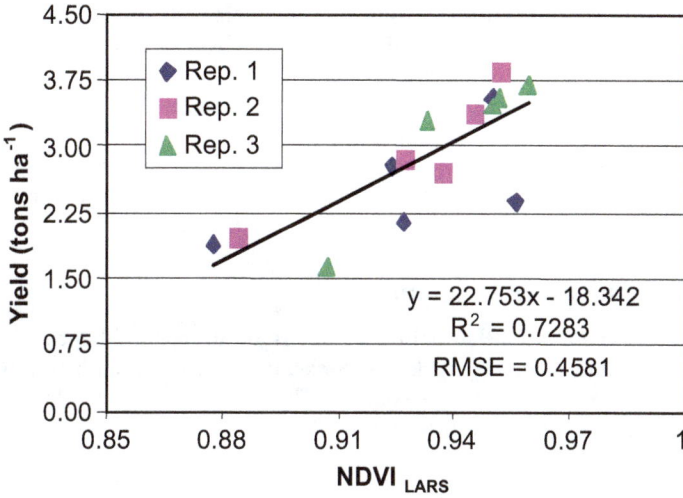

Fig. 7. Estimation of rice yield with NDVI$_{LARS}$ values.

4.3 Estimation of protein content

Protein content is one of the major food nutrients to determine quality of the food-grain. It could be measured as the total available nitrogen content in the food stuff (Kennedy, 1995). The rice was powdered and sieved before testing for total nitrogen with standard method. The linear model of total nitrogen against NDVI$_{LARS}$ (with r^2 = 0.591, Figure 8) showed positive relationships, and would be useful to the farmers, as they can get idea of quality of rice grain well in advance, at booting stage (from the image taken during booting stage).

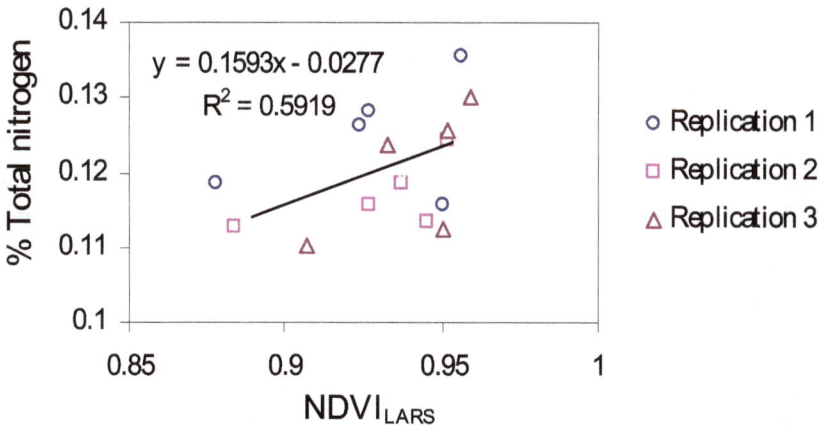

Fig. 8. Estimation of protein content with NDVI$_{LARS}$ values.

4.4 Discussions

The NDVI value, calculated from the LARS images collected for 65 days old crop, was suitable to estimate the total biomass (r2 = 0.760) and rice crop yield (r2 = 0.728). The protein content estimated through NDVI value was marginally suitable, capable to provide overall rice crop quality before-hand. The LARS system is suitable for real and near-real time crop parameter estimation, monitoring and evaluations. The NIR sensors can be substituted with any professional camera or cheap digital camera to optimize cost involved. The overall efficiency of the LARS system will be dependent on the sensors mounted on the helicopter. A skilled labor can easily handle the whole system with least supervision. The LARS not only replaces the satellite based image processing system but also ground level spectrophotometer, chlorophyll content measuring equipments. With little time, the system will be ready for taking images, for instance, just after rainfall.

5. Field application of LARS systems

Sugiura et al. (2007) mounted a thermal band camera on unmanned helicopter platform to estimate soil water status in paddy fields and correlation was obtained between the thermal image temperature and soil moisture content. The coefficient of determination (r^2) for moisture content and temperature model at 10.00 a.m. and 3.00 p.m. were 0·69 and 0·64 respectively (Figure 9).

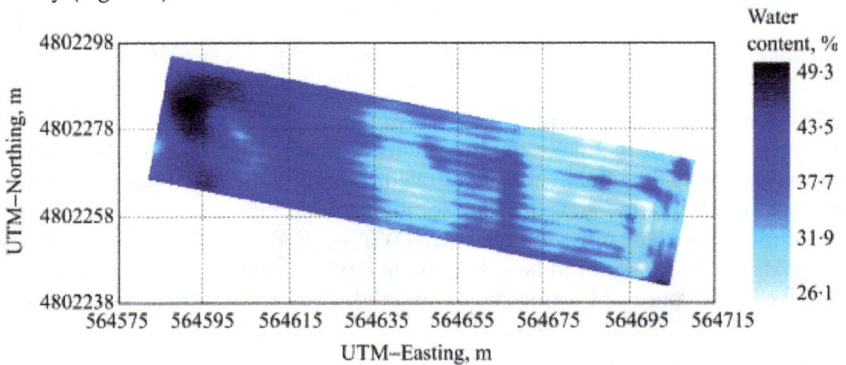

Fig. 9. Soil moisture content estimation with LARS images (Sugiura et al., 2007)

The r^2 between moisture content and difference in temperature was 0·42. The development was intended assisting in proper irrigation scheduling and monitoring water stressed situations for rain-fed cropping. Ishii et al. (2005) developed a system that can generate a map regarding crop status obtained by mounting an imaging sensor on an unmanned helicopter. They achieved an accuracy of 38 cm using RTK GPS receiver and GDS unit. The maps are accurate enough to be used for variable rate nutrients and pesticides application for the farmland.

Lenthe et al. (2007) used unmanned helicopter based IR thermography imaging system as a tool for monitoring the microclimatic conditions promoting incidence and severity of diseases within wheat fields with a high spatial resolution. Zhou et al. (2009) used R44 helicopter for aerial electrostatic spraying system. The results of the studies showed that electrostatic spraying with helicopters could produce uniform and fine droplets with better droplet adhesion and distribution, higher depositing efficiency, lower environmental contamination; lower pesticide application rate and aerial spray can improve efficiency for prevention and cure pests in agriculture and forestry.

6. Conclusions

A radio-controlled helicopter-based LARS system can be used to acquire multispectral images over a rice canopy to estimate rice yield. The study indicated that the LARS platform could substitute for satellite-based and costly airborne remote sensing system. Images are obtained successfully by the multispectral camera mounted on the radio-controlled helicopter at a height of 20 m over rice plots. Rice yield and total biomass were found to be significantly different at the 0.05 and 0.1 significance levels, respectively, under different N treatment regimes. The relationship between $NDVI_{LARS}$ and $NDVI_{SPECTRO}$ ($r^2 = 0.897$, RMSE $= 0.012$) shows the applicability of LARS sensor-based images for estimating NDVI values, which varied over the five levels of applied N. A linear regression model shows a good fit ($r^2 = 0.728$, RMSE $= 0.458$ ton ha^{-1}) for estimating total biomass for rice using LARS image-based NDVI values. A linear model ($r^2 = 0.760$, RMSE $= 0.598$ ton ha^{-1}) indicates that rice yield could also be predicted with NDVI values derived from LARS images. The protein content can be positively estimated well in advance to actual crop harvesting. The regression model procedure outlined herein can be followed for larger rice fields by recording crop input rates and acquiring LARS images.

7. References

Abdullah, N., Y. W. Ho, and S. Jalaludin. 1992. Microbial colonization and digestion of feed materials in cattle and buffaloes: II. Rice straw and palm press fibre. *Asian-Australasian J. Animal Sci.* 5(2): 329-335.

Aber, J. S., K. Aaviksoo, E. Karofeld, and S. W. Aber. 2002. Patterns in Estonian bogs as depicted in color kite aerial photographs. *Suo (Mires and Peat)* 53(1): 1-15.

Alchanatis, V., Z. Schmilovitch, and M. Meron. 2005. In-field assessment of single-leaf nitrogen status by spectral reflectance measurements. *Precision Agric.* 6(1): 25-39.

Alvaro, F., L. F. García del Moral, and C. Royo. 2007. Usefulness of remote sensing for the assessment of growth traits in individual cereal plants grown in the field. *Intl. J. Remote Sensing* 28(11): 2497-2512.

Amoroso, L., and R. Arrowsmith. 2000. Balloon photography of brush fire scars east of Carefree, Arizona. Tempe, Ariz.: Arizona State University, Department of Geological Sciences.

Auernhammer, H., M. Demmel, F. X. Maidl, U. Shmidnalter, T. Schneider, and P. Wagner. 1999. An on-farm communication system for precision farming with nitrogen real-time application. ASAE Paper No. 991150. St. Joseph, Mich.: ASAE.

Buick, R. 2002. GPS guidance: Making an informed decision. In Proc. of 6th International Conference on Precision Agriculture 1979-2004. CD-ROM, P.C.Robert et al. Madison,Wisc, ASA, CSSA, and SSA.

Bausch, W., and J. A. Delgado. 2005. Impact of residual soil nitrate on in-season nitrogen applications to irrigated corn based on remotely sensed assessments of crop nitrogen status. *Precision Agric.* 6(6): 509-519.

Beerwinkle, K. R. 2001. An automatic capture-detection, time-logging instrumentation system for boll weevil pheromone traps. *Applied Eng. in Agric.* 17(6): 893-898.

Biermacher, J. T., F. M. Epplin, B. W. Brorsen, J. B. Solie, and W. R. Raun. 2006. Maximum benefit of a precise nitrogen application system for wheat. *Precision Agric.* 7(3): 193-204.

Chang, J., D. E. Clay, K. Dalsted, S. Clay, and M. O'Neill. 2003. Corn (*Zea mays* L.) yield prediction using multispectral and multidate reflectance. *Agron. J.* 95(6): 1447-1453.

Chang, K. W., Y. Shen, and J. C. Lo. 2005. Predicting rice yield using canopy reflectance measured at booting stage. *Agronomy Journal,* 97(3): 872-878.

Chang, Y. K., Q. U. Zaman, A. W. Schumann, and D. C. Percival. 2011. Performance tests of g-ratio index and color co-occurrence matrix based machine vision algorithms in the wild blueberry fields. ASABE Paper No 09558 (ASAE, St. Joseph, MI, USA)

Chen, S., C. W. Huang, C. C. Huang, C. K. Yang, T. H Wu, Y. Z. Tsai, and P. L. Miao. 2003. Determination of nitrogen content in rice crop using multi-spectral imaging. ASAE Paper No. 031132. St. Joseph, Mich.: ASAE.Daughtry, C. S. T., C. L. Walthall, M. S. Kim, E. B. de Colstoun, and J. E. McMurtrey III. 2000. Estimating corn leaf chlorophyll concentration from leaf and canopy reflectance. *Remote Sensing of Environment,* 74(2): 229-23.

de Matthaeis, P., P. Ferrazzoli, G. Schiavon and D. Solimini.1995. Crop type identification and biomass estimation by SAR. *Geoscience and Remote Sensing Symposium, 1995. IGARSS '95. 'Quantitative Remote Sensing for Science and Applications', International.*

Doraiswamy, P. C., S. Moulin, P. W. Cook, and A. Stern. 2003. Crop yield assessment from remote sensing. *Photogram. Eng. and Remote Sensing,* 69(6): 665-674.

Eisenbiss, H. 2004. A mini unmanned aerial vehicle (UAV): System overview and image acquisition. In *Proc. Intl. Workshop on Processing and Visualization Using High-Resolution Imagery.* International Society for Photogrammetry and Remote Sensing (ISPRS).

Fablo, M., and R. Felix. 2001. Analysis of GAC NDVI data for cropland identification and yield forecasting in Mediterranean African countries. *Photogram. Eng. and Remote Sensing* 67(5): 593-602.

FAOSTAT. 2007. *Agricultural Statistics Yearbook: 2006.* Rome, Italy: Food and Agriculture Organization of the United Nations.Fadel, M. 2004. Performance assessment of VRT-based granular fertilizer broadcasting systems. *Agricultural Engineering International: the CIGR Journal of Scientific Research and Development,* Manuscript PM 03 001, Vol.6.Field crop report. 1998. Field crop report for rice cropping in Thailand. Bangkok, Thailand: Field Crop Research Station.

Fukagawa, T., K. Ishii, N. Noguchi, and H. Terao. 2003. Detecting crop growth by a multispectral imaging sensor. ASAE Paper No. 033125. St. Joseph, Mich.: ASAE.

Gebhardt, S., J. Schellberg, R. Lock, and W. Kuhbauch. 2006. Identification of broad-leaved dock (*Rumex obtusifolius* L.) on grassland by means of digital image processing. *Precision Agriculture,* 7(3): 165-178.

Gitelson, A. A., Y. J. Kaufman and M. N. Merzlyak. 1996. Use of a green channel in remote sensing of global vegetation from EOS-MODIS, *Rem. Sens. Environ.,* 58: 289-298.

Hayward, R. C., D. Gebre-Egziabher, and J. D. Powell. 1998. GPS based attitude for aircraft: www. waas. standford.edu.

Herwitz, S. R., L. F. Johnson, S. E. Dunagan, and R. G. Higgins. 2004. Imaging from an unmanned aerial vehicle: Agricultural surveillance and decision support. *Computers and Electronics in Agric.* 44(1): 49-61.

Huang, Y., Y. Lan, W. C. Hoffmann, and B. Fritz. 2008. Development of an unmanned aerial vehicle-based remote sensing system for site-specific management in precision agriculture. In *Proc. 9th Intl. Conf. on Precision Agriculture.* Denver, Colo.: Colorado State University.

Hunt, E. R., C. L. Walthall, and C. S. T. Daughtry. 2005. High-resolution multispectral digital photography using unmanned airborne vehicles. In *Proc. 20th Biennial Workshop on Aerial Photography, Videography, and High-Resolution Digital Imagery for Resource Assessment.* Bethesda, Md.: American Society for Photogrammetry and Remote Sensing (ASPRS).

IRRI. 2006. *Atlas of Rice and World Rice Statistics.* Manila, Philippines: International Rice Research Institute.

Ishii, K., N., Noguchi and R. Sugiura. 2005. Remote-sensing technology for vegetation monitoring using an unmanned helicopter, *Biosystems Engineering,* 90 (4), 369-379.

Jensen, A., B. Lorenzen, H. Spelling-Ostergaard, and E. Kloster-Hvelplund. 1990. Radiometric estimation of biomass and N content of barley grown at different N levels. *Intl. J. Remote Sensing* 11(10): 1809-1820.

Johnson, R. A., and G. K. Bhattacharyya. 2001. *Statistics: Principles and Methods.* 4th ed. John Wiley and Sons.

Jones, C. L., N. O. Maness, M. L. Stone, and R. Jayasekara. 2004. Chlorophyll estimation using multi-spectral reflectance and height sensing. ASAE/CSAE Paper No. 043081. St. Joseph, Mich.: ASAE.

Kahabka, J. E., H. M. V. Es, E. J. McClenahan, and W. J. Cox. 2004. Spatial analysis of maize response to nitrogen fertilizer in central New York. *Precision Agric.* 5(5): 463-476.

Kennedy, P. M. 1995. Intake and digestion in swamp buffaloes and cattle: 3. Comparison with four forage diets, and with rice straw supplemented with energy and protein. *J. Agric. Sci. Cambridge* 124(2): 265-275.

Kim, Y., and J. F. Reid. 2006. Modeling and calibration of a multi-spectral imaging sensor for in-field crop nitrogen assessment. *Applied Eng. in Agric.* 22(6): 935-941.

Lamb, D. W., and R. B. Brown. 2001. Remote sensing and mapping of weeds in crops. *J. Agric. Eng. Res.* 78(2): 117-125.

Lee, W. S., and S. Searcy. 2000. Multispectral sensor for detecting nitrogen in corn plants. ASAE Paper No. 001010. St. Joseph, Mich.: ASAE.

Lee, Y. S., C. M. Yang, and A. H. Chang. 2002. Changes of nitrogen and chlorophyll contents and reflectance spectral characteristics to the application of nitrogen fertilizer in rice plants. *J. Agric. Res. China,* 51(1): 1-14.

Lenthe, J.H., E.C. Oerke and H.W. Dehne. 2007. Digital infrared thermography for monitoring canopy health of wheat. *Precision Agriculture,* 8(1):15-26.

MacArthur, D., J. K. Schueller, and C. D. Crane. 2005. Remotely piloted mini-helicopter imaging of citrus. ASAE Paper No. 051055. St. Joseph, Mich.: ASABE.

MacArthur, D., J. K. Schueller, W. S. Lee, C. D. Crane, E. Z. MacArthur, and L. R. Parsons. 2006. Remotely piloted helicopter citrus yield map estimation. ASABE Paper No. 063096. St. Joseph, Mich.: ASABE.

McNairn, H., D. Wood, Q. H. J. Gwyn, R. J. Brown, and F. Charbonneau. 1998. Mapping tillage and crop residue management practices with RADARSAT. *Canadian J. Remote Sensing* 24(1): 110-115.

Min, M., W. S. Lee, T. F. Burks, J. D. Jordan, A. W. Schumann, J. K. Schueller, and H. Xie. 2008. Design of a hyperspectral nitrogen sensing system for orange leaves. *Computers and Electronics in Agric.* 63(2): 215-226.

Markwell, J., J. C. Osterman, and J. L. Mitchell. 1995. Calibration of the Minolta SPAD-502 leaf chlorophyll meter", *Photosynth. Res.*, 46, 467-472.

Noh, H. K., Q. Zhang, and S. Han. 2004. Sensor-based variable-rate application of nitrogen by using multi-spectral image sensor. ASAE Paper No. 041133. St. Joseph, Mich.: ASAE.

Okamoto, H., T. Murata, T. Kataoka, and S-I Hata. 2007. Plant classification for weed detection using hyperspectral imaging with wavelet analysis. *Weed Biol. and Mgmt.* 7(1): 31-37.Pordesimo, L.O., W.C. Edens, and S. Sokhansanj. 2004. Distribution of aboveground biomass in corn stover. *Biomass and Bioenergy* 26:337-343.

Rama Rao, N., M. Kapoor, N. Sharma, and K. Venkateswarlu. 2007. Yield prediction and waterlogging assessment for tea plantation land using satellite image-based techniques. *Intl. J. Remote Sensing* 28(7): 1561-1576.

Reese, G.A., R.L. Bayn, and N.E. West. 1980. Evaluation of double-sampling estimators of subalpine herbage production. *Journal of Range Management* 33:300-306.

Reyniers, M., and E. Vrindts. 2006. Measuring wheat nitrogen status from space and ground-based platform. *Intl. J. Remote Sensing* 27(3): 549-567.

Rouse, J. W., R. H. Haas, J. A. Shell, and D. W. Deering. 1973. Monitoring vegetation systems in the Great Plains with ERTS-1. In *Proc. 3rd Earth Resources Technology Satellite Symp.*, 1: 309-317. NASA SP-351. Washington, D.C.: NASA.

Seang, T. P. and J-P Mund. 2006. Geo-referenced balloon digital aerial photo technique: A low-cost high-resolution option for developing countries. In *Proc. Map Asia 2006: 5th Annual Conf. on Geographic Information Technology and Application*. GIS Development Pvt. Ltd., Noida, India.

Stafford, J. V. 2000. Implementing precision agriculture in the 21st century. *J. Agric. Eng. Res.* 76(3): 267-275.

Stombaugh, T., A. Simpson, J. Jacobs and T. Mueller. 2003. A low cost platform for obtaining remote sensed imagery. In: *Precision Agriculture*, Edited by J. Stafford and A. Werner, pp.665-676.

Sugiura, R., N. Noguchi, K. Ishii, and H. Terao. 2002. The development of remote sensing system using unmanned helicopter. In *Proc. Automation Technology for Off-Road Equipment*, 120-128. ASAE Paper No. 701P0502. Q. Zhang, ed. St. Joseph, Mich.: ASAE.

Sugiura, R., K. Ishii, and N. Noguchi. 2004. Remote sensing technology for field information using an unmanned helicopter. In *Proc. Automation Technology for Off-Road Equipment*. ASAE Paper No. 701P1004. St. Joseph, Mich.: ASAE.Sujiura, R., N. Naguchi and K. Ishii. 2007. Correction of low-altitude thermal images applied to estimating of soil water status, *Biosystems Engineering*, 96(3), 301-313.

Tennakoon, S. B., V. V. N. Murty, and A. Eiumnoh. 1992. Estimation of cropped area and grain yield of rice using remote sensing data. *Intl. J. Remote Sensing* 13(3): 427-439.

Thoma, D., S. Gupta, and M. Bauer. 2004. Evaluation of optical remote sensing models for crop residue cover assessment. *J. Soil and Water Cons. Soc.* 59(5): 224-233.

Thomasson, J. A., R. Sui, and D. C. Akins. 2000. Spectral changes in picked cotton leaves with time. In: *Proc. 5th Intl. Conf. on Precision Agriculture & Other Resources Mgmt.* Ames, Iowa: USA.

Thomoson, S. J., J. E., Hanks, and G. F. Sassenrath-Cole. 2002. Continuous georeferencing for video-based remote sensing on agricultural air craft. *Trans. of ASAE*, 45(40): 1177-1189.

Thomson, S. J., and D. G. Sullivan. 2006. Crop status monitoring using multispectral and thermal imaging systems for accessible aerial platforms. ASABE Paper No. 061179. St. Joseph, Mich.: ASABE.

Trotter, T. F., P.S. Frazier, M. G. Trotter and D. W. Lamb. 2008. Objective biomass assessment using and active plant sensor (crop circle)- Preliminary experience on a variety of agricultural landscapes, Report from Precision Agriculture Research Group, The University of New England, Australia.

Tumbo, S. D., D. G. Wagner, and P. H. Heinemann. 2001. On-the-go sensing of chlorophyll status in corn. ASAE Paper No. 011175. St. Joseph, Mich.: ASAE.

Wood, C. W., D. W. Reeves, and D. G. Himelrick. 1993. Relationships between chlorophyll meter readings and leaf chlorophyll concentration, N status, and crop yield: A review. *Proc Agron. Soc. New Zealand* 23: 1-9.

Xiang, H., and L. Tian. 2006. Development of autonomous unmanned helicopter-based agricultural remote sensing system. ASABE Paper No. 063097. St. Joseph, Mich.: ASABE.

Xiang, H., and L. Tian. 2007a. An autonomous helicopter system for aerial image collection. ASABE Paper No. 071136. St. Joseph, Mich.: ASABE.

Xiang, H., and L. Tian. 2007b. Autonomous aerial image georeferencing for an UAV-based data collection platform using integrated navigation system. ASABE Paper No. 073046. St. Joseph, Mich.: ASABE.

Zaman, Q. U., K. C. Swain, A. W. Schumann, and D. C. Percival. 2010. Automated, low-costyield mapping of wild blueberry fruit. Applied Engineering in Agriculture. 26(2): 225-232.

Zaman, Q. U., A. W. Schumann, and S. Shibusawa. 2006. Impact of variable rate fertilizationon nitrate leaching in citrus orchards. 8th Int. Precision Agriculture Conf. Minnesota. July 24-26, 2006.

Zaman, Q., A. W. Schumann, and K. H. Hostler. 2006. Estimation of citrus fruit yield usingultrasonically sensed tree size. *Applied Eng. in Agric.* 22(1): 39-44.Zaman, Q. U., and A. W. Schumann. 2006. Nutrient management zones for citrus based on variation in soil properties and tree performance. Precision Agriculture 7(1):45-63.

Zhang, F., B. Wu, and C. Liu. 2003. Using time series of SPOT VGT NDVI for yield forecasting. In *Proc. Geoscience and Remote Sensing Symp.*, 1: 386-388. Piscataway, N.J.: IEEE.

Zhang, M., Z. Qin, and X. Liu. 2005. Remote sensed spectral imagery to detect late blight in field tomatoes. *Precision Agric.* 6(6): 489-508.

Zhang, N., and R. K. Taylor. 2001. Applications of a field-level geographic information system (FIS) in precision agriculture. *Applied Eng. in Agric.* 17(6): 885-892.

Zhou, H., Y. Ru, C. Shu, J. Zheng, and H. Zhu. 2009. Design and experiments of aerial electrostatic spraying system assembled in helicopter. ASABE Paper No. 097378, Annual Meeting at Reno, Nevada, June 21-24.

12

Mapping Aboveground and Foliage Biomass Over the Porcupine Caribou Habitat in Northern Yukon and Alaska Using Landsat and JERS-1/SAR Data

Wenjun Chen, Weirong Chen, Junhua Li, Yu Zhang, Robert Fraser,
Ian Olthof, Sylvain G. Leblanc and Zhaohua Chen
Canada Centre for Remote Sensing, Natural Resources Canada
Canada

1. Introduction

The linkage between caribou and the aboriginal people in the North America has existed for thousands of years. Caribou have played a critical role in the economy, culture, and way of life of the aboriginal people (Hall, 1989; Madsen, 2001). Currently, there are 60 major migratary tundra caribou herds circum-arctic, of which 30 are located in North America, including the Porcupine caribou herd in northern Yukon, Canada and northern Alaska, USA (Russell et al., 1992; Russell et al., 1993; Russell & McNeil, 2002; Russell et al., 2002; Griffith et al., 2002). The Porcupine caribou herd has been at the centre of debate between wildlife habitat conservation and industrial development in the Arctic because of the potential oil drilling in the Arctic National Wildlife Refuge (ANWR) 1002 area, which happens to largely overlap with the calving ground of the Porcupine caribou herd (Griffith et al., 2002; Kaiser, 2002; National Research Council, 2003; Heuer, 2006).

One of the objectives of the Canadian International Polar Year (IPY) project entitled "Climate Change Impacts on Canadian Arctic Tundra Ecosystems (CiCAT): Interdisciplinary and Multi-scale Assessments" was to assess the impact of climate change on caribou habitats over Canada's north, in close collaboration with the CircumArctic Ranfiger Monitoring and Assessment network (CARMA) (http://www.rangifer.net /carma/). Because of the vastness and remoteness of the arctic landmass, inherent logistic difficulty, and high cost of conducting field measurements, an approach that is solely based on field inventory is clearly impractical for monitoring and assessing the impact of climate on caribou habitats. Satellite remote sensing can monitor land surfaces from space repeatedly and consistently over large areas. Therefore, remote sensing provides a powerful tool for monitoring and assessing the impact of climate on caribou habitats, when calibrated and validated against the field measurements and other independent data.

In this study, we report the development of baseline maps of aboveground and foliage biomass over the Porcupine caribou habitat in northern Yukon and Alaska, using Landsat and JERS-1/SAR data. Specifically, we will (1) describe aboveground and foliage biomass measurement, (2) establish and validate relationships between measurements and remote sensing indices, and (3) map aboveground and foliage biomass for the Porcupine caribou habitat.

2. Data sources and methods

2.1 Field measurements of aboveground and foliage biomass

Field measurements were made in Yukon and North Western Territories (Fig. 1). Aboveground biomass was measured at 43 sites in the summer of 2004 along the Dempster Highway, which goes through the winter and summer ranges of the Porcupine caribou habitat (Fig. 1). Foliage biomass was measured at 10 non-treed sites along the Dempster Highway in the summer of 2006, and again in the summer of 2008 at 11 non-treed sites in the Ivvavik National Park. The Ivvavik National Park is located at northern tip of the Yukon and overlaps with the calving ground and summer range of the Porcupine caribou herd. Details of measurement procedure, calculation method, and results are described as follows.

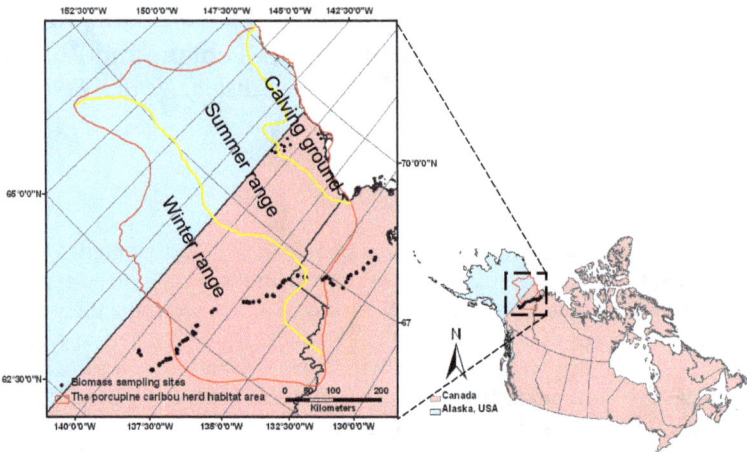

Fig. 1. Locations of the Porcupine caribou calving ground, summer range, and winter range, as well as that of the aboveground and foliage biomass measurement sites along the Dempster Highway transect during the summers of 2004 and 2006, and in the Ivvavik National Park during the summer of 2008.

We selected sites that were representative of local vegetation conditions, relatively homogenous, and of size at least 3 × 3 Landsat pixels (i.e., > 90 m × 90 m), to ensure that the field measurements can be reliably correlated with Landsat-scale remote sensing indices. In the north where the growing season is very short, biophysical parameters (e.g., foliage biomass) vary significantly both during the beginning and end of the growing season, but are relatively stable and achieve their maximums during the middle of the growing season. Therefore, we measured aboveground biomass and foliage biomass during middle of growing season (e.g., July 18-27, 2004, July 19-27, 2006, and July 15-26, 2008).

We used a systematic approach to sample and measure the aboveground biomass: layer by layer from top to the ground. If a site was sparsely treed woodland, we first measured tree biomass with a variable sampling plot scheme using a prism (Halliwell & Apps, 1997). The basic principle of the variable sampling plot scheme is that each tree is selected using a circular plot with a radius r that is proportional to its diameter at the breast height (DBH). The constant of proportionality is called the plot radius factor (PRF), and thus for tree i we can write:

$$PRF = r_i\,/\,DBH_i \quad or \quad r_i = PRF \times DBH_i \tag{1}$$

with DBH is in centimeters and plot radius is in meters, PRF is in m cm^{-1}. A tree is included or "in" if $DBH_i > r_i/PRF$, and excluded or "out" otherwise. A difficulty may arise as whether to include or exclude a tree if it's at the edge, namely, $DBH_i \approx r_i/PRF$. If this is the case, the distance from the plot centre to the tree is measured. Using the tree's DBH and the appropriate PRF, this distance can then be compared to the radius of the plot as calculated by equation 1, and the tree place "in" or "out" of the plot after-the-fact. The measurements included tree height, diameter at breast height (DBH), and stand density. The heights of trees are recorded using a laser height measurement instrument (Impulse Forest Pro, Lasertech, Clarkston, MI, USA). Trees for height measurement are selected on basis of the point plot scheme. The value of DBH was measured using a tape. The stocking and biomass of the trees were then calculated on basis of allometric equations.

If tree regeneration was present, we measured the biomass of the seedlings using a fixed circular plot of radius of 3.99 m. The measurements included counting the number of stems per tree species, selecting an average-sized seedling to estimate its ground stem diameter, height, and sampling one or two average-sized seedling to measure its fresh and oven-dry weight.

For the understory layers, namely, high shrub, low shrub, herbs/graminoids, moss, and lichen, we harvested samples at five 1 m × 1 m plots to measure the total aboveground biomass: four at four directions and one random (Fig. 2). All the aboveground biomass within the plots were cut and collected into plastic bags and their green biomass weighed. Out of the five plots, we selected one representative plot to conduct an intensive sampling. For this plot, we measured the average height, visually estimate the ground cover percentage, and then cut and weighed the fresh biomass, sequentially, for tall shrub, low shrubs, herbs/graminoids, moss, and lichen. The ratio of green to oven-dry biomass was measured by bringing a sample of the aboveground biomass from the plot back to laboratory.

If a site had no trees, then we applied the aforementioned understory procedure. Note that the plot number and design of a specific site were determined on basis of site conditions: the more heterogeneous and sparse the vegetation is at a site, the more plots are needed (Chen et al., 2009b; Chen et al., 2010b). The procedure was modified for tall shrubs if the tall shrubs (higher than 50 cm) were clustered instead of homogeneous distribution. Our field experience indicated that it was often the case for tall shrubs. In this case, we measured the biomass of tall shrubs using a fixed circular plot of radius = 3.99 m. The measurements included identifying shrub species, counting number of clusters of shrub in the plot, and selecting two average-sized clusters of shrub for more extensive sampling. All aboveground biomass of the two clusters of shrub were cut and weighed for total green biomass. A fraction of the biomass was brought back to measure the ratio of oven-dry to fresh weight. If there are more than one tall shrub species, we repeated the measurements for each of them. The procedure for low shrub, herbs/graminoids, moss, and lichen remain the same.

For foliage biomass measurement during the summers of 2006 and 2008, we selected only non-treed sites. At each sites, five 1 m × 1 m plots were sampled (Fig. 3). Fig. 3 shows a photograph of the field sampling at a coastal plain site in the Ivvavik National Park during the summer of 2008. At each plot, all plants were harvested, sorted into dead and live, different species, and leaves and stem, and recorded for their fresh weights. A fraction of the harvested biomass for all components were brought back to the laboratory, oven-dried, and weighed. Oven-dry foliage biomass of vascular plants were than calculated using the ratio of oven-dry to fresh foliage biomass and the corresponding fresh weight records.

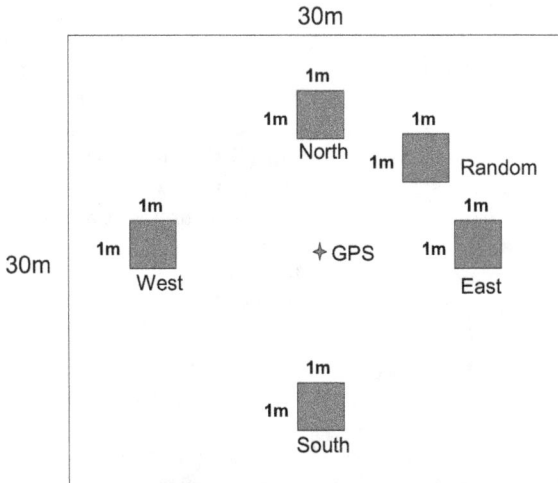

Fig. 2. Plot design for measuring aboveground and foliage biomass at non-forested sites.

The tree oven-dry aboveground biomass at a woodland site was calculated using Canada-wide tree biomass equations by Evert (1985) and the Weibull tree-size distribution function within a stand by Chen (2004).

In order to determine the tree-size distribution for a woodland site, the Parameter Prediction Method (PPM) based on basal area was implemented (Chen, 2004). With measured average DBH, height, and stand density (trees ha^{-1}) as inputs, the tree-size distribution function gives 15 tree-size classes for each site, and output mean DBH, height, and number of trees for each class. Here, we used trembling aspen functions for hardwood species, jack pine functions for all pine species and pine-mixture, black spruce functions for all spruce and fir species, and mixed-stand function for mixed softwoods/hardwoods.

The Canada-wide biomass equations calculate the oven-dry biomass of stem, bark, and crown of a single tree for 18 Canadian tree species (Evert, 1985), with a given DBH and height. Using the information of tree-size distribution and the biomass equations, we first calculated each biomass component (stem, bark, and crown), added these components to get total aboveground biomass, and then multiplied with the number of trees in each tree-size distribution class to get the total tree biomass per hectare, B_t, in t ha^{-1}.

The aboveground oven-dry biomass of clustered high shrub and regenerating trees at a site, B_s in t ha^{-1}, was calculated by

$$B_s = \frac{(b_{s1} + b_{s2})}{2} \frac{N_c}{5} R_s \qquad (2)$$

where $(b_{s1} + b_{s2})/2$ is the mean green biomass of the two clusters of shrub sampled or that of regenerating trees (kg per cluster or regenerating tree), N_c is the number of clusters of shrub or regenerating trees in the 3.99 m-radius plot, and R_s is the ratio of oven-dry to fresh biomass for shrub or the regenerating trees.

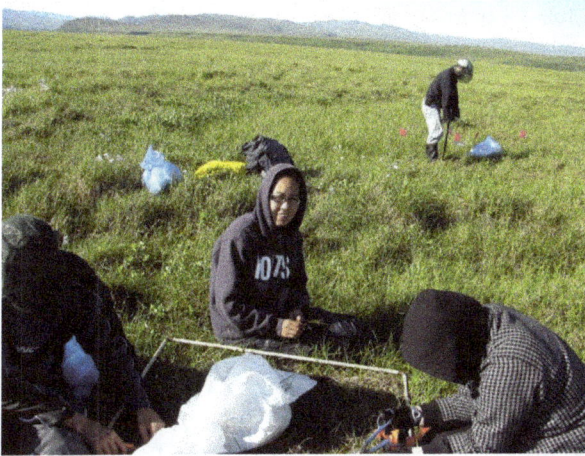

Fig. 3. Photograph showing biomass sampling at a coastal plain site in the Ivvavik National Park, July 25, 2008. Front are 3 northern students from Aklavik, NWT (left to right: Alexander Gordon, Jayneta Pascal & Kayla Arey), and at the background Weiroing Chen of CCRS was sampling root biomass (photo by Wenjun Chen).

The aboveground oven-dry biomass of graminoids, herbs and shrubs in the lower layers, B_g in t ha^{-1}, was calculated by

$$B_g = \frac{(b_{g1} + b_{g2} + b_{g3} + b_{g4} + b_{g5})}{5} 10 R_g \qquad (3)$$

where b_{g1}, b_{g2}, b_{g3}, b_{g4}, and b_{g5} are, respectively, the aboveground green biomass of the graminoids/herbs and shrubs mixture for each of the five plots in a site (kg m^{-2}), and R_g is the ratio of oven-dry to fresh biomass for the mixture of graminoids/herbs and shrubs. The oven-dry biomass of lichen, B_l in t ha^{-1}, was calculated by

$$B_l = \frac{(p_{l1} + p_{l2} + p_{l3} + p_{l4} + p_{l5})}{5} \frac{(b_{l1} + b_{l2})}{2} 1000 R_l \qquad (4)$$

where P_{l1}, P_{l2}, P_{l3}, P_{l4}, and P_{l5} are, respectively, the percentage of lich ground cover at the 5 plots, b_{l1} and b_{l2} are, respectively, the fresh lichen biomass of each of the two 10-cm by 10-cm samples collected from one of the plots (kg m^{-2}), and R_l is the ratio of oven-dry to fresh biomass for lichen. The oven-dry biomass of live moss, B_m in t ha^{-1}, was calculated by

$$B_m = \frac{(p_{m1} + p_{m2} + p_{m3} + p_{m4} + p_{m5})}{5} \frac{(b_{m1} + b_{m2})}{2} 1000 R_m D_m \qquad (5)$$

where P_{m1}, P_{m2}, P_{m3}, P_{m4}, and P_{m5} are, respectively, the percentage of moss ground cover at the 5 plots, b_{m1} and b_{m2} are, respectively, the aboveground green biomass of each of the two 1-cm depth, 10 cm by 10 cm square moss samples collected from one of the plots (kg m^{-2}), D_m is the depth of alive moss in cm, and R_m is the ratio of oven-dry to green biomass for moss. The aboveground biomass for a site, B_a, was thus the summation of all these components where proper.

The calculation of foliage biomass, B_f in unit g m^{-2}, was similar to that of aboveground biomass, except that only the foliage component of vascular plants was included.

2.2 Remote sensing data sources and processing

Nearly clear-sky Landsat TM/ETM+ level 1G orthoimagery was downloaded from the United Stated Geological Survey (USGS) website (http://earthexplorer.usgs.gov) and from the Centre for Topographic Information (CTI) of Natural Resources Canada, available through Geogratis (http://geogratis.cgdi.gc.ca/). 23 scenes are needed to cover the entire Porcupine caribou habitat. Most of them were acquired within the middle growing season (July 10 to August 15) during 1999 – 2003. Only Bands 3 (0.63–0.69 μm), 4 (0.75–0.90 μm), and 5 (1.55–1.75 μm) were used in the study, because Bands 1-3 are highly correlated, as are Bands 5 and 7. Prior to further analysis, all scenes, if necessary, were re-projected to Lambert Conformal Conic (LCC) projection with 95° W, 49° N as the true origin and 49° N and 77° N as standard parallels. Surface reflectance was derived by radiometrically normalizing each scene to 250 m resolution clear sky MODIS imagery, which was a 10-day composite acquired during July 21-31, 2001 with matching bands calibrated to a strip of atmospherically corrected Landsat TM/ETM+ images. A Landsat mosaic of surface reflectance was then generated with further normalizations for individual scenes if substantial discrepancy exists. All radiometric normalization and calibration equations were developed using a Scattergram Controlled Regression method (Elvidge et al., 1995; Yuan & Elvidge, 1996; Chen et al., 2010a). Furthermore, clouds and forested areas were removed and masked, and the mosaic was clipped based on the Porcupine caribou habitat boundary.

The JERS-1/SAR datasets were extracted from the North America JERS mosaics (acquired in the summer of 1998) to cover the Porcupine caribou habitat (Kyle McDonald, JPL, personal communication). In addition, data of 1:50000 DEM covering the habitat were also obtained for orthorectifying the JERS mosaic. The North America JERS summer mosaics were not precisely geo-referenced, because topographic distortions were not removed during the mosaic process due to the lack of adequate DEM (Sheng & Alsdorf, 2005). Visual inspection showed that the offset between the JERS mosaic and the Landsat images may be 300 m ~ 5000 m. The positioning error was expected to be much larger in mountainous regions.

In this study, we employed a SAR image simulation method from DEM data to correct topographic distortions (Sheng & Alsdof, 2005). Briefly speaking, this method includes four steps: (1) simulating a SAR image in an azimuth-range projection from the DEM according to imaging geometry of real SAR image; (2) collecting ground control points that tie the uncorrected SAR image to the simulated SAR image; (3) warping the real SAR image to the simulated SAR image using a polynomial function fitted from the ground control points; and (4) projecting the warped real SAR image back to the DEM map coordinate system. The method requires three types of inputs: individual scene of SAR imagery, DEM data, and SAR imaging geometry parameters (i.e. sensor altitude H, minimum look angle θ, the orbital azimuth angle β). Since the year-day file, which contains the date of specific JERS image's acquisition, was delivered with the JERS mosaics data, we could extract individual path image in the JERS mosaic. The path image was used as an input in place of individual scene of JERS. In addition, since the imaging geometry information of the individual path in the JERS mosaic was unavailable, general JERS imaging geometry parameters were used in the SAR simulation method. For the purpose of image matching for ground control point selection, a correct geometry was preferred but is not necessary (Sheng and Alsdof, 2005;

Linder and Meuser; 1993). In this study, we used H = 568000 meter, θ =35 degree, and β =190 degree in SAR image simulation. Finally, the Digital Number (*DN*) values of JERS mosaic could be converted to backscatter coefficients σ (in db format) by the equation:

$$\sigma = 20 \times \log_{10} DN - 48.54 \tag{6}$$

The co-registered and geo-referenced Landsat TM/ETM+ and JERS-1 mosaics were then compared with measurements of aboveground and foliage biomass, with their best-fit relationships being applied back to the mosaics to produce maps of aboveground and foliage biomass over the Porcupine caribou habitat, as outlined in Fig. 4.

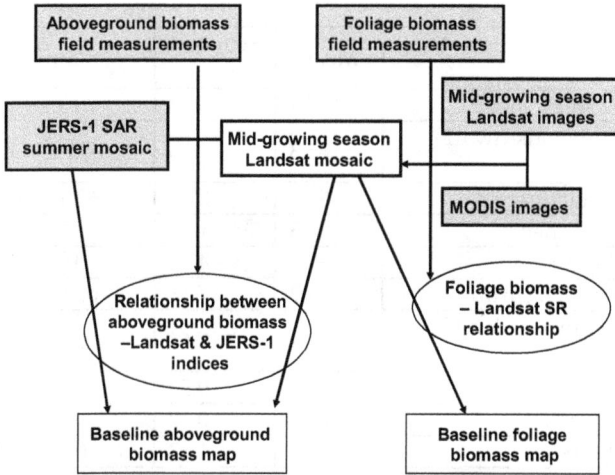

Fig. 4. Flow chart describing the procedures of using field measurements and remote sensing data to produce aboveground and foliage biomass baseline maps.

3. Results and discussions

3.1 Measured aboveground and foliage biomass over the Porcupine caribou habitat

Table 1 summarizes the mean value, standard deviation, and range of aboveground biomass measured at sites within and around the Porcupine caribou winter and summer range along the Dempster Highway, Yukon, during the summer of 2004. Average values of measured aboveground biomass of sparsely treed woodland, low-high shrub lands, and mixed graminoids-dwarf shrub-herb lands were, respectively, 57.3, 11.1, and 2.3 t ha-1. Within each vegetation type, the ranges of measured aboveground biomass were very large. The standard deviations of measured aboveground biomass among sites were often larger than their corresponding mean values, especially for low-high shrubs and mixed graminoids-dwarf shrub-herb, and when all types of vegetation were considered. The measured aboveground biomass ranged from 10 to 100 t ha-1 for sparsely treed woodlands, from 1 to 100 t ha-1 for the low-high shrub sites, from 0.5 to 10 t ha-1 for mixed graminoids-dwarf shrub-herb sites. These measurements indicate that there are significant overlaps in the ranges of aboveground biomass between sparsely treed woodlands and low-high shrub sites, and between low-high shrub sites and mixed graminoids-dwarf shrub-herb sites.

Dominant vegetation type	Mean (t ha^{-1})	Standard deviation (t ha^{-1})	Range (t ha^{-1})	Number of sites
Sparse woodland	57.3	34.0	9.1 – 94.5	8
Low-high shrub	11.1	11.4	2.0 – 47.2	19
Graminoids-dwarf shrub	2.3	2.2	0.4 – 8.3	16
All types	16.4	25.6	0.4 – 94.5	43

Table 1. Mean value, standard deviation, and range of aboveground biomass measured at sites within and around the Porcupine caribou winter and summer range along the Dempster Highway, Yukon, during the summer of 2004.

	Dominant vegetation type	Mean	Standard deviation	Range	Number of sites
Foliage biomass (g m^{-2})	Low-high shrub	135.4	36.3	95.3 – 198.4	6
	Graminoids-dwarf shrub	65.3	16.3	37.9 – 92.3	9
	Coastal tussock	87.9	22.2	63.1 – 106.1	3
	Rock lichen	11.3	10.2	0.0 – 20.0	3
	All types	80.8	47.2	0.0 – 198.4	21
Aboveground biomass (t h^{-1})	Low-high shrub	7.02	7.58	0.36 – 18.16	6
	Graminoids-dwarf shrub	3.11	2.64	0.32 – 8.55	9
	Coastal tussock	0.77	0.53	0.21 – 1.27	3
	Rock lichen	0.65	0.57	0.0 – 1.06	3
	All types	3.54	4.83	0.0 – 18.16	21

Table 2. Mean value, standard deviation, and range of foliage biomass measured at sites within and around the Porcupine caribou winter and summer ranges along the Dempster Highway, Yukon during the summer of 2006, as well as at sites within and around the Porcupine caribou calving ground and summer range inside the Ivvavik National Park, Yukon during the summer of 2008. Also shown are statistics for the corresponding aboveground biomass measurements.

Similarly, the foliage biomass values measured at sites within a specific dominant vegetation type also varied significantly (Table 2). For example, foliage biomass ranges from 95.3 to 198.4 g m^{-2} for low-high shrub sites, 37.9–92.3 g m^{-2} for mixed graminoids-dwarf shrub-herb sites, 63.1–106.1 g m^{-2} for coastal plain tussock sites, and 0.0–20.0 g m^{-2} for hill-top rock lichen sites. Consequently, assigning aboveground or foliage biomass value to a site according to its vegetation type can result in substantial error (Gould et al., 2003; Walker et al., 2003).

3.2 Relationships between aboveground biomass and remote sensing indices
In this study, we investigated the applicability of both optical and radar data as well as their combinations. To obtain the best regression model for estimating aboveground biomass, robust multiple regressions were conducted between field-measured aboveground biomass and various remote sensing-derived variables such as TM spectral reflectance, vegetation

indices, and JESR-1/SAR backscatter coefficients. We used 3 × 3 pixels (i.e., 90 m by 90 m) averaged value in place of single pixel value in order to reduce the effect of erroneous spectral features, e.g., features of adjacent pixels may have been assigned to some field plots of the data due to errors in image registration and the location of sample plots. The Landsat images were re-sampled to 100 m resolution for matching the resolution of the North America JERS summer mosaic.

For the sites along the Dempster Highway, we found that strong correlations exist between $\ln(B_a)$ and remote sensing signals (Table 3). When all types were mixed, the strongest correlation was found against the L-band JERS-1/SAR backscatter, followed by the Landat B4/B5 (Fig. 5). The Landsat bands 3 and 5 show strong negative relationships for the sparse woodlands and all types mixed, but not for shrub and graminoids lands.

		B3	B4	B5	B4/B5	SR	NDVI	SWVI	JERS
Sparse woodland	r	-0.77	-0.18	-0.78	0.40	0.53	0.21	0.44	0.68
	r^2	0.59	0.03	0.61	0.16	0.28	0.04	0.19	0.46
Low-high-Shrub	r	-0.17	0.04	-0.39	0.59	0.47	0.26	0.59	0.42
	r^2	0.03	0.00	0.15	0.35	0.22	0.07	0.35	0.18
Graminoid s-dwarf shrub	r	-0.13	0.22	0.01	0.21	0.26	0.31	0.18	0.64
	r^2	0.02	0.05	0.00	0.04	0.07	0.10	0.03	0.41
All types	r	-0.65	-0.02	-0.68	0.70	0.63	0.53	0.68	0.73
	r^2	0.42	0.00	0.46	0.49	0.40	0.28	0.46	0.53

Table 3. Correlation coefficient (r) and coefficient of determination (r^2) between $\ln(B_a)$ and remote sensing indices for mixed graminoids-dwarf shrub-herb, low-high shrub, sparse woodlands, and all types for aboveground biomass measurements along the Dempster Highway in 2004. Remote sensing indices include Landsat red band reflectance (B3), near infrared band reflectance (B4), shortwave infrared band reflectance (B5), ratio of B4/B5, simple ratio (SR = B4/B3), normalized differential vegetation index (NDVI = (B4 - B3)/(B4 + B3)), shortwave vegetation index (SWVI = (B4 - B5)/(B4 + B5)), and L-band JERS-1/SAR backscatter coefficient (JERS).

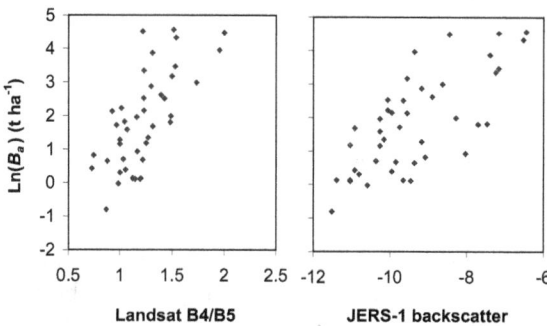

Fig. 5. Scatter plots between $\ln(B_a)$ and Landsat B4/B5 as well as JERS-1/SAR backscatter coefficients.

Many vegetation indices have been developed and applied for estimating aboveground biomass (Anderson & Hanson, 1992; Anderson et al., 1993; Mutanga & Skidmore, 2004; Lu 2005). In Table 3, we examined 4 vegetation indices: ratio of Landsat B4/B5, simple ratio (SR = B4/B3), normalized differential vegetation index (NDVI = (B4 - B3)/(B4 + B3)), shortwave vegetation index (SWVI = (B4 - B5)/(B4 + B5)). In general, vegetation indices can partially reduce the impacts on reflectance caused by environmental conditions and shadows, thus improve correlation between AGB and vegetation indices, especially in those sites with complex vegetation stand structures (Lu et al., 2004). However as shown in Table 3, not all vegetation indices are significantly correlated with aboveground biomass. Consequently, various degrees of success had been obtained in estimating aboveground biomass using Landsat vegetation indices (Sader et al., 1989; Lee & Nakane, 1996; Nelson et al., 2000; Steininger, 2000; Foody et al., 2003; Phua & Saito, 2003; Zheng et al., 2004). For example, Nelson et al. (2000) found that aboveground biomass cannot be reliably estimated using Landsat data without the inclusion of secondary forest age. Steininger (2000) explored the ability of Landsat data for estimating aboveground biomass of tropical secondary forests and found that saturation was a problem for advanced successional forests.

Similarly, different degrees of success had been obtained in previous studies of using radar data for estimating aboveground biomass, with longer-wavelength L-band and P-band SAR data proven to be more valuable (Sader, 1987; Le Toan et al., 1992; Rauste et al., 1994; Ranson et al., 1997; Luckman et al., 1997; Kurvonen et al., 1999; Kuplich et al., 2000; Tsolmon et al., 2002; Sun et al., 2002; Castel et al., 2002; Santos et al., 2002). For example, Kuplich et al. (2000) used JERS-1/SAR data for aboveground biomass estimation of regenerating forests. Sun et al. (2002) found that multi-polarization L-band SAR data were useful for estimating aboveground biomass of forest stands in mountainous areas. Castel et al. (2002) identified the significant relationships between the backscatter coefficient of JERS-1/SAR data and the stand biomass of a pine plantation. Santos et al. (2002) used JERS-1/SAR data to analyse the relationships between backscatter signals and biomass of forest and savanna formations. The significant correlation between aboveground biomass and JERS-1/SAR backscatter coefficient, as shown in Table 3 and Fig. 4, indicates that longer-wavelength L-band SAR data are also valuable in the Arctic. Nevertheless, the saturation problem is also common in estimating aboveground biomass using radar data (Luckman et al., 1997; Balzter, 2001, Lucas et al., 2004; Kasischke et al., 2004). For example, Luckman et al. (1997) found that the longer-wavelength L-band SAR image was more suitable to discriminate different levels of forest biomass up to a certain threshold than shorter-wavelength C-band SAR data.

To take advantages of the ability of Landsat vegetation indices and JERS-1/SAR backscatter coefficient, we used both data for estimating aboveground biomass. The best fit relationship between aboveground biomass and JERS-1/SAR backscatter coefficient σ as well as Landsat B4/B5 for all types mixed in the Dempster Highway study area is given by:

$$\ln(B_a) = 2.3759(B4/B5)+0.5542\sigma+4.0948, \qquad (7)$$

with a coefficient of determination r^2 = 0.72, and standard estimation error (SEE) = 0.78. Because the logarithmic equations could introduce a systematic bias when used for back calculating biomass, it has now become fairly widely recognized that a correction factor is necessary to counteract this bias (Sprugel, 1983). The correction factor (CF) can be calculated by using the formula:

$$CF = e^{SEE^2/2} \qquad\qquad (8)$$

For equation (7), the $CF = 1.35$. The 1:1 comparison of estimated aboveground biomass using CF-corrected equation (7) and the measured values for the sites along the Dempster Highway is shown in Fig. 6, with the slope = 0.95, $r^2 = 0.72$, and $SEE = 13.5$ t h⁻¹ over measured aboveground biomass range from 0.4 to 94.5 t h⁻¹. These results agreed well with the finding of Moghaddam et al. (2002) that the estimation accuracy of forest biomass was significantly improved when radar and optical data were used in combination, compared to estimates using a single data type alone.

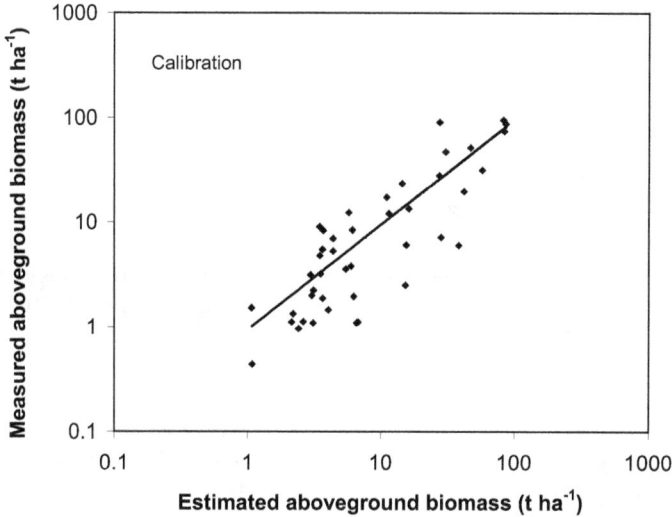

Fig. 6. A 1:1 comparison between estimated and measured aboveground biomass values for the calibration sites along the Dempster Highway transect measured during the summer of 2004. For clarity over the low biomass range, the results are shown on the log-log scale.

3.3 Validation of aboveground biomass relationship with remote sensing indices

In a previous study (Chen et al., 2009a), we validated equation (7) using aboveground biomass measurements at 33 sites around Yellowknife, Northwest Territories, and Lupin Gold Mine, Nunavut Territories. The 1:1 comparison of estimated aboveground biomass and the measurements for the Yellowknife and Lupin Gold Mine study area indicates that the equation holds up very well, with $r^2 = 0.81$, slope =1.17, and $SEE = 9.67$ t h⁻¹ over measured aboveground biomass range from 0.9 to 103.3 t h⁻¹.

In this study, we further validated the relationship using measurements collected over the same study area at later dates. We used aboveground biomass measured at 21 sites along the Dempster Highway during the summer of 2006 in the Ivvavik National Park during the summer of 2008, where foliage biomass was the main measurement target, to validate the above relationship. Fig. 7 shows the validation result, with the slope = 0.91, $r^2 = 0.90$, and $SEE = 1.5$ t h⁻¹ over measured aboveground biomass range from 0 to 18.16 t h⁻¹.

These validation results suggest that the aboveground biomass can be reliably estimated using Landsat B4/B5 and JERS-1/SAR backscatter coefficient.

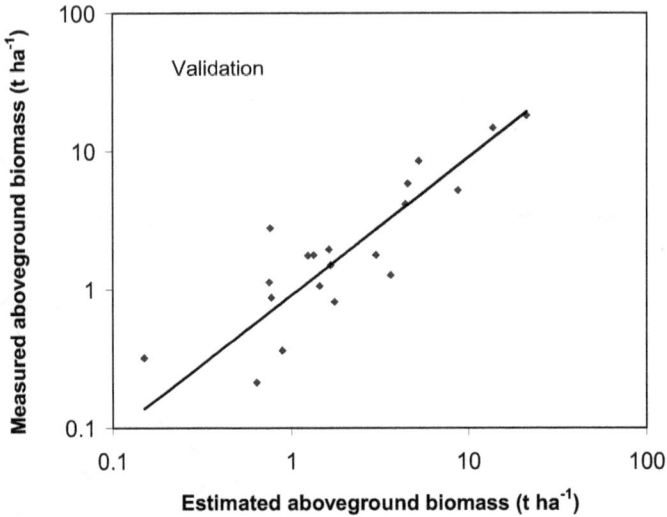

Fig. 7. A 1:1 comparison between estimated and measured aboveground biomass values for the validation sites along the Dempster Highway transect measured during the summer of 2006 and in the Ivvavik National Park during the summer of 2008. For clarity over the low biomass range, the results are shown on the log-log scale.

3.4. Relationships between foliage biomass and remote sensing indices

As shown in Table 4, the Landsat-based *SR* can explain 81% variations in measured foliage biomass within the Porcupine caribou habitat. The explanation power of other Landsat-based indices in the descending order is 76% for the ratio of B4/B5, 71% for SWVI, and 67% for NDVI. Single band reflectance of Landsat has lower power of explanation.

	B3	B4	B5	SR	NDVI	B4/B5	SWVI
r	-0.73	0.52	-0.54	0.90	0.82	0.87	-0.84
r^2	0.53	0.27	0.29	0.81	0.67	0.76	0.71
SEE	31.4	39.1	38.6	20.6	26.4	21.9	25.2

Table 4. Correlation coefficients (r), coefficient of determination (r^2), and standard estimation error (*SEE*, in g m^{-2}) between foliage biomass (B_f) and remote sensing signals for foliage biomass measured at the 21 sites along the Dempster Highway in 2006 and in the Ivvavik National Park in 2008. Remote sensing indices include Landsat red band reflectance (B3), near-infrared band reflectance (B4), shortwave infrared band reflectance (B5), ratio of B4/B5, simple ratio (SR = B4/B3), normalized differential vegetation index (NDVI = (B4 - B3)/(B4 + B3)), and shortwave vegetation index (SWVI = (B4 - B5)/(B4 + B5)).

In comparison with results for aboveground biomass shown in Table 3, Landsat-based vegetation indices have a much improved power of explanation for foliage biomass (Table 4). This is in good agreement with the fact that optical sensors mainly capture canopy information, thus the optical sensor data are more suitable for estimation of canopy parameters such as foliage biomass than aboveground biomass, as demon stared by previous studies (Franklin & Hiernaux, 1991; Hall et al., 1995; Chen & Cihlar 1996; Turner et al., 1999; Brown et al., 2000; Chen et al., 2002; Wylie et al., 2002; Phua & Saito, 2003; Laidler & Treitz, 2003; Lu, 2004). The best fit relationship between foliage biomass (B_f, g m^{-2}) and Landsat-based simple ratio (SR) over the Porcupine caribou habitat is given as follows

$$B_f = 16.62SR - 1.1906,$$ (9)

with $r^2 = 0.81$, $SEE = 20.6$ g m^{-2}, $F = 81$, $P = 2.7 \times 10^{-8}$, and $n = 21$ (Fig. 8).

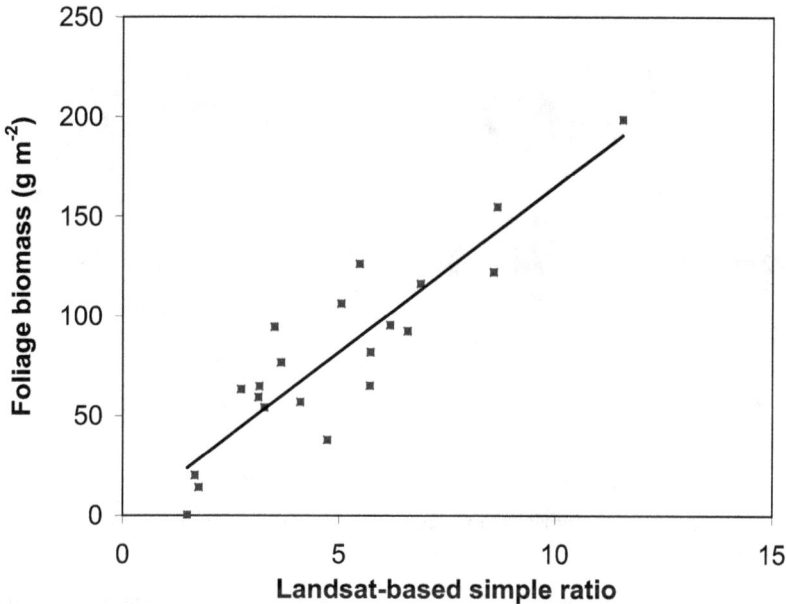

Fig. 8. Relationship between Landsat simple ratio and foliage biomass measured at sites along the Dempster Highway during the summer of 2006 and in the Ivvavik National Park during the summer of 2008.

Because we had only a relative small sample size of foliage biomass over the Porcupine caribou habitat, we didn't leave a fraction of the foliage biomass measurement points as validation. Nevertheless, we did find a similar relationship between Landsat-based simple ration and foliage biomass for the Bathurst caribou habitat located in Northwest Territory, Nunavut Territory, and northern Saskatchewan (Chen et al., 2011), with $r^2 = 0.86$, $SEE = 26.3$ g m^{-2}, $F = 158$, $P = 2.6 \times 10^{-12}$, and $n = 27$.

3.5 Baseline maps of aboveground and foliage biomass

Applying equations (7) and (8) to the co-registered and geo-referenced Landsat and JERS-1/SAR mosaics data over the Porcupine caribou habitat, we produced aboveground biomass for the Porcupine caribou habitat. Fig. 9 shows aboveground biomass distribution in circa 2000 over the Porcupine caribou habitat.

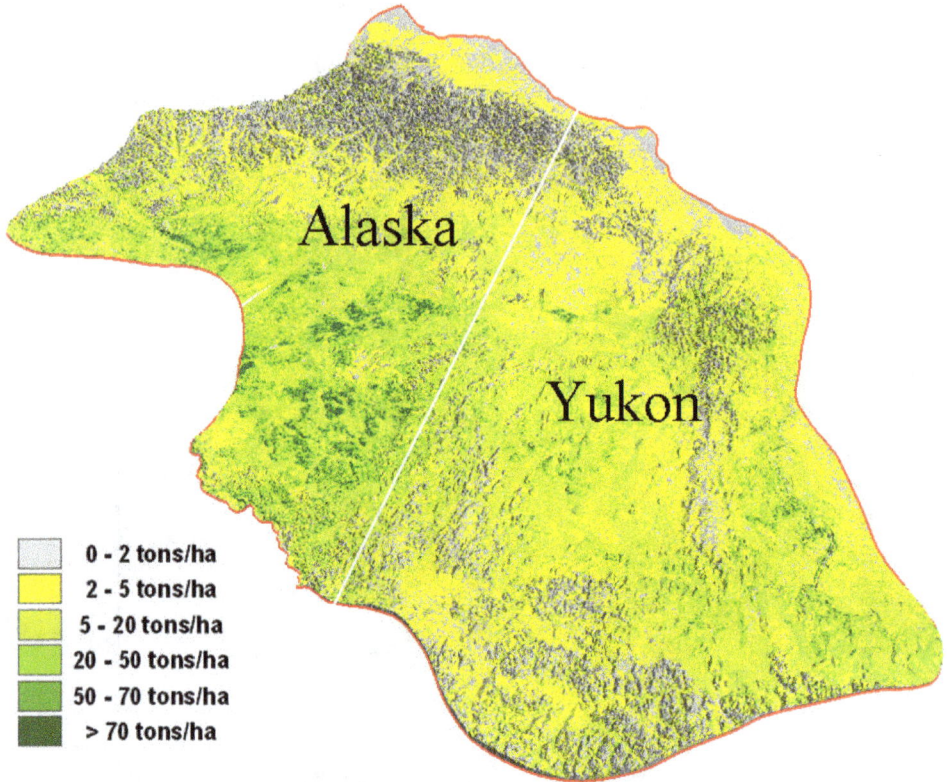

Fig. 9. Aboveground biomass distribution over the Porcupine caribou habitat circa 2000. The 3-d effect was generated using the DEM data.

The amount of aboveground biomass was less than 5 t ha^{-1} for most area of the Porcupine caribou calving ground located in northern coastal area in both Alaska and Yukon. Along the coastal line and at high mountain ridges, the aboveground biomass further decreased to less than 2 t ha^{-1}. For the summer and winter ranges of the Porcupine caribou habitat to the south, the values of aboveground biomass generally increased, with a large percentage of the areas having aboveground biomass in the range of 5-20 t ha^{-1}, as well as a significant fraction of the area of 2-5 t ha^{-1} aboveground biomass. The areas of aboveground biomass 2-5 t ha^{-1} appeared to be dominated by graminoids and herbs, while those of 5-20 t ha^{-1} aboveground biomass were mainly shrub land. The Porcupine summer and winter ranges also had a small fraction of area that had more than 20 t ha^{-1} aboveground biomass, likely associated with treed woodland. Areas of 0-2 t ha^{-1} aboveground biomass also occur in the summer and winter ranges of the Porcupine caribou habitat, mostly happened at high mountain ridges.

The information on foliage biomass can be more useful to local caribou management boards and researchers because it is directly related forage availability (Russell et al., 1993; Russell & McNeil, 2002; Russell et al., 2002). Using equation (9), we developed circa 2000 foliage biomass distribution over the Porcupine caribou habitat (Fig. 10).

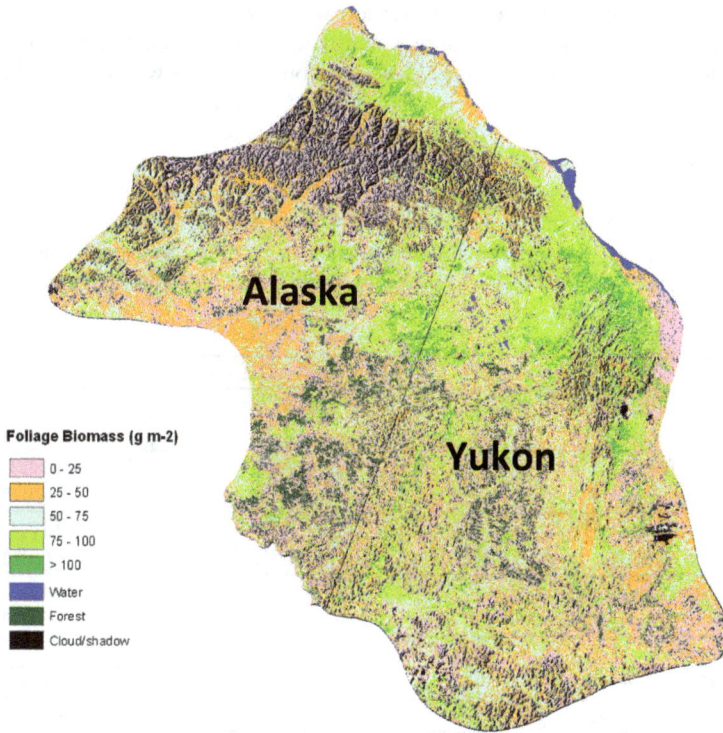

Fig. 10. Foliage biomass distribution over the Porcupine caribou habitat cirac 2000. The 3-d
effect was generated using the DEM data. Areas covered by water, forest, and cloud were
blocked out.

Fig. 10 shows the circa 2000 foliage biomass distribution of non-forest land areas in the
Porcupine caribou habitat. Because of lacking foliage biomass measurements over forested
land, foliage biomass was not estimated over forest areas in the Porcupine caribou habitat
and areas covered forests were blocked out in Fig. 10. We also blocked out water and cloud-
cover areas in Fig. 10. As shown in Fig. 10, most of the northern coastal areas in Alaska and
Yukon, over which the calving ground of the Porcupine caribou habitat is located, had a
quite high foliage biomass from 50 to > 100 g m-2. This is a contrast to the distribution of the
aboveground biomass. Over these same coastal areas the aboveground biomass was quite
low. Both aboveground biomass and foliage biomass were found to be lower at high
mountain ridges, and coastal beaches.

The foliage biomass map reveals that on average the amount of seasonal peak foliage
biomass in the calving ground of the Porcupine caribou herd was similar to that in the
summer range, and was even higher if only the concentrated calving ground over the
Northern Alaska and Yukon coastal plain is concerned (Table 5). To the contrast, the
average seasonal peak foliage biomass in the calving ground of the Bathurst caribou herd
was much lower than that in the summer range during the same time period (Chen et al.,
2011). This result agree well with the ground survey by Griffith et al. (2001), which
suggested that on June 14, during 1998-1999, the total available forage was 63 g m-2 for the

Bathurst caribou calving ground, in comparison with 460 g m^{-2} for the Porcupine caribou calving ground. Of the 63 g m^{-2} forage over the Bathurst caribou calving ground, 33 g m^{-2} was lichen and moss, 14 g m^{-2} was stand dead, and only 16 g m^{-2} was vascular biomass. On the contrast for the Porcupine caribou calving ground, 50 g m^{-2} was lichen and moss, 250 g m^{-2} was stand dead, and 160 g m^{-2} was vascular biomass. The live biomass over the Porcupine calving ground was 10 time that over the Bathurst calving ground. In addition, the high value of stand dead over the Porcupine calving ground also suggest the high foliage biomass value over previous years. The difference in calving ground foliage biomass collaborates well with the fact that cows of the Bathurst herd leave calving ground soon after giving birth while those of Porcupine herd stay for a much longer period (Griffith et al., 2001).

	Porcupine	Bathurst	Data sources
Measured foliage biomass, calving ground (g m^{-2})	63 - 106	17 - 94	This study, Chen et al. (2011)
Area-averaged foliage biomass, calving ground (g m^{-2})	78	24	This study, Chen et al. (2011)
Measured foliage biomass, summer range (g m^{-2})	14 – 198	17 - 267	This study, Chen et al. (2011)
Area-averaged foliage biomass, summer range (g m^{-2})	69	43	This study, Chen et al. (2011)
Milk production (l d^{-1})	2.02	1.09 – 1.79	Griffith et al. (2001)
Calve growth rate (g d^{-2})	493	150 - 407	Griffith et al. (2001)
Cow body weight range (kg)	83 - 96	66 - 78	Griffith et al. (2001)

Table 5. Comparison of measured and area-averaged foliage biomass values between Porcupine caribou habitat and Bathurst caribou habitat. The area-averaged values were calculated from the circa 2000 baseline foliage biomass maps. Field measurement of foliage biomass over the Bathurst caribou habitat was conducted during the summer of 2005 (Chen et al., 2011). Also included in the comparison are cow milk production, calve growth rate, and cow body weight.

The difference in calving ground forage availability, in turn, results in difference in caribou biological measures (Table 5). For example, the body weights of Porcupine cows are about 20% higher than that of Bathurst herd (Griffith et al., 2001). The Bathurst cow milk production was 1.09 – 1.79 liter per day, against 2.02 liter per day by an average Porcupine caribou cow, with the corresponding calve growth rate being 150 -407 gram per day for the Bathurst herd against 493 gram per day for that of the Porcupine herd (Table 5).

4. Conclusions

Aboveground biomass was measured at 43 sites in the summer of 2004 along the Dempster Highway, which goes through the winter and summer ranges of the Porcupine caribou habitat. The measured aboveground biomass ranged 10-100 t ha^{-1} for sparsely forested woodlands, 1-100 t ha^{-1} for the low-high shrub sites, and 0.5-10 t ha^{-1} for mixed graminoids-dwarf shrub-herb sites.

Foliage biomass was measured at 10 non-forested sites along the Dempster Highway in the summer of 2006, and again in the summer of 2008 at 11 non-forested sites in the Ivvavik National Park located at northern tip of Yukon, which overlaps with the calving ground and summer range of the Porcupine caribou herd. The measured foliage biomass ranged 0.95-2 t ha^{-1} for low-high shrub sites, 0.38-0.92 t ha^{-1} for mixed graminoids-dwarf shrub-herb sites, 0.63-1.06 t ha^{-1} for coastal tussock sites, and < 0.2 t ha^{-1} for hill-top rock lichen sites.

When all data points were pooled together, the best relationship between aboveground biomass and remote sensing data was found to be with the combination of JERS-1 backscatter and Landsat B4/B5, with r^2 = 0.72. Validation using aboveground biomass measurements at foliage biomass measurement sites gives similar result, with the slope = 0.91, r^2 = 0.90, and SEE = 1.5 t h^{-1} over measured aboveground biomass range from 0 to 18.16 t h^{-1}. Similar validation results were obtained using aboveground biomass measurements at 33 sites around Yellowknife, Northwest Territories, and Lupin Gold Mine, Nunavut Territories in a previous study (Chen et al., 2009a), with r^2 = 0.81, slope =1.17, and SEE = 9.67 t h^{-1} over measured aboveground biomass range from 0.9 to 103.3 t h^{-1}.

For the foliage biomass, the Landsat-based simple ration gives the best fit, with r^2 = 0.81, SEE = 20.6 g m^{-2}, F= 81, P = 2.7×10^{-8}, and n = 21.

Appling these relationships to the mosaic of Landsat and JERS-1 images covering the entire Porcupine caribou habitat, we produced baseline maps of aboveground biomass and foliage biomass for the Porcupine caribou habitat. The foliage biomass map reveals that on average the amount of seasonal peak foliage biomass in the calving ground of the Porcupine caribou herd was similar to that in the summer range, and was even higher if only the concentrated calving ground over the Northern Alaska and Yukon coastal plain is concerned. To the contrast, the average seasonal peak foliage biomass in the calving ground of the Bathurst caribou herd was much lower than that in the summer range during the same time period. The difference in calving ground foliage biomass collaborates well with the fact that cows of the Bathurst herd leave calving ground soon after giving birth while those of Porcupine herd stay for a much longer period, which in turn partially explain why the Porcupine caribou calves can have several times higher growth rate and the body weights of Porcupine cows are about 20% higher than that of Bathurst herd.

It should be emphasized, however, that assessment of the impacts of climate change on the Porcupine caribou habitat requires integration of various information sources from remote sensing products, climate records, and other relevant tempo-spatial data. Furthermore, there are many other factors besides the habitat, such as harvest, predators, diseases/parasites, industrial development, extreme weather events, climate change, and pollution, may influence the abundance of a caribou herd. Therefore, this work is just one of the first steps towards informed and proper decision-making that balances the needs of caribou habitat protection and industrial development under a global change environment.

5. Acknowledgment

This study is financially supported by grants from the Canadian IPY program through a project entitled "Climate Change Impacts on Canadian Arctic Tundra Ecosystems (CiCAT): Interdisciplinary and Multi-scale Assessments", Canadian Space Agency's Government Related Initiatives Program (GRIP) through a project entitled "ParkSPACE: Towards an Operational Satellite-based System for Monitoring Ecological Integrity of Arctic National Parks", and from the Earth Sciences Sector, NRCan's "Climate Change Geosciences

Program". Colleagues from CARMA (e.g., Drs. Don Russell, Wendy Nixon, and Brad Griffith) have provided great insights about research priorities and directions on caribou habitats. Parks Canada Agency (e.g., Dr. Don McLennan, Paul Dixon, et al.) organized 2008 field work in the Ivvavik National Park, arranged licence application, and provided logistics supports including helicopter riding and camping facility. David Jones and 3 northern students from Aklavik, NWT (Alexander Gordon, Jayneta Pascal, Kayla Arey) participated in and contributed his through knowledge of northern vegetation to the field measurement campaign in the summers of 2004, 2006, or 2008. Drs. Kyle McDonald of the Jet Propel Laboratory/NASA and Mahta Moghaddam of University of Michigan provided the JERS coverage over Canada's north. The Aurora Research institute and the Gwich'in Renewable Resource Board, Inuvik, NWT helped us in obtaining research licenses. Steve Wolfe of the Geological Survey Canada arranged the use of the laboratory space and biomass drying ovens. Drs. Yinsuo Zhang and Nadia Rochdi provided critical reviewed of the manuscript. The authors wish to thank all for providing their assistances.

6. References

Anderson, G.L. & Hanson, J.D. (1992). Evaluating handheld radiometer derived vegetation indices for estimating above ground biomass. *Geocarto International*, Vol. 7, pp. 71–78.

Anderson, G.L.; Hanson, J.D. & Haas, R.H. (1993) Evaluating Landsat Thematic Mapper derived vegetation indices for estimating aboveground biomass on semiarid rangelands. *Remote Sensing of Environment*, Vol. 45, pp. 165–175.

Balzter, H. (2001) Forest mapping and monitoring with interferometric synthetic aperture radar (InSAR). *Progress in Physical Geography*, Vol. 25, pp. 159–177.

Brown, L.; Chen, J.M.; Leblanc, S.G. & Cihlar, J. (2000). A shortwave infrared modification to the simple ratio for LAI retrieval in boreal forests: an image and model analysis. *Remote Sensing of Environment*, Vol. 71, pp. 16–25.

Castel, T.; Guerra, F.; Caraglio, Y. & Houllier, F. (2002) Retrieval biomass of a large Venezuelan pine plantation using JERS-1 SAR data: analysis of forest structure impact on radar signature. *Remote Sensing of Environment*, Vol. 79, pp. 30–41.

Chen, J.M. & Cihlar, J. (1996). Retrieving leaf area index of boreal conifer forests using Landsat TM images. *Remote Sensing of Environment*, Vol. 55, pp. 153–162.

Chen, J.M.; Pavlic, G., Brown, L.; Cihlar, J.; Leblanc, S.G.; White, H.P.; Hall, R.J.; Peddle, D.R.; King, D.J.; Trofymow, J.A.; Swift, E.; Van der sanden, J. & Pellikka, P.K.E. (2002). Derivation and validation of Canada-wide coarse resolution leaf area index maps using high-resolution satellite imagery and ground measurements. *Remote Sensing of Environment*, Vol. 80, pp. 165–184.

Chen, W. (2004). Tree size distribution functions of four boreal forest types for biomass mapping. *Forest Science*, Vol. 50, pp. 436-449.

Chen, W.; Blain, D.; Li, J.; Keohler, K.; Fraser, R.; Zhang, Y.; Leblanc, S.G.; Olthof, I.; Wang, J. & McGovern, M. (2009a). Biomass measurements and relationships with Landsat-7/ETM+ and JERS-1/SAR data over Canada's western sub-arctic and low arctic. *International Journal of Remote Sensing*, Vol. 30, No. 9, pp. 2355 – 2376.

Chen, W.; Li, J.; Zhang, Y.; Zhou, F.; Koehler, K.; Leblanc, S.; Fraser, R.; Olthof, I.; Zhang, Y. & Wang, J. (2009b). Relating biomass and leaf area index to non-destructive measurements for monitoring changes in arctic vegetation. *Arctic*, Vol. 62, No. 3, pp. 281–294

Chen, W.; Russell, D.E.; Gunn, A.; Croft, B.; Li, J.; Chen, W.R.; Zhang, Y.; Koehler, K.; Olthof, I.; Fraser, R.H.; Leblanc, S.G.; Henry, G.R.; White, R.G. & Finstad, G.L. (2011). Migratory tundra caribou habitat quality indicators and relationships with climate: calving ground and summer range. *Climatic change* (submitted).

Chen, W.R.; Chen W. & Li, J. (2010a). Comparison of surface reflectance derived by relative radiometric normalization versus atmospheric correction for generating large-scale Landsat mosaics. *Remote Sensing Letters*, Vol. 1, No. 2, pp. 103 – 109.

Chen, Z.; Chen, W.; Leblanc, S.G. & Henry, G. (2010b). Digital photograph analysis for measuring percent plant cover in the Arctic. *Arctic* Vol. 63, No. 3, pp. 315-326.

Evert, F. (1985). *Systems of equations for estimating oven-dry mass of 18 Canadian tree species.* Canadian Forestry Service, Petawawa National Forestry Institute.

Foody, G.M.; Boyd, D.S. & Culter, M.E.J. (2003). Predictive relations of tropical forest biomass from Landsat TM data and their transferability between regions. *Remote Sensing of Environment*, Vol. 85, pp. 463 – 474.

Franklin, J. & Hiernaux, P.Y.H. (1991). Estimating foliage and woody biomass in Sahelian and Sudanian woodlands using a remote sensing model. *International Journal of Remote Sensing*, Vol. 12, pp. 1387–1404.

Gould, W.A.; Raynolds, M. & Walker, D.A. (2003). Vegetation, plant biomass, and net primary productivity patterns in the Canadian Arctic. *Journal of Geophysical Research*, 108(D2), 8167, doi:10.1029/2001JD000948.

Griffith, B.; Douglas, D.; Walsh, N.; Young, D.; McCabe, T.; Russell, D.E.; White, R.G.; Cameron, R. & Whitten K (2002). The Porcupine caribou herd. In: *Arctic Refuge coastal plain terrestrial wildlife research summaries.* D. Douglas; P. Reynolds & E. Rhode (Eds.). U.S. Geological Survey, Biological Science Report BSR-2002-0001.

Griffith, B.; Gunn, A.; Russell, D.E.; Johnstone. J.; Kielland, K.; Wolfe, S. & Douglas, D.C. (2001). *Bathurst caribou calving ground studies: Influence of nutrition and human activity on calving ground location.* Final report submitted to West Kitikmeot Slave Study Society, Yellowknife, NWT, Canada, pp 90.

Hall, E. (Ed.) (1989) *People & Caribou in the Northwest Territories.* Government of Northwest Territories, Department of Renewable Resources, Yellowknife, NWT, Canada.

Hall, F.G.; Shimabukura, Y.E. & Huemmrich, K.F. (1995). Remote sensing of forest biophysical structure using mixture decomposition and geometric reflectance models. *Ecological Applications*, Vol. 5, pp. 993–1013.

Halliwell, D.H. & Apps, M.J. (1997). *BOREAS biometry and auxiliary sites: over story and under story data.* Natural Resources Canada, Canadian Forest Service, Northern Forest Centre, Edmonton, Alberta, Canada, pp. 244.

Heiskanen, J. (2006). Estimating aboveground tree biomass and leaf area index in a mountain birch forest using ASTER satellite data. *International Journal of Remote Sensing*, Vol. 27, No. 6, pp. 1135–1158.

Heuer, K. (2006). *Being Caribou: Five Months on Foot with an Arctic Herd.* McClelland & Stewart Ltd., Toronto, ON, Canada, pp. 235.

Kaiser, J. (2002). Caribou study fuels debate on drilling in Arctic Refuge. *Science*, Vol. 296, pp. 444-445.

Kasischke, E.S.; Goetz, S.; Hansen, M.C.; Ustin, S.L.; Ozdogan, M.; Woodcock, C.E. & Rogan, J. (2004). Temperate and boreal forests. In: *Remote Sensing for Natural Resource Management and Environmental Monitoring*, S.L. Ustin (Ed.), pp. 147–238, John Wiley & Sons Hoboken, New Jose, USA.

Kuplich, T.M.; Salvatori, V. & Curran, P.J. (2000). JERS-1/SAR backscatter and its relationship with biomass of regenerating forests. *International Journal of Remote Sensing*, Vol. 21, pp. 2513 – 2518.

Kurvonen, L.; Pulliainen, J. & Hallikainen, M. (1999). Retrieval of biomass in boreal forests from multitemporal ERS-1 and JERS-1 SAR data. *IEEE Transactions on Geoscience and Remote Sensing*, Vol. 37, pp. 198–205.

Laidler, G.J. & Treitz, P. (2003). Biophysical remote sensing of arctic environments. *Progress in Physical Geography*, Vol. 27, pp. 44-68.

Le Toan, L.; Beaudoin, A.; Riom, J. & Guyon, D. (1992). Relating forest biomass to SAR data. *IEEE Transactions on Geoscience and Remote Sensing*, Vol. 30, pp. 403 – 411.

Lee, N.J. & Nakane, K. (1996). Forest vegetation classification and biomass estimation based on Landsat-TM data in a mountainous region of west Japan. In: *The Use of Remote Sensing in The Modeling of Forest Productivity*, H.L. Gholz, K. Nakane & H. Shimoda (Eds.), pp. 159 – 171, Kluwer Academic Publishers, Dordrecht, The Netherlands.

Linder, W. & Meuser, H.F. (1993). Automatic tiepointing in SAR images. In: *SAR Geocoding: Data and Systems*, G. Schreier (Ed.), pp. 202 – 212, Karlsruhe, Wichmann, Germany.

Lu, D. (2005). Aboveground biomass estimation using Landsat TM data in the Brazilian Amazon Basin. *International Journal of Remote Sensing*, Vol. 26, pp. 2509–2525.

Lu, D.; Mausel, P.; Brondizio, E. & Moran, E. (2004) Relationships between forest stand parameters and Landsat Thematic Mapper spectral responses in the Brazilian Amazon basin. *Forest Ecology and Management*, Vol. 198, pp. 149–167.

Luckman, A.; Baker, J.R.; Kuplich, T.M.; Yanasse, C.C.F. & Frery, A.C. (1997). A study of the relationship between radar backscatter and regenerating forest biomass for space borne SAR instrument. *Remote Sensing of Environment*, Vol. 60, pp. 1–13.

Lucus, R.M.; Held, A.A.; Phinn, S.R. & Saatchi, S. (2004). Tropical forests. In: *Remote Sensing for Natural Resource Management and Environmental Monitoring*, S.L. Ustin (Ed.), pp. 239–315, John Wiley & Sons, Hoboken, New Jose, USA.

Madsen, K. (2001). *Project Caribou: An Educator's Guide to Wild Caribou of North America*. Department of Renewable Resources, Government of Yukon, Whitehorse, Yukou Territories, Canada.

Moghaddam, M.; Dungan, J.L. & Acker, S. (2002). Forest variable estimation from fusion of SAR and multispectral optical data. *IEEE Transactions on Geoscience and Remote Sensing*, Vol. 40, pp. 2176 – 2187.

Mutanga, O. & Skidmore, A.K. (2004). Narrow band vegetation indices overcome the saturation problem in biomass estimation. *International Journal of Remote Sensing*, Vol. 25, pp. 3999–4014.

National Research Council, (2003). *Cumulative Environmental Effects of Oil and Gas Activities on Alaska's North Slope*. The National Academies Press, Washington DC, USA.

Nelson, R.F.; Kimes, D.S.; Salas, W.A. & Routhier, M. (2000). Secondary forest age and tropical forest biomass estimation using Thematic Mapper imagery. *Bioscience*, Vol. 50, pp. 419–431.

Olthof, I.; Pouliot, D.; Fernandes, R. & Latifovic, R. (2005). Landsat-7 ETM+ radiometric normalization comparison for northern mapping applications. *Remote Sensing of Environment*, Vol. 95, pp. 388-398.

Phua, M. & Saito, H. (2003). Estimation of biomass of a mountainous tropical forest using Landsat TM data. *Canadian Journal of Remote Sensing*, Vol. 29, pp. 429 – 440.

Ranson, K.; Sun, G.; Lang, R.H.; Chauhan, N.S.; Cacciola, R.J. & Kilic, O. (1997). Mapping of boreal forest biomass from spaceborne synthetic aperture radar. *Journal of Geophysical Research*, Vol. 102, pp. 29599 – 29610.

Rauste, Y.; Hame, T.; Pulliainen, J.; Heiska, K. & Hallikainen, M. (1994). Radar-based forest biomass estimation. *International Journal of Remote Sensing*, Vol. 15, pp. 2797 – 2808.

Russell, D.E. & McNeil, P. (2002). *Summer ecology of the Porcupine caribou herd*. Porcupine Caribou Management Board, Whitehorse, Yukon Territories, Canada

Russell, D.E.; Kofinas, G. & Griffith, B. (2002). *Barren-Ground Caribou Calving Ground Workshop: Report of Proceedings*. Technical Report Series No. 390, Canadian Wildlife Service, Ottawa, Ontario, Canada.

Russell, D.E.; Martell, A.M. & Nixon, W.A.C. (1993). Range ecology of the Porcupine caribou herd in Canada. *Rangifer*, special issue No. 8.

Russell, D.E.; Whitten, K.R.; Farnell, R. & van de Wetering, D. (1992). *Movement and distribution of the Porcupine caribou herd, 1970-1990*. Technical Report Series No 138, Environment Canada, Canadian Wildlife Service, Whitehorse, Yukon Territoies, Canada.

Sader, S.A. (1987). Forest biomass, canopy structure, and species composition relationships with multipolarization L-band synthetic aperture radar data. *Photogrammetric Engineering and Remote Sensing*, Vol. 53, pp. 193–202.

Sader, S.A.; Waide, R.B.; Lawrence, W.T. & Joyce, A.T. (1989). Tropical forest biomass and successional age class relationships to a vegetation index derived from Landsat-TM data. *Remote Sensing of Environment*, Vol. 28, pp. 143–156.

Santos, J.R.; Pardi Lacruz, M.S.; Araujo, L.S. & Keil, M. (2002). Savanna and tropical rainforest biomass estimation and spatialization using JERS-1 data. *International Journal of Remote Sensing*, Vol. 23, pp. 1217–1229.

Serreze, M.C.; Walsh, J.E.; Chapin III, F.S.; Osterkamp, T.; Dyurgerov, M.; Romanovsky, V.; Oechel, W.C.; Morison, J.; Zhang, T. & Barry, R.G. (2000). Observational evidence of recent change in the northern high-latitude environment. *Climatic Change*, Vol. 46, pp. 159-207.

Sheng, Y. & Alsdorf, D.E. (2005). Automated georeferencing and orthorectification of Amazon basin-wide SAR mosaics using SRTM DEM data. *IEEE Transactions on Geoscience and Remote Sensing*, Vol. 43, pp. 1929 – 1940.

Sprugel, D.G. (1983). Correcting for bias in log-transformed allometric equations. *Ecology*, Vol. 64, pp. 209-210.

Steininger, M.K. (2000). Satellite estimation of tropical secondary forest aboveground biomass data from Brazil and Bolivia. *International Journal of Remote Sensing*, Vol. 21, pp. 1139–1157.

Sturm, M.; Racine, C. & Tape, K. (2001). Increasing shrub abundance in the Arctic. *Nature*, Vol. 411, pp. 546– 547.

Sun, G.; Ranson, K.J. & Kharuk, V.I. (2002). Radiometric slope correction for forest biomass estimation from SAR data in the western Sayani Mountains, Siberia. *Remote Sensing of Environment*, Vol. 79, pp. 279–287.

Theau, J.; Peddle, D.R. & Duguay, C.R. (2005). Mapping lichen in a caribou habitat of Northern Quebec, Canada, using an enhancement-classification method and spectral mixture analysis. *Remote Sensing of Environment*, Vol. 94, pp. 232–243.

Tsolmon, R.; Tateishi, R., & Tetuko, J.S.S. (2002). A method to estimate forest biomass and its application to monitor Mongplian Taiga using JERS-1 SAR data. *International Journal of Remote Sensing*, Vol. 23, pp. 4971 – 4978.

Turner, D.P.; Cohen, W.B.; Kennedy, R.E.; Fassnacht, K.S. & Briggs, J.M. (1999). Relationships between leaf area index and Landsat TM spectral vegetation indices across three temperate zone sites. *Remote Sensing of Environment*, Vol. 70, pp. 52–68.

Walker, D.A.; Epstein, H.E.; Jia, G.J.; Balser, A.; Copass, C.; Edwards, E.L.; Gould, W.A.; Hollingsworth, J.; Knudson, J.; Maier, H.A.; Moody, A. & Raynolds, M.K. (2003). Phytomass, LAI, and NDVI in northern Alaska: Relationships to summer warmth, soil pH, plant functional types and extrapolation to the circumpolar Arctic. *Journal of Geophysical Research*, 108 (D2), 8169, doi: 10.1029/ 2001JD000986.

Wylie, B.K.; Meyer, D.J.; Tieszer, L.L. & Mannel, S. (2002). Satellite mapping of surface biophysical parameters at the biome scale over the North American grasslands, a case study. *Remote Sensing of Environment*, Vol. 79, pp. 266–278.

Zheng, D.; Rademacher, J.; Chen, J.; Crow, T.; Bresee, M.; Moine, J.L. & Ryu, S.R. (2004). Estimating aboveground biomass using Landsat 7 ETM+data across a managed landscape in northern Wisconsin, USA. *Remote Sensing of Environment*, Vol. 93, pp. 402 – 411.

13

Using Remote Sensing to Estimate a Renewable Resource: Forest Residual Biomass

A. García-Martín, J. de la Riva, F. Pérez-Cabello and R. Montorio
Department of Geography and Spatial Management, University of Zaragoza, Zaragoza,
Spain

1. Introduction

Regarding the three objectives of European Union (EU) energy policy (secure supply, competitiveness and environmental protection) the EU Commission published the Communication entitled 'White Paper: Energy for the Future - Renewable Sources of Energy' (EU Commission, 1997; Mourelatou & Smith, 2004). This document, which was the starting point for the European promotion and development of renewable energy, stated the objective that 12% of energy production in 2010 would come from renewable sources. In Spain, this objective was recognized by the government in 1999 in the Plan to the Promotion of Renewable Energies (PPRE). To achieve this objective, the PPRE focused on increasing the use of biomass, identifying forest residues as one of the biomass sources. Specifically, this document established an increase in the use of forest residual biomass to 450 000 tonnes of petroleum equivalent per year (TPE/year) (IDAE, 1999).

Among the different sources of renewable energy, this chapter focuses on forest residual biomass (FRB). This term refers to branches, foliage, and unmerchantable stem tops that are commercially unsuitable in terms of the timber obtained from regular operations in forest management or in timber exploitation (Esteban et al., 2004; IDAE, 1999; Velázquez, 2006). Following the 'complete-tree concept,' the term 'branches' include the wood and bark of live and dead branches; 'foliage' refers to all leaves-needles, new shoots and reproductive organs; and the 'unmerchantable stem top' is the upper section of the stem that is left unutilized in logging operations due to its small diameter and high degree of branching (the bottom stem diameter of this unmerchantable top usually ranges from 5 to 10 cm) (Hakkila & Parikka, 2002). The treatments commonly applied to these residues in Spain include controlled burning, stacking within the forest and, less commonly, splintering to improve their incorporation into the soil (IDAE, 2005a, 2007; Velázquez, 2006).

However, this biomass segment can also be used as a source of energy in heating applications (fuel for domestic or industrial stoves and boilers) and in the generation of electricity (replacing fossil fuels in power stations) (Asikainen et al., 2002; IDAE, 2005b), with the majority of the residue currently being utilized for the latter in Spain (IDAE, 2007). The benefits associated with this energy-related use can be divided into two types: environmental and socio-economic. The first benefits are generated in the production phase, as the recovery and elimination of FRB reduces the risk of forest fires and their severity (IDAE, 2005a; Velázquez, 2006), and also because the implementation of silviculture can improve the health of forests (Eriksson et al., 2002; Raison, 2002). Environmental benefits are also obtained in the application phase. Considering only the combustion process, the energy produced is almost neutral in terms of carbon dioxide, as the amount of CO_2 emitted to the atmosphere due to FRB combustion is the same than the fixed in the FRB formation (net primary production). In addition, sulphur and chlorine emissions are very low. Consequently, taking into account only the combustion process, the generation of energy from FRB does not contribute to an increase in greenhouse gasses (Hakkila & Parikka, 2002; IDAE, 2005a). The social and economic benefits can be separated into two categories: (i) those that occur at the national scale, associated with a reduction in the use of non-renewable fossil fuels (Domínguez, 2002; Eriksson et al., 2002; IDAE, 2005a), and (ii) those that occur at the local scale, as the utilization of a waste product leads to increasing harvests, transportation, and utilization of forest residues in power stations, which also leads to an increase in rural employment. These considerations are important for rural areas in which the level of unemployment and depopulation is a public policy issue, and where increased employment can help to support a population that can maintain the natural environment (Borsboom et al., 2002; Domínguez, 2002; IDAE, 2005a, 2007).

Despite the benefits outlined above, previous studies conducted before the subsequent revision of PPRE, the Plan to Renewable Energies 2005–2010 (PRE), showed that the use of FRB had a long way to go in achieving the anticipated objectives. In the PRE analysis, the lack of specific methodologies to assess the regional-scale quantity of FRB was identified as one of the main problems (IDAE, 2005a). This is a fundamental concern because power stations that use FRB require knowledge of the amount of resource available (Esteban et al., 2004; IDAE, 2005b, 2007; Velázquez, 2006). To overcome this problem, a methodology is needed that quantifies the potential of this forest resource in a given area. This is an important matter considering that the new Plan to Renewable Energies in Spain 2011-2020 (currently being written using the provisions of Directive 2009/28/EC) set a final energy consumption target of 20% derived from renewable energy, doubling the previous mark in the 2010 Plan (MITYC, 2010).

Several studies have demonstrated the effectiveness of satellite images in estimating forest variables, including biomass. These studies were carried out using both passive (optical) and active (radar and lidar) sensors. Focusing on optical sensors, different experiments have been conducted using multispectral or hyperspectral coarse spatial resolution data (i.e. Anaya et al., 2009; Muukkonen & Heiskanen, 2007) and fine spatial resolution data (i.e. Gonzalez et al., 2010; Proisy et al., 2007); although the most frequent experiments have been performed using medium spatial resolution data, mainly using Landsat Thematic Mapper (TM) or Enhanced Thematic Mapper Plus (ETM+) data (Lu, 2006; Powell et al., 2010). Radar data has also been used extensively, establishing significant correlations between biomass and radar backscatter at C-band (Kurvonen et al., 2002) and L-band (Austin et al., 2003; Lucas et al., 2010), with the latter being more consistent and robust. Recently, forest biomass assessment research has focused on using polarimetric synthetic aperture radar interferometry (PolInSAR) and laser scanning systems (lidar) data, mostly by means of airborne sensors. Koch (2010) offers a complete review of the recent advances and future developments in this issue.

Images derived from remote sensing register continuous and complete information across a landscape and such images can be obtained at frequent intervals. These characteristics help to overcome some of the problems associated with inventory methods exclusively based on field work, interpolation techniques and GIS (Franklin, 2001; Lu, 2006; Salvador & Pons, 1998a).

Thus, remotely sensed data have not only facilitated an increase in the speed, cost efficiency, precision, and timeliness of inventories, but they have also allowed the construction of maps of forest attributes with spatial resolutions and accuracies that were not feasible even a few years ago (McRoberts & Tomppo, 2007). These advances have made remotely sensed data the primary source for biomass estimation (Lu, 2006). However, no studies have yet presented a technique for biomass estimation that is consistent, reproducible and entirely applicable at regional scales (Muukkonen and Heiskanen, 2005; Powell et al., 2010), especially in forests located in areas of irregular topography and characterized by heterogeneity in species composition, complex stand structures and environmental conditions (Brown et al., 1999; Lu & Batistiella, 2005; Lu, 2006; Mallinis et al., 2004;). Those characteristics are inherent features of Mediterranean forests (Shoshany, 2000), where not many studies have been carried out.

The objective of this chapter is to explain a methodology developed to estimate the amount of FRB potentially suitable for renewable energy production in the pine forests of Mediterranean areas at regional scale, using satellite images and forest inventory data. It is intended, therefore, to eliminate a major barrier to the use of this renewable source of energy. In turn, by using a plain methodology, it is intended that the method developed can be adopted by decision makers and land managers for both forest management and regional planning, considering that energy planning is a major component of land management.

For this study, we used FRB data obtained from allometric equations applied to the Second Spanish National Forest Inventory (NFI-2) (dependent variable) and spectral data from Landsat 5 TM imagery (independent variable). In order to avoid the effects of forest heterogeneity in the establishment of accurate predictive models, different methods were tested to extract the spectral data from the Landsat images.

2. Materials and methods

The methodology applied in this study was developed in the framework of geographic information technologies (Remote Sensing and Geographical Information Systems, GIS) as information sources and tools for forest management. The software Erdas Imagine was used to process the Landsat images, ARC-GIS was used for the management of ancillary information, and SPSS was employed for statistical analyses.

The methodology proposed and developed in this study was divided into five phases or steps. Firstly, FRB data was obtained by means of field work and pre-existing forest data. Secondly, suitable satellite images were selected and different treatments were applied in order to convert the data to a suitable format for quantitative analysis and to guarantee the validity of the results. Thirdly, three different methods were used to relate field data and radiometric information in order to overcome the difficulties involved in estimating forest parameters in heterogeneous Mediterranean forests. The fourth phase focused on developing and running regression models to map FRB in the study area, and to evaluate and compare the results at pixel level obtained using the applied extraction methods. Finally, the best model was selected for application to a recent image to quantify the amount of FRB at present time in the study area.

2.1 Study area

The study area is the Province of Teruel, which is located in the northeastern Iberian Peninsula (Figure 1). This province was selected in the context of the LIGNOSTRUM project

(2002-2005), of which the present research was a component. The aim of the Spanish government-financed LIGNOSTRUM project was to increase the use of agricultural and forest residues as energy resources. There were two main reasons for selecting this territory as a setting to measure the benefits of energy use of forest residues: (i) the presence of large agricultural and forest areas; and (ii) the existence of economically disadvantaged rural areas.

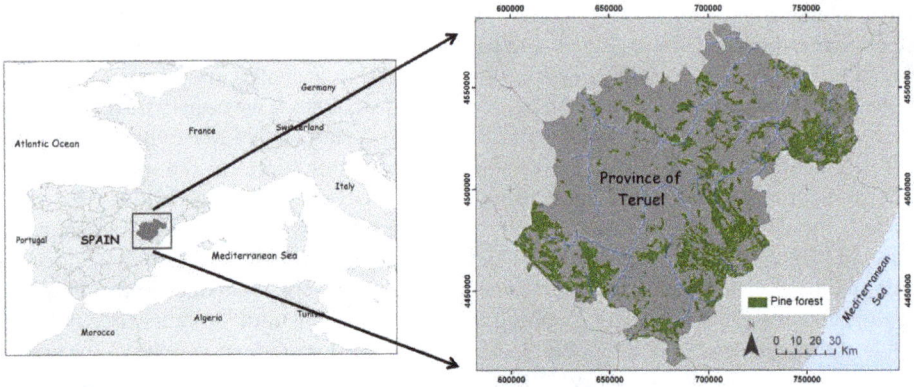

Fig. 1. Location of the study area

There are three characteristics of this province that mean difficulties to achieve the objective of estimate FRB using satellite images: (i) the size of the province (14,804 km²), (ii) the concentration of forest resources in the mountain areas (as the topography modifies the reflectance values of equal ground covers), and (iii) the Mediterranean characteristics of these forests (presence of multiple species with a complex spatial structure resulting from multiple interrelationships between different ecological factors and a history of human activity extending back for many centuries). These peculiarities determined the types of images that were used, the treatments applied to the images and the methods applied to extract the radiometric data. Among all types of forests present in the study area, only pine forests were selected for study since they represent slightly over 71% of the total forest area of the province and they have the greatest potential for FRB generation from harvest (IDAE, 2007; MMA, 1996).

2.2 FRB data

Aboveground biomass (AGB) is normally estimated using allometric equations that relate it to tree data, usually diameter at breast height (DBH) and height (Ketterings et al., 2001; Montero et al., 2005; Pilli et al., 2006; Wang, 2006; Zianis & Mencuccini, 2004). These equations show a large variation related to tree species, geographic location (Ketterings et al., 2001; Montero et al., 2005), and other factors such as climate and tree population age (Zianis & Mencuccini, 2004). Few biomass equations have been developed for Spain (Montero et al., 2005), with even fewer focusing on residual biomass (in some cases, crown biomass has been assessed). Consequently, it was necessary to develop specific FRB regressions for each pine species in the study area for application to pre-existing tree dimensional data.

At the start of the LIGNOSTRUM project in 2002 the most recent, accurate, and complete information on the dimensional parameters of trees in the forests of the study area was found in the Second National Forest Inventory (NFI-2), which was completed in 1994 (MMA, 1996). The Spanish NFI is carried out every 10 years as a systematic field sampling of permanent plots. These plots are located in the corners of the UTM 1x1 km grid of the 1:50,000 National Topographic Cartography that are within forested areas. The placement of plots in the field was performed using georeferenced 1:30,000 aerial photographs and topographic cartography. Plots have a circular shape with radii ranging from 5 m to 25 m depending on the DBH of the trees; only trees with a DBH greater than 7.5 cm were considered. The NFI-2 fieldwork in Teruel Province was undertaken from March to August 1994, sampling a total of 2083 plots. The DBH (two perpendicular diameters measured at a height of 1.30 m) and total height (accurate to within half a meter) were measured for all trees. A detailed description of NFI-2 is provided in the summary report of the inventory (MMA, 1996).

Taking into account these characteristics of NFI-2, we devised a species-based sampling strategy to obtain FBR regressions. We located all the sample areas in forests managed by the Teruel Environment Service owing to the destructive nature of this sampling. Fieldwork was carried out from November 2003 to June 2004. For each species, the distribution of sampling points was proportional to the number of trees, the volume of wood, and the stand basal area documented in NFI-2 data. We considered a normal diameter range from 7.5 to 40 cm because these are the minimum diameters recorded in NFI-2 data and they are the most frequently observed in the tables of production of the study area.

In the field, we used a scale with an accuracy of 250 grams to obtain the wet weight of the residual biomass. In addition, we took two size measurements for each sampled tree: DBH (accuracy of 5 mm) and height (centimeter accuracy). Finally, to avoid the influence of the variations in the degree of wetness in the samples, we collected samples of leaves and branches in the different trees in order to calculate the dry weight by applying the method described by Joosten et al. (2004).

The total number of sampled trees was 186 (30 of *Pinus sylvestris*, 59 of *Pinus halepensis*, 57 of *Pinus nigra*, and 40 of *Pinus pinaster*). We obtained a regression equation for each species, but two in the case of *Pinus pinaster* regarding the tree origin (natural or reforested) (Table 1).

Species	N° of sampled trees	Models	R^2_a	Std. Desv. (kg)	Weight (X-k)
P. sylvestris	30	$FRB=0.064 \cdot DBH^{3.3}/H^{1.5}$	0.974	12.298	-
P. nigra	57	$FRB=338.416 \cdot e^{-35.116/DBH}$	0.910	18.836	$1/DBH$
P. halepensis	59	$FRB=0.067\ DBH3/H$	0.969	13.063	$1/(DBH*(1+ORI))$
P. pinaster natural	28	$FRB=1.101 \cdot 10^{-3} \cdot DBH^4/H$	0.973	6.004	-
P. pinaster reforested	12	$FRB=1.97 \cdot 10^{-4} \cdot DBH^{3.823} \cdot H^{0.337}$	0.974	12.175	-

Table 1. Regression equations obtained by species, R^2 adjusted, standard error associated and weight expression to accomplish the assumption of residual homocedasticity. Source: Alonso et al., (2005), where FRB is forest residual biomass, DBH is tree diameter at breast height, H is tree height, and ORI is the tree origin (0=reforested; 1=natural)

In applying the obtained FRB equations to the NFI-2 plots, we only chose those in which pines were among the range used in the equations, thus obtaining 617 plots. These residual biomass data were linked to a table with all of the other NFI-2 variables, including their locations in UTM coordinates. To avoid complexity associated with the mixture of tree species in the spectral data of the Landsat image, we selected only monospecies pine plots (Salvador & Pons, 1998b). In addition, we removed from the dataset plots located in an area affected by a wildfire that occurred in 1994 and those affected by cloud shadows in the southeast corner of the Landsat image. Following this selection process, a total of 482 plots were finally obtained. We created a point map with these plots to read the Landsat TM spectral information. The estimated residual biomass in selected plots ranged from 0.107 to 64.720 tons/ha. This large range is related to the high degree of heterogeneity of Mediterranean forests.

2.3 Image pre-processing and development of new spectral indices

Among the different types of remote sensing data available to achieve the objective of this research, Landsat images were selected because they are one of the most common in forestry-related applications and estimates of aboveground biomass (AGB) at regional-local scales (i.e. Fazakas et al., 1999; Foody et al., 2001, 2003; Gasparri et al., 2010; Hall et al., 2006; Labrecque et al., 2003, 2006; Lu, 2005; Lu & Batistiella, 2005; Lu et al., 2004; Mäkelä & Pekkarinen, 2004; Mallinis et al., 2004; Meng et al., 2009; Powell et al., 2010; Roy & Ravan, 1996; Salvador & Pons, 1998a, 1998b; Steininger, 2000; Tangki & Chappell, 2008; Wulder et al., 2008; Zheng et al., 2004, 2007). In addition, taking into account the research objectives, there were two other important reasons for using Landsat images. Firstly, essentially all of the study area was represented in one scene, an advantage over other approaches in which errors are introduced owing to inconsistencies in radiometric corrections, necessary when multiple scenes are used (Chuvieco, 2002; Foody et al., 2003). Secondly, the methodology would be useful and replicable in other areas with similar characteristics because of the systematic and frequent coverage of Landsat images over time as well as their ease of distribution at no cost to users via the Internet (Chander et al., 2009).

The study used two Landsat 5 TM images recorded on 10 June 1994 and 5 July 2008 (Path 199/Row 32). The earlier image was selected on the basis of temporal coincidence with NFI-2 field work, so it was useful in performing the methodology, whereas the later image was chosen for a recent FRB estimation and inventory in the study area.

We applied three pre-processing techniques to both images to obtain radiometric variables consistent and suitable to be related to field data, and to guarantee the applicability of the estimation model carried out with the 1994 image to the 2008 image: (i) geometric correction, (ii) radiometric correction, and (iii) transformations and ratios. The importance of accurate geometric rectification in the present research arises from the fact that the image data were linked with ground data. Both Landsat TM images were geometrically rectified into a local UTM projection using a second-order polynomial model. The rectification model incorporated a Digital Elevation Model (DEM) with 25 m spatial resolution. Ground Control Points (GCPs) were taken from high-resolution ortho-photographs (pixel size = 1 x 1 m). A total of 125 GCPs were used to re-project each image with an estimated Root Mean Squared Error (RMSE) of 0.60 and 0.48 pixels, respectively. A nearest-neighbor resampling technique was used to minimize changes in the radiometric values of the ground data, with the pixel re-projected to 25 m. The radiometric correction addresses atmospheric and topographic effects in the surface reflectivity registered in both Landsat images. In this process, the relative data on surface reflectivity given in digital numbers (DN) is transformed into an absolute form in reflectance values. In areas of rough terrain, as in the present study area, a good topographic normalization is required to compensate for the varying solar illumination associated with the irregular shape of the terrain. To accomplish this, first, the default transmittance method, proposed in Chavez (1996), was applied to eliminate the atmospheric effects present in both Landsat images. Second, conversion from DN values to reflectance measurements was carried out using the Minnaert Correction Method (Colby, 1991). Different transformations (determine in previous work focusing on forest parameter estimation) were applied to the 1994 image to increase the amount of spectral information, including different vegetation indices. Figure 2 shows the location of the 482 NFI-2 plots over a false colour band combination 7/4/3 (RGB) of the study area.

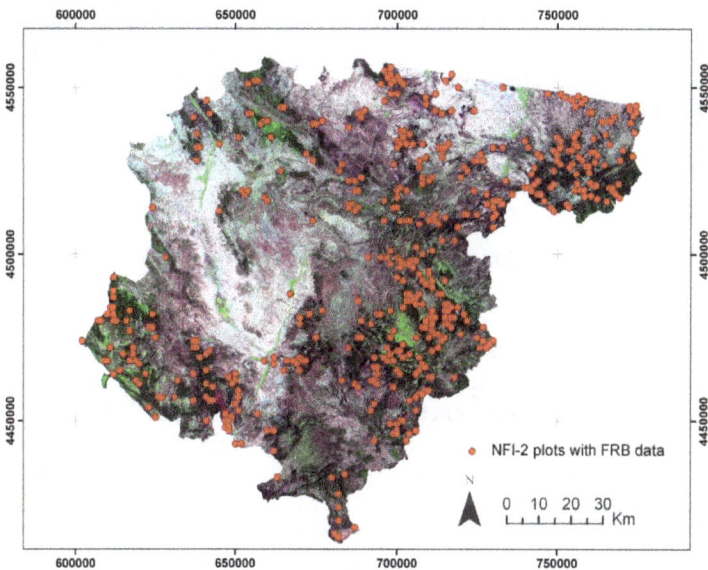

Fig. 2. Location of the 482 NFI-2 plots over a false colour band combination 7/4/3 (RGB) of the study area

2.4 Linking FRB ground data to 1994 radiometric data

As mentioned previously, heterogeneity is an inherent characteristic of Mediterranean forests and is one of the main factors that complicates forest parameter estimation in this area using remote sensing images. This is mainly because spatial heterogeneity is registered in the satellite image, resulting in a large spectral variability in the spectral response of the areas covered by these forests. This introduces difficulties in building accurate predictive models. As a result, two NFI-2 plots with equal amounts of FRB can show different reflectance values because of the presence of other landscape elements (i.e. firebreaks, trails), the position of the FRB plot on the border of another land cover type (i.e. scrublands, farmlands) or because of high variability within the forested area (presence of different tree species, ages). In addition, other problems detected in previous research in Mediterranean forests are inaccuracies in the localization of inventory field plots, small plot sizes, and the small number of plots used in the analysis (Mallinis et al., 2004; Maselli & Chiesi, 2006; Salvador & Pons, 1998a, 1998b; Shoshany, 2000; Vázquez de la Cueva, 2005).

The present study tested three different methods to extract the radiometric data in order to overcome the problems outlined above and achieve accurate FRB regression models: (i) fixed pixel windows or kernels, (ii) visual analysis, and (iii) spectral segmentation.

2.4.1 Fixed pixel windows

The use of kernels or fixed pixel windows greater than one pixel is the most common method to extract radiometric information in works focused on forest parameter estimation by means of satellite images using field plots data (i.e. Foody et al., 2001; Hall et al., 2006; Labrecque et al., 2003; Lu et al., 2004; Lu, 2005; and Lu & Batistiella, 2005 Roy & Ravan, 1996). In our case, taking into account the high variability of Mediterranean forests, a 3 x 3 pixel window centered on each plot was used to extract the mean value and standard deviation for all radiometric variables derived from the 1994 image. Following the methodology used in Labrecque et al. (2003), the variability in each plot was calculated over the six spectral reflectance bands using Pearson's coefficient of variation (CV) (1):

$$CV = \frac{S}{|\overline{X}|} \qquad\qquad (1)$$

where S is the standard deviation calculated for the 3 x 3 pixel window centered on each plot and X is the average calculated for the same window.

Thus, information is obtained regarding the degree of spectral heterogeneity within the immediate surroundings of each plot: plots with high CV have a high degree of spatial heterogeneity, while plots with low CV are the most homogeneous. With the aim of determining the influence of spectral heterogeneity on estimates of FRB, the CV values were used to divide the sample into 10 homogeneity levels. Initially, we calculated the thresholds needed to delimit the 10 CV percentiles in each reflectance band and, after that, these thresholds were used to create 10 groups of plots as follows: the first group contained all of the plots, the second contained only those plots whose CV values were lower than that of percentile 9 in all TM bands, the third contained only those plots whose CV values were lower than that of percentile 8 in all of the TM bands, and so on, until the 10 groups were delineated. As a result, the first of the delineated groups contains the plots with the highest degree of spectral heterogeneity, while the last contains the plots that are the most homogeneous. For the rest of this chapter, these plot groups are named using the CV percentile employed in creating them. Figure 3 shows the differences regarding the degree of spectral heterogeneity within the immediate vicinity among one plot classified in group CV9 and another classified in group CV4.

2.4.2 Visual analysis

In the second method to extract the radiometric data from Landsat image, we use visual image analysis to delimit homogeneous FRB units larger than the NFI-2 plots. Different studies have demonstrated that biomass can be estimated more accurately if the ground data is referenced over areas larger than the pixel level using similar units in forest and spectral characteristics (Hyyppä & Hyyppä 2001; Muukkonen & Heiskanen, 2005; Zheng, 2004). In addition, this method avoids some problems related to the use of fixed windows, as they do not completely eliminate errors in image registration and location of the sample plots, may intersect several stands with different spectral and forest characteristics, and not fit well with the limited spatial and spectral resolution of Landsat data (Mäkelä & Pekkarinen, 2001; Mäkelä & Pekkarinen, 2004; Muukkonen & Heiskanen, 2005; Pekkarinen, 2002).

FRB cartography with stands larger than NFI-2 plots did not exist in the study area; however, the interpretation of high-resolution aerial photographs has shown to be useful in forest-inventory applications such as stratification, volume estimation, and measurements of forest characteristics (Lu, 2006; Mäkelä & Pekkarinen, 2004). In this context, the high-resolution ortho-photographs used in geometric rectification were applied to extend the plot areas to larger sizes with visually similar composition and forest structure. This was performed using a "heads-up," on-screen digitizing technique within a GIS application. The selected NFI-2 plots, with their respective radii size, were displayed over the composite digital aerial photograph with 1 m spatial resolution. Where possible, larger homogeneous areas containing the in situ plots were then defined. Aragon 1:50,000 digital forest cartography was used to guarantee the pure composition of the new areas. To avoid errors in the results, only homogeneous areas with a high degree of similarity between observations from the aerial photographs of the NFI-2 plots were selected. As a result, a total of 131 areas were selected. In addition, when extracting mean values, only those pixels located in the core areas of the delimited stand were selected. Pixels located on the borders were avoided, as these can reflect properties of landscape elements in the immediate vicinity of the delimited stands (Figure 4).

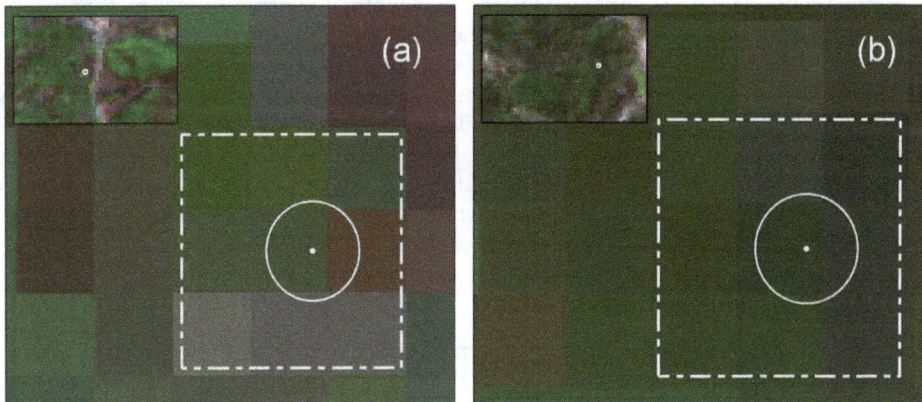

Fig. 3. (a) 3 x 3 pixel window centred on a plot with high heterogeneity (high CV); (b) 3 x 3 pixel window centred on a plot with low heterogeneity (low CV)

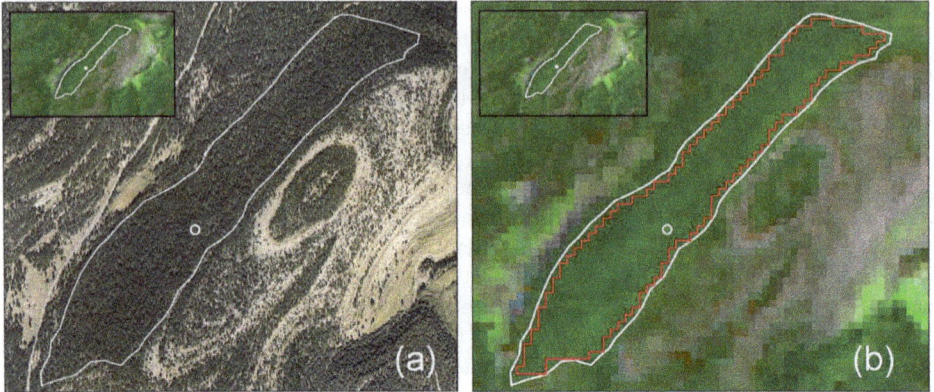

Fig. 4. (a) Homogeneous area delimited using high-resolution aerial photographs; (b) Core pixels of the homogeneous area used to extract spectral data from the Landsat TM image

2.4.3 Spectral segmentation

The main problems in constructing predictive models related to the extraction of radiometric data using high-resolution aerial photographs of delimited forest area are: (i) human errors in the establishment of limits during the visual image interpretation process, (ii) assumption of a FRB constant value for the entire homogeneous area based on a point location that is not representative of larger forest stands (Pekkarinen, 2002).

Image segmentation can be applied to delimit homogeneous spectral features as a reference for spectral data extraction (Hall et al., 2006; Mäkelä & Pekkarinen, 2001; Pekkarinen, 2002). In this study, the segmentation algorithm RGB clustering incorporated in the software Erdas Imagine was applied. The essential parameters that control the results in this method are the RGB composition (which is used to apply the segmentation), the stretch applied to each band in the composition, and the number of bins into which each band is divided. Four different segmentations were performed, modifying the stretch (2 or 4 standard deviations) and the bins (7-6-6 and 4-3-3) using a RGB composition TM7-TM4-TM3 to model the heterogeneity of the analyzed forest (Table 2). In addition, in order to remove pixels within the immediate vicinity that were not attributable to the IFN-2 plot, the obtained image segments containing field plots were then intersected with 3 x 3 pixel windows, resulting in a mask of homogeneous pixels. Thus, the mean pixel value for each image band was then computed from the pixels that belong to the same spectral category as that of the central pixel of the NFI-2 plot (Figure 5).

Segmentation	Number of Std. dev. in each band	Number of bins		
		Red	Green	Blue
S1	2	7	6	6
S2	4	7	6	6
S3	2	4	3	3
S4	4	4	3	3

Table 2. Parameters considered in the four different segmentations performed using RGB composition TM7-TM4-TM3

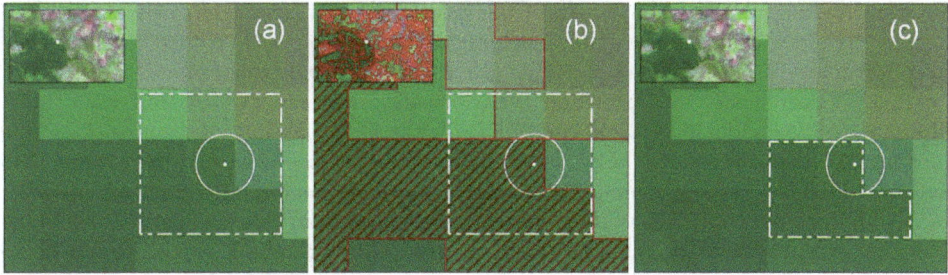

Fig. 5. Procedure to extract radiometric data combining the use of segmentation and 3 x 3 pixel windows with restrictions

2.5 FRB mapping and validation method

We used Pearson's coefficient to explore the feasibility of building accurate and representative predictive models using the plot groups derived from the three extraction methods. After this selection, each of the three groups was divided into two samples: 80% of the sample was used to carry out the predictive model and the remaining 20% was used to validate it. This sample division in each group was made randomly to guarantee the execution of the estimation equations and the validation processes. Moreover, to assess the robustness of the models, the sample division was done five times, completing the respective estimation model and its validation each time.

The multiple linear regression model was performed, using the option "stepwise" to include only the significant variables ($p < 0.05$). In addition, performance was verified for all of the principles assumed for this type of regression at the model and variable level. To evaluate the performance of models, the coefficient of determination (R^2) was used, and the Root Mean Square Error (RMSE) and the relative RMSE ($RMSE_r$) were calculated using 20% of the sample previously reserved for the validation. Finally, the best model conducted with each of the extraction methods was applied to the 1994 Landsat data in order to obtain the FRB estimation cartography. The three derived cartographies were validated using plots not included in the groups considered in the regression models, and the RMSE and $RMSE_r$ were calculated to evaluate the results.

2.6 Model application and inventory

The results obtained with the June 1994 Landsat image (selected on the basis of its temporal coincidence with NFI-2 fieldwork) at the regression model level and in the cartography validation (in terms of R^2 and RMSE and $RMSE_r$, respectively) were analyzed. Then, the most suitable estimation model was selected for application to the July 2008 Landsat image. As a result, current information was obtained about the potential quantity of this energy resource and its spatial distribution in the study area.

3. Results

3.1 Models run using fixed pixel windows to extract the radiometric information

Table 3 shows the correlation coefficients tendency of the original bands and some of the variables derived from them correlated to FRB in the first nine groups delimited using the CV. As expected, higher correlations were obtained with increasing spectral homogeneity and a decreasing number of plots. All of the groups showed significant correlations, with the majority of the considered variables yielding p-values of generally $p < 0.05$, with the one exception being the first CV group, as this group contained only three plots. In all groups, the highest coefficients of correlation were obtained for variables related to wetness information (TM band 5 –TM5-, TM band 7 –TM7-, the third principal component -PC3-, the third tasseled cap component -TC3-, the moisture stress index –MSI-, and the sum of TM5 and TM7 -MID57-), reaching similar coefficients than the Normalized Difference Vegetation Index (NDVI). The coefficients of these variables increased from values of 0.450–0.460 in the group of percentile 10 to more than 0.850 in the group of percentile 3 with all results being statistically significant.

	CV10	CV9	CV8	CV7	CV6	CV5	CV4	CV3	CV2
N° plots	482	381	285	208	149	111	68	36	14
TM1	-0.435*	-0.493*	-0.500*	-0.529*	-0.550*	-0.513*	-0.542*	-0.708*	-0.647*
TM2	-0.409*	-0.468*	-0.470*	-0.484*	-0.499*	-0.454*	-0.495*	-0.713*	-0.638*
TM3	-0.413*	-0.474*	-0.477*	-0.492*	-0.512*	-0.464*	-0.512*	-0.734*	-0.673*
TM4	-0.199*	-0.257*	-0.232*	-0.189*	-0.163*	-0.110	-0.213	-0.267	-0.573*
TM5	-0.451*	-0.524*	-0.521*	-0.552*	-0.576*	-0.552*	-0.641*	-0.793*	-0.791*
TM7	-0.452*	-0.521*	-0.523*	-0.562*	-0.603*	-0.571*	-0.639*	-0.788*	-0.780*
PC1	-0.429*	-0.498*	-0.493*	-0.509*	-0.528*	-0.489*	-0.560*	-0.737*	-0.743*
PC2	-0.032	0.012	-0.037	-0.087	-0.153	-0.194*	-0.115	-0.213	0.410
PC3	0.421*	0.474*	0.508*	0.590*	0.656*	0.663*	0.754*	0.869*	0.853*
TC1	-0.414*	-0.482*	-0.475*	-0.483*	-0.493*	-0.451*	-0.521*	-0.707*	-0.721*
TC2	0.310*	0.334*	0.371*	0.436*	0.494*	0.506*	0.546*	0.664*	0.425
TC3	0.453*	0.525*	0.541*	0.603*	0.654*	0.645*	0.750*	0.852*	0.852*
NDVI	0.457*	0.510*	0.525*	0.587*	0.634*	0.605*	0.684*	0.807*	0.737*
SAVI	0.455*	0.507*	0.523*	0.585*	0.632*	0.603*	0.682*	0.805*	0.735*
MSI	-0.458*	-0.518*	-0.540*	-0.618*	-0.669*	-0.674*	-0.772*	-0.883*	-0.864*
MID57	-0.454*	-0.525*	-0.525*	-0.558*	-0.590*	-0.562*	-0.641*	-0.792*	-0.787*

Table 3. Pearson's coefficient of correlation (R) between spectral variables and plots at the level of 3 x 3 pixel windows (* correlation is significant at the 0.05 level); TM, Thematic Mapper; PC, principal component analysis; TC, tasseled cap transform; NDVI, normalized difference vegetation index; SAVI, soil adjusted vegetation index; MSI, moisture stress index; MID57, sum of middle infrared wavelengths

These results show that the degree of spectral heterogeneity determines the feasibility of building accurate predictive models. Thus, if the groups with more plots are used, the models will have low prediction capacity. By contrast, if the groups with lower numbers of plots are used, the models will have higher predictive capacity, but, in turn, the models will be biased to the sample, being not representative of all of the forests of Teruel Province. Therefore, we selected the CV4 group because, among the groups characterized by an elevated homogeneity of plots (allowing execution of models with high R^2), it is the one with the most number of plots (68). Scatter plots comparing FRB and the spectral variables clearly reveal nonlinear relationships. As a result, the most suitable standard transformation was applied to the independent and dependent variables in order to guarantee the linearity principle in the multiple linear regression model (logarithmic –ln-, square root –sq-, exponential –exp- and inverse –inv-) (Hair et al., 1998). Table 4 shows linear regression models performed with each one of the five random divisions of its sample. All of them include only one variable, which is always related to water content in vegetation. The lack of a second variable in these models is due to the high degree of auto-correlations between the spectral variables. This avoids violation of one of the principles of multiple linear regression models. Even so, the R^2 values achieved were higher than 0.7. Consequently, they are suitable for FRB estimation in our study area. The results obtained from validation using 20% of the sample showed that the model run with MID57 (TM5+TM7) is the most suitable, because the $RMSE_r$ was only 26.67%, significantly lower than the others. As a result, it was selected to obtain FRB cartography with the 1994 Landsat image, using the pine forest areas in the Aragon 1:50,000 cartography as a mask (Figure 6).

Model	Variable	R^2_a	β_0	β_1	RMSE (ton/ha)	$RMSE_r$ (%)
N_1	ln_MID57	0.711	18.879	-4.663	4.843	26.67
N_2	ln_TM5	0.713	17.933	-5.073	4.591	43.38
N_3	ln_TM5	0.750	17.900	-5.053	4.767	51.91
N_4	ln_TM7	0.735	12.603	-3.906	6.144	42.92
N_5	ln_TM5	0.750	18.253	-5.204	4.792	34.93

Table 4. Linear regression models obtained from 3 x 3 pixel windows

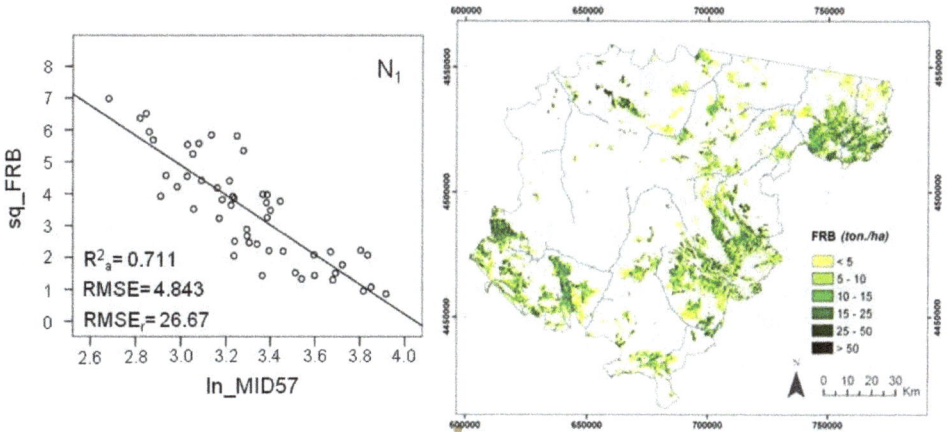

Fig. 6. The regression model selected using fixed pixel windows and the FRB map produced applying it

3.2 Models completed using visual analysis to extract the radiometric information

Correlation analysis completed using homogeneous areas derived from visual analysis showed that spectral variables related to wetness yielded the highest Pearson coefficients. Among them, TM5, TM7, TC3, MSI and MID57 were identified as the best predictor variables, with an R > 0.68. In addition, vegetation indices were again the second group of variables that showed high correlation levels, with NDVI showing the strongest correlation

(Table 5). Finally, note that scatter plots comparing FRB and spectral variables also showed nonlinear relationships; therefore both independent and dependent variables were transformed using standard transformation before introducing them in the SPSS software for the linear regression analysis.

Variable	R	Variable	R
TM1	-0.633*	PC3	0.678*
TM2	-0.628*	TC1	-0.648*
TM3	-0.638*	TC2	0.537*
TM4	-0.330*	TC3	0.708*
TM5	-0.693*	NDVI	0.652*
TM7	-0.688*	SAVI	0.651*
PC1	-0.665*	MSI	-0.708*
PC2	-0.174*	MID57	-0.692*

Table 5. Pearson's coefficient of correlation (R) between spectral variables and homogeneous areas delimited by visual analysis (* correlation is significant at the 0.05 level)

Consequently, it is possible to run estimation models using visual homogeneous areas with similar accuracy level as with fixed windows because the correlation coefficients are comparable. Therefore, the main difference in the analysis made using visual homogeneous data comes from the number of plots included, which is nearly double that of the group of percentile 4 (131 versus 68). As a result, the probability of constructing an over-fitted model is reduced (Hair et al., 1998). Therefore, this method to extract radiometric data appears to be more suitable to perform regression models for FRB estimation, as these models could be more representative of all environmental characteristics of Teruel pine forests.

Only one of the five models successfully included more than one independent variable (Table 6), owing to the high auto-correlation between them. The variable chosen in each regression attempt depends on sample divisions; but in all cases, this variable was related to wetness. The model run with the sample N_3 was selected to produce a map (Figure 7), as it showed good conciliation between its R^2 and its $RMSE_r$. In addition, this model allowed direct comparison to the previous extraction method because both models use the same radiometric variable (MID57).

Model	Variable	R^2_a	β_0	β_1	β_2	RMSE (ton/ha)	$RMSE_r$ (%)
N_1	ln_MID57	0.610	16.822	-3.960	-	12.207	64.95
N_2	ln_TM5	0.562	16.625	-4.541	-	7.497	41.78
N_3	ln_MID57	0.595	17.675	-4.191	-	8.839	59.48
N_4	MSI, inv_TM4	0.579	6.649	-5.909	45.999	8.079	54.32
N_5	ln_TM5	0.558	17.492	-4.838	-	11.238	56.68

Table 6. Linear regression models obtained using homogeneous areas delimited by visual analysis

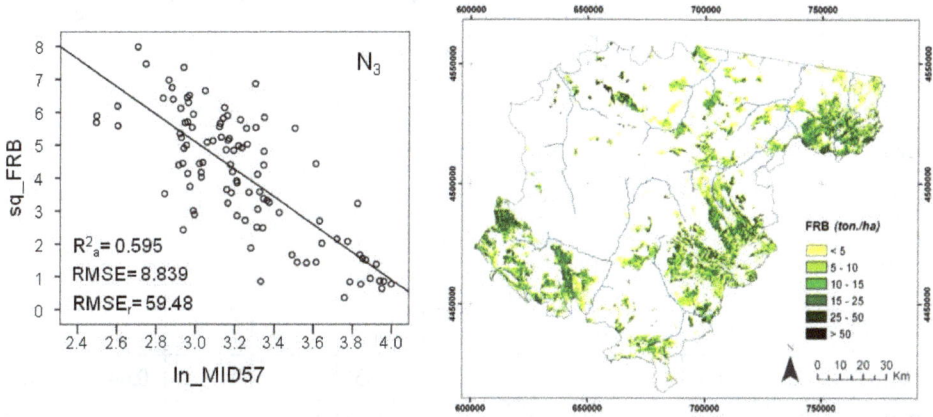

Fig. 7. The regression model selected using homogeneous areas delimited by visual analysis and the FRB map produced applying it

3.3 Models carried out using spectral segmentation to extract the radiometric information

The correlation analysis using the data derived from the third extraction method was only performed using the plot groups CV7 and CV6 for two reasons: (i) to directly reject in the analysis the plots with high probability of containing radiometric data related to different spectral features than the FRB data; and (ii) to identify a regression model with similar R^2 values than were derived from the use of fixed 3 x 3 windows, but using a higher number of plots, making it more representative of the entire study area. In addition, only NFI-2 plots inside areas classified as pine forest in the Aragon 1:50,000 forest map were considered for this analysis. As is shown in Table 7, the radiometric variables related to wetness were again the most correlated. In addition, it was shown that the correlation coefficients increase as the homogeneity in the sample increases, independently of the performed segmentation. Focusing on these results, only the S4 segmentation in group CV6 shows higher regression coefficients than using a fixed 3x3 window without restrictions. Concretely, the maximum R value S4 was reached using the variables TM7 and MID57 (R>0.7).

Nonlinear relationships between the dependent variable (FRB) and the independent variables (radiometric data) were again revealed; thus transformations were applied. Table 8 shows the regression models that were obtained using plots within group CV6. Four models were performed using a wetness radiometric variable. This type of variable was also selected initially in the only one model that used more than one variable. The R^2 coefficients were situated between 0.53 and 0.59 and only one model indicated an $RMSE_r$ lower than 40% in its validation. Owing to the limited difference in terms of R^2, this model with the lowest $RMSE_r$ was selected to derive FRB cartography for the study area (Figure 8).

	Segmentation S1		Segmentation S2		Segmentation S3		Segmentation S4		Fixed 3x3 without restrictions	
	CV7	CV6	CV7	CV6	CV7	CV6	CV7	CV6	CV7	CV6
N° plots	182	130	182	130	182	130	182	130	182	130
TM1	-0.595*	-0.616*	-0.600*	-0.621*	-0.599*	-0.617*	-0.606*	-0.628*	-0.598*	-0.630*
TM2	-0.585*	-0.617*	-0.580*	-0.613*	-0.573*	-0.609*	-0.579*	-0.619*	-0.585*	-0.632*
TM3	-0.594*	-0.638*	-0.591*	-0.640*	-0.588*	-0.629*	-0.597*	-0.641*	-0.602*	-0.657*
TM4	-0.224*	-0.182*	-0.243*	-0.210*	-0.220*	-0.192*	-0.224*	-0.188*	-0.240*	-0.205*
TM5	-0.622*	-0.678*	-0.642*	-0.696*	-0.635*	-0.684*	-0.638*	-0.696*	-0.650*	-0.698*
TM7	-0.605*	-0.683*	-0.625*	-0.696*	-0.621*	-0.685*	-0.632*	-0.704*	-0.641*	-0.701*
PC1	-0.592*	-0.653*	-0.608*	-0.663*	-0.601*	-0.651*	-0.606*	-0.664*	-0.615*	-0.668*
PC2	-0.059	-0.131	-0.063	-0.147	-0.084	-0.156	-0.082	-0.154	-0.082	-0.159
PC3	-0.563*	-0.621*	0.579*	0.634*	0.573*	0.621*	0.591*	0.636*	0.605*	0.649*
TC1	-0.578*	-0.629*	-0.588*	-0.637*	-0.581*	-0.624*	-0.584*	-0.636*	-0.595*	-0.643*
TC2	0.437*	0.477*	0.443*	0.484*	0.445*	0.479*	0.453*	0.484*	0.449*	0.491*
TC3	0.609*	0.670*	0.629*	0.694*	0.622*	0.678*	0.636*	0.696*	0.654*	0.700*
NDVI	0.626*	0.665*	0.627*	0.660*	0.626*	0.655*	0.634*	0.657*	0.630*	0.665*
SAVI	0.624*	0.663*	0.625*	0.657*	0.624*	0.653*	0.631*	0.655*	0.628*	0.663*
MSI	-0.592*	-0.640*	-0.629*	-0.667*	-0.617*	-0.652*	-0.633*	-0.667*	-0.643*	-0.676*
MID57	-0.618*	-0.684*	-0.638*	-0.699*	-0.632*	-0.687*	-0.639*	-0.702*	-0.648*	-0.701*

Table 7. Pearson's coefficient of correlation (R) between spectral variables and data obtained using segmentation and 3x3 pixel windows with restrictions and without restrictions (* correlation is significant at the 0.05 level)

Model	Variable	R^2_a	β_0	β_1	β_2	RMSE (ton/ha)	$RMSE_r$ (%)
N_1	TC3, ln_TM1	0.547	7.746	0.303	-1.087	9.064	44.95
N_2	ln_MID57	0.535	16.409	-3.888	-	5.657	35.58
N_3	ln_TM7	0.560	11.395	-3.411	-	8.840	50.07
N_4	ln_MID57	0.596	16.563	-3.928	-	9.688	66.67
N_5	ln_MID57	0.533	16.217	-3.846	-	8.231	51.01

Table 8. Linear regression models obtained using segmentation and 3 x 3 pixel windows with restrictions

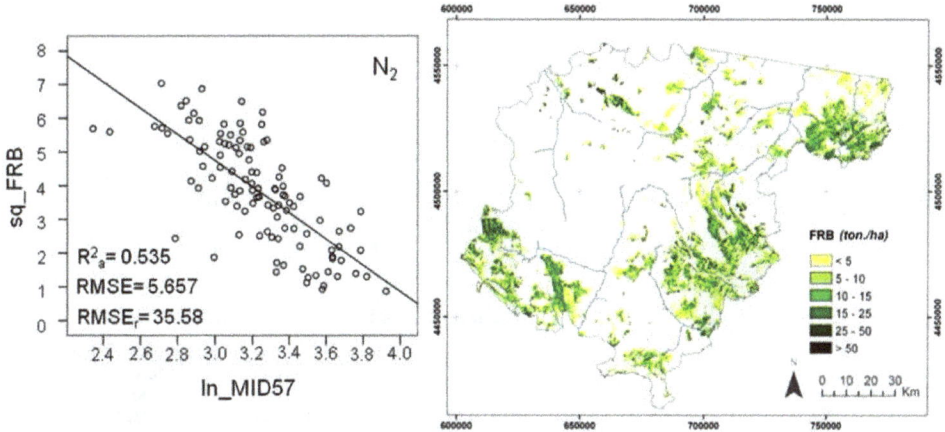

Fig. 8. The regression model selected using segmentation and 3 x 3 pixel windows with restrictions and the FRB map produced applying it

3.4 FRB cartography validation

The accuracy assessments of every regression equation (RMSE and $RMSE_r$) were completed using plots that showed the same homogeneity criteria as was used to run the model. However, since the selected estimation models have been applied to each one of the Landsat pixels located in forested areas in Teruel Province, the degree of success in these must also be evaluated at that scale.

To accomplish this, the NFI-2 plots excluded from the estimation models and their validation were considered. In order to guarantee the results, those plots that were affected by inaccuracies in their field location and/or by the radiometric response of different landscape elements located in their immediate vicinity, were removed from the validation sample. Consequently, group CV8 was used since it includes a high number of plots, which ensures that the validation results were not biased by using only the ideal plots.

As it can be seen in figure 9, the results show few differences between the three maps. Those obtained from 3 x 3 fixed windows yieded a $RMSE_r$ of 64.26%, while spectral homogeneous forest areas had $RMSE_r$ values of 66, 71% and 65.06%, respectively. These results at pixel level can be considered tolerable for the study area considering previous experiments using similar methodologies for boreal environments less affected by heterogeneity than Mediterranean forests. Thus, Tokola et al. (1996), Tokola & Heikkila (1997), Mäkkelä & Pekkarien (2001) and Katila & Tomppo (2001) reported $RMSE_r$ to estimate forest parameters such us timber volume or total volume from about 65% to more than 100%. In this respect, it is important to emphasize that estimation error in cartography derived from satellite images decreases with an increase in the size of the area used to validate it. For example, Fazakas et al. (1999) showed a $RMSE_r$ of 66.5% at pixel level, but when using an aggregation area of 598 ha, the $RMSE_r$ was 8.7%. However, it was not possible to carry out a similar analysis in our study area because no other FRB data were available at any scale.

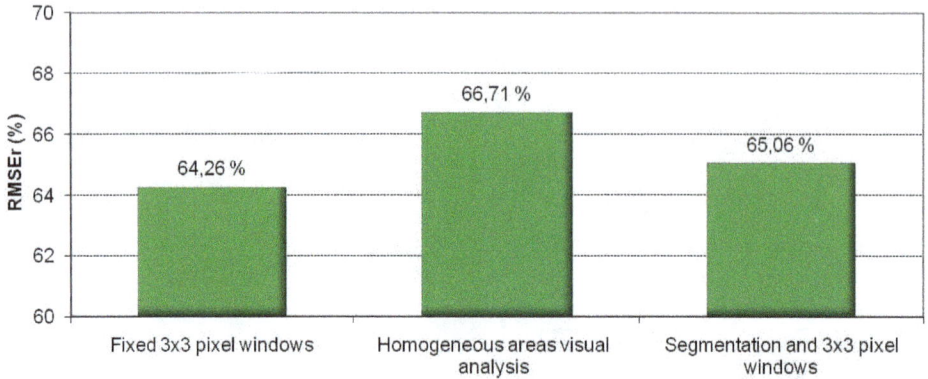

Fig. 9. RMSE$_r$ differences between FRB maps obtained using the three extraction methods

3.5 Inventory of potentially available FRB

Previous research has shown the utility of wetness variables obtained from Landsat TM images to estimate FRB, regardless of the image date in the summer period. This is because these kinds of variables yield high and similar estimation FRB models using June, July and August scenes. In addition, statistical differences were not found in the moisture content of the four pines studied over the 3 month period (García-Martín et al., 2008a).

In relation to results in previous sections, the model selected using the extraction of 3 x 3 fixed windows was applied to the MID57 neocanal image derived from the July 2008 Landsat image in order to inventory the FRB. This model was finally selected because it had higher predictive power (R^2 of 0.711 versus 0.595 on the model selected in the visual analysis method and 0.535 in the segmentation method) and allows the development of maps with the lowest estimation error. This last point shows that the limited set of plots used in the 3 x 3 fixed windows method is as representative for FRB estimation as those used in the other two methods, although they were composed of nearly double the number of plots.

Figure 10 shows the estimation cartography obtained for the entire study area. This cartography allows calculation of the total amount of FRB resource at the provincial level, which amounts to 5,449,252 tons. In addition, with the high spatial resolution (25 x 25 m), the cartography precisely reveals the richest regions and FRB distribution within them.

This makes this cartography especially suitable for determining optimal areas, taking into account other spatial variables that also determine the technical and economic feasibility in the harvest of this renewable energy resource, for example: (i) slope, which influences the possibility of using machinery and its efficiency; (ii) distance to forest tracks, which determines a portion of the transport costs; and (iii) area of forest stands, which is related to the necessary displacement during the working day. These three spatial factors and the quantity of FRB derived from remote sensing data at 25 m resolution can be integrated into a Geographical Information System (GIS) to identify areas more suitable for harvest of FRB, with attention to principles of sustainable ecological forest management (Pascual et al., 2007; García-Martín et al., 2011).

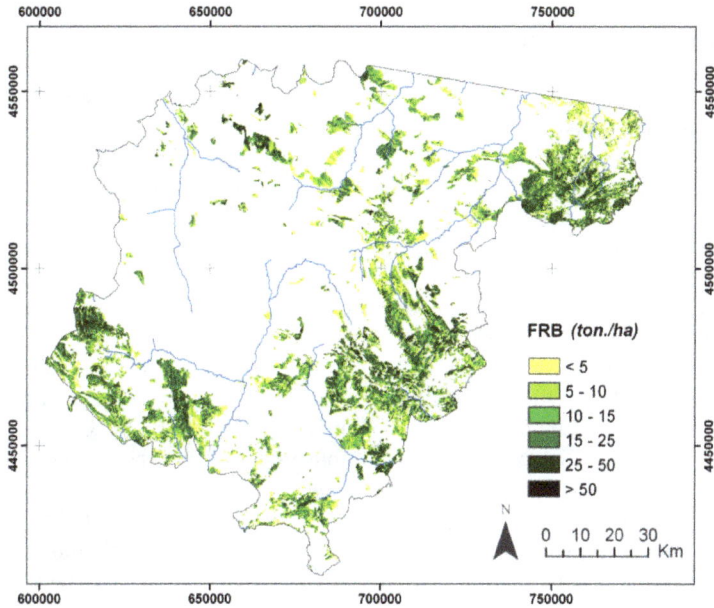

Fig. 10. FRB in the study area in July 2008

4. Conclusions

This study demonstrates the utility of Landsat TM images and forest inventory data in estimating FRB in Mediterranean areas. The methodology employed provides a continuous and complete estimation of FRB that overcomes problems associated with the limitations of point inventory. This information can be very useful in determining the most suitable areas in which to install power stations that make use of this resource; the lack of this type of information is one of the main problems currently facing the industry. As a result, an increase in the use of this kind of biomass resource will help to achieve the stated objectives of renewable energy production in Spain. Moreover, this is especially important considering two facts: (i) the current socio-economic context, with an increase in the price of petrol due to international instability, and public concerns regarding nuclear power, owing to Fukushima incident; and (ii) biomass is currently the only renewable energy that can be used as a strategic source of energy, as it is always available independent of weather conditions and can be stored easily prior to use (Jarabo, 1999).

The three sources of information typically used for biomass estimation (data from field sampling, satellite imagery and ancillary data) (Lu, 2006) have been carefully integrated, bearing in mind the specific characteristics of the Mediterranean environment. Thus, to reduce the problem that heterogeneity introduced into estimation models, three different methods were tested to extract radiometric data from a TM image. In agreement with previous studies focused on AGB estimation (i.e. Foody, 2001; Lu et al., 2004; Lu, 2005; Lu & Batistiella, 2005; Steininger, 2000) and with previous analyses carried out with our data (García-Martín et al., 2006, 2008a, 2008b), the derived spectral variables related to wetness showed the strongest correlations to FRB, independent of the method used to extract the radiometric data. Analyses undertaken revealed statistically significant correlations with the three methods, but the use of 3 x 3 pixel windows and CV were indicated as the most appropriate to isolate a homogeneous group of plots that allow the most accurate regression equations (R^2>0.7). In addition, despite the fact that these models were performed using half of the plots than those derived from the other two extraction methods, it was representative of the entire study area, as was shown in the validation results obtained at pixel level ($RMSE_r$ was similar in the cartographies obtained from the three methods). The visual analysis method did not yield such good results probably because of inaccuracies in the delineation of the homogeneous areas and because of the assumption of a constant value of BRF for the entire area from one precise location. The segmentation method did not improve the results either. This fact can be related to the lack of success to model effectively the spatial distribution pattern of BRF in the study area using RGB clustering. Finally, it is important to note that any of the methodologies tested to extract radiometric data successfully completed multivariate regressions models. This was because of the high auto-correlation among the well-correlated radiometric variables with FRB (wetness variables and vegetation indices). The use of these variables together will run a model affected by collinearity.

Validation cartography was based at the pixel level using plots that had not been previously used in the regression equation, removing only those with higher probability of error. This approach ensured the independence of the validation. $RMSE_r$ obtained in the three cartographies (above 65%) were better or at the same level than those produced in previous experiments conducted in boreal environments to estimate forest parameters using Landsat images. This is a positive result considering the higher complexity of Mediterranean forests. The remaining errors in the estimates of FRB can be related to the factors highlighted in García-Martín et al. (2008a,b): (i) inaccuracies in the fieldwork undertaken to establish the allometric equations and statistical analyses of the equations; (ii) problems involved in relating NFI plots to satellite data, mainly related to inaccuracies in the plot field placement (the three methodologies applied in the present study in linking ground data and remotely sensed data helped to reduce this problem, but it is still possible that errors were present in the final sample data); (iii) inaccuracies related to heterogeneity of the sample, because different pine species were considered and they are distributed in different regions of the study area; and (iv) limitations related to the spectral, radiometric, and spatial resolution of the TM in highly heterogeneous environments. However, despite these limitations, the size of the scenes and ease of distribution at no cost, make Landsat TM the most suitable in terms of achieving the objective of developing a useful methodology for estimations of FRB at regional-scales.

Finally, in order to improve outcomes, different lines of research should be considered. Firstly, different segmentation methods other than RGB clustering should be explored to determine if they can better model the forest spatial pattern and, as a result, obtain more accurate estimation models. Concerning this, the eCognition and the Definiens Developer segmentation procedures offer the possibility of considering additional features together with spectral data, based on similarities in shape and size. Secondly, focusing only on Landsat data, it is necessary to explore the use of other statistical methods that allow highly auto-correlated dependent variables to be considered jointly. In addition, it would be useful to explore the capability of hyperespectral sensors to identify narrow spectral bands highly correlated to FRB and poorly correlated between them and to examine the capability of SAR data from different sensors. This will allow the execution of better predictive multivariate regression models not affected by collinearity. Lastly, it would be useful to integrate data from physical variables that can be related to FRB quantity and distribution such as elevation, slope and aspect, as well as diverse biophysical parameters such as soil, lithology and climate.

5. Acknowledgment

This research has been supported by a grant provided by the Ministry of Science and Technology (AP2003-3097) and the project LIGNOSTRUM (AGL2002-03917-AGR-FOR), financed also by the Spanish Ministry of Science and Technology.

6. References

Alonso, E.; Asín, J. & Pascual, J. (2005). Biomasa residual forestal: regresiones para las especies del género Pinus existentes en la provincia de Teruel, *La ciencia forestal: respuestas para la sostenibilidad. 4° Congreso Forestal Español*, CD-ROM, Zaragoza, Spain, September 26-30, 2005

Anaya, J.A.; Chuvieco, E. & Palacios-Orueta, A. (2009). Aboveground biomass assessment in Colombia: A remote sensing approach. *Forest Ecology and Management*, Vol. 257, pp. 1237-1246

Asikainen, A.; Björheden, R. & Nousiainen, I. (2002). Cost of wood energy, In: *Bioenergy from Sustainable Forestry: Guiding Principles and Practice*, J. Richardson; R. Björheden; P. Hakkila, A.T. Lowe & C.T. Smith, (Eds.), 125-157, Kluwer Academic Publishers, Dordrecht, Netherlands

Austin, J.M.; Mackey, B.G. & Van Niel, K.P. (2003). Estimating forest biomass using satellite radar: an explanatory study in a temperature Australian Eucalyptus forest. *Forest Ecology and Management*, Vol. 176, pp. 575-583

Borsboom, N.W.J.; Hektor, B.; McCallum, B. & Remedio, E. (2002). Social implications of forest energy production, In: *Bioenergy from Sustainable Forestry: Guiding Principles and Practice*, J. Richardson; R. Björheden; P. Hakkila, A.T. Lowe & C.T. Smith, (Eds.), 265-297, Kluwer Academic Publishers, Dordrecht, Netherlands

Brown, S.L.; Schroeder, P. & Kern, J.S. (1999). Spatial distribution of biomass in forest of the eastern USA. *Forest Ecology and Management*, Vol. 123, pp. 81-90

Chander, G; Markham, B.L. & Helder, D.L. (2009). Summary of current radiometric calibration coefficients for Landsat MSS, TM, ETM+, and EO-1 ALI sensors. *Remote Sensing of Environment*, Vol. 113, pp. 893-903

Chavez, P.S. (1996). Image-based atmospheric corrections: Revisited and improved. *Photogrammetric Engineering and Remote Sensing*, Vol. 62, pp. 1025–1036

Chuvieco, E. (2002). *Teledetección ambiental. La observación de la tierra desde el espacio*, Ariel, Barcelona, Spain.

Colby, J.D. (1991). Topographic normalization in rugged terrain. *Photogrammetric Engineering and Remote Sensing*, Vol. 57, pp. 531-537.

Domínguez, J. (2002). *Los sistemas de información geográfica en la planificación e integración de energías renovables*, CIEMAT, Madrid, Spain

Eriksson, H.M.; Hall, J.P. & Helynen, S. (2002). Rationale for forest energy production, In: *Bioenergy from Sustainable Forestry: Guiding Principles and Practice*, J. Richardson; R.

Björheden; P. Hakkila, A.T. Lowe & C.T. Smith, (Eds.), 1-17, Kluwer Academic Publishers, Dordrecht, Netherlands

Esteban, L.S.; Pérez, P.; Ciria, P. & Carrasco, J.E. (2004). *Evaluación de los recursos de biomasa forestal en la provincia de Soria. Análisis de alternativas para su aprovechamiento energético*, CIEMAT, Madrid, Spain

EU Commission (1997). *Energy for the Future: Renewable Sources of Energy. White Paper for a Community Strategy and Action Plan*, European Commission, COM (97) 599 final, Brussels, Belgium

Fazakas, Z.; Nilsson, M. & Olsson, H. (1999). Regional forest biomass and wood volume estimation using satellite data and ancillary data. *Agricultural and Forest Meteorology*, Vol. 98-99, pp. 417-425.

Foody, G.M.; Cutler, M.E.; McMorrow, J.; Pelz, D.; Tangki, H.; Boyd, D.S. & Douglas, I. (2001). Mapping the biomass of Bornean tropical rain forest from remotely sensed data. *Global Ecology and Biogeography*, Vol. 10, pp. 379-387

Foody, G.M.; Boyd, D.S. & Cutler, M. (2003). Predictive relations of tropical forest biomass from Landsat TM data and their transferability between regions. *Remote Sensing of Environment*, Vol. 85, pp. 463-474

Franklin, S.E. (2001). *Remote sensing for sustainable forest management*, Taylor & Francis-CRC Press, New York, USA

García-Martín, A.; Pérez-Cabello, F. & de la Riva Fernández, J. (2006). Evaluación de los recursos de biomasa residual forestal mediante imágenes del satélite Landsat y SIG. *GeoFocus*, Vol. 6, pp. 205-230.

García-Martín, A.; de la Riva Fernández, J.; Pérez-Cabello, F.; Montorio-Llovería, R.; García Galindo, D. & Pascual Puigdevall, J. (2008a). Evaluation of the effect of temporality on forest residual biomass estimation using summer Landsat TM imagery, *Proceedings of the 16th European Biomass Conference and Exhibition*, pp. 254-262, Valencia, Spain, June 2-6, 2008

García-Martín, A., Pérez-Cabello, F.; de la Riva Fernández, J. & Montorio Llovería, R. (2008b). Estimation of crown biomass of Pinus spp. from Landsat TM and its effect on burn severity in a Spanish fire scar. *IEEE Journal of Selected Topics in Applied Earth Observations and Remote Sensing* Vol. 1, pp. 254-265

García-Martín, A.; García-Galindo, D.; Pascual, J.; de la Riva, J.; Pérez-Cabello, F. & Montorio, R. (2011). Determinación de zonas adecuadas para la extracción de biomasa residual forestal en la provincia de Teruel mediante SIG y Teledetección. *GeoFocus*, Vol. 11, pp. 19-50.

Gasparri, N.I.; Parmuchi, M.G.; Bono, J., Karszenbaum, H. & Montenegro, C.L. (2010). Assessing multi-temporal Landsat 7 ETM+ images for estimating above-ground biomass in subtropical dry forests of Argentina. *Journal of Arid Environments*, Vol. 74, pp. 1262-1270

Gonzalez, P.; Asner, G.P.; Battles, J.J.; Lefsky, M.A.; Waring, K.M. & Palace, M. (2010). Forest carbon densities and uncertainties from Lidar, QuickBird, and field measurements in California. *Remote Sensing of Environment*, Vol. 114, Issue 7, pp. 1561-1575

Hair, J.F.; Anderson, R.E.; Tatham, R.L. & Black, W.C. (1998). *Multivariate data análisis*, Prentice-Hall (5th ed.), Upper Saddle River, New Jersey, USA

Hakkila, P. & Parikka, M. (2002). Fuel resources from the forest, In: *Bioenergy from Sustainable Forestry: Guiding Principles and Practice*, J. Richardson; R. Björheden; P. Hakkila, A.T.

Lowe & C.T. Smith, (Eds.), 19-48, Kluwer Academic Publishers, Dordrecht, Netherlands

Hall, R.J.; Skakun, R.S.; Arsenault, E.J. & Case, B.S. (2006). Modeling forest stand structure attributes using Landsat ETM+ data: application to mapping of aboveground biomass and stand volume. *Forest Ecology and Management*, Vol. 225, pp. 378-390

Hyyppä, H.J. & Hyyppä, J.M. (2001). Effects of stand size on the accuracy of remote sensing-based forest inventory. *IEEE Transactions on Geosciences and Remote Sensing*, Vol. 39, pp. 2613-2621

IDAE (1999). *Plan de fomento de las energías renovables en España*, Instituto para la Diversificación y Ahorro de la Energía, Ministerio de Ciencia y Tecnología, Madrid, Spain

IDAE (2005a). *Plan de Energías Renovables en España 2005-2010*, Instituto para la Diversificación y Ahorro de la Energía, Ministerio de Industria, Turismo y Comercio, Madrid, Spain

IDAE (2005b). *Resumen del Plan de Energías Renovables en España 2005-2010*, Instituto para la Diversificación y Ahorro de la Energía, Ministerio de Industria, Turismo y Comercio, Madrid, Spain

IDAE (2007). *Energía de la biomasa*, Instituto para la Diversificación y Ahorro de la Energía, Ministerio de Industria, Turismo y Comercio, Madrid, Spain

Jarabo, F. (1999). *La energía de la biomasa*, S.A.P.T. Publicaciones Técnicas, Madrid, Spain

Joosten, R.; Schumacher, J.; Wirth, C. & Sculte, A. (2004). Evaluating tree carbon predictions for beech (Fagus sylvatica L.) in western Germany. *Forest Ecology and Management*, Vol. 189, pp. 87-96

Katila, M. & Tomppo, E. (2001). Selecting estimation parameters for the Finnish multisource national forest inventory. *Remote Sensing of Environment*, Vol. 76, pp. 16–32

Ketterings, Q.M.; Coe, R.; van Noordwijk, M.; Ambagau, Y. & Palm, C.A. (2001). Reducing uncertainty in the use of allometric biomass equations for predicting above-ground biomass in mixed secondary forests. *Forest Ecology and Management*, Vol. 146, pp. 199-209

Koch, B. (2010). Status and future of laser scanning, synthetic aperture radar and hyperspectral remote sensing data for forest biomass assessment. *ISPRS Journal of Photogrammetry and Remote Sensing*, Vol. 65, pp. 581-590

Kurvonen, L., Pulliainen, J. y Hallikainen, M. (2002). Active and passive microwave remote sensing of boreal forest. *Acta Astronautica*, Vol. 51, pp. 707-713

Labrecque, S.; Fournier, R.A.; Luther, J.E. & Piercey, D.E. (2003). A comparison of three approaches to map forest biomass from Landsat-TM and inventory data in Western Newfoundland, *Proceedings of 25th Canadian Symposium on Remote Sensing*, CD-Rom, Montreal, Canada, October 14-16, 2003

Labrecque, S.; Fournier, R.A.; Luther, J.E. & Piercey, D.E. (2006). A comparison of four methods to map biomass from Landsat-TM and inventory data in Western Newfoundland. *Forest Ecology and Management*, Vol. 226, pp. 129-144

Lu, D. (2005). Aboveground biomass estimation using Landsat TM data in the Brazilian Amazon. *International Journal of Remote Sensing*, Vol. 26, pp. 2509-2525

Lu, D. (2006). The potential and challenge of remote sensing-based biomass estimation. *International Journal of Remote Sensing*, Vol. 27, pp. 1297-1328

Lu, D. & Batistella, M. (2005). Exploring TM image texture and its relationships with biomass estimation in Rondônia, Brazilian Amazon. *Acta Amazonica*, Vol. 35, pp. 249-257

Lu, D.; Mausel, P.; Brondízio, E. & Moran, E. (2004). Relationships between forest stand parameters and Landsat TM spectral responses in the Brazilian Amazon Basin. *Forest Ecology and Management*, Vol. 198, pp. 149-167

Lucas, R.; Armston, J.; Fairfax, R.; Fensham, R.; Accad, A.; Carreiras, J.; Kelley, J.; Bunting, P.; Clewley, D.; Bray, S.; Metcalfe, D.; Dwyer, J.; Bowen, M.; Eyre, T.; Laidlaw, M. & Shimada, M. (2010). An Evaluation of the ALOS PALSAR L-Band Backscatter – Above Ground Biomass Relationship Queensland, Australia: Impacts of Surface Moisture Condition and Vegetation Structure. *IEEE Journal of Selected Topics in Applied Earth Observations and Remote Sensing*, Vol. 3, pp. 576 – 593

Mäkelä, H. & Pekkarinen, A. (2001). Estimation of timber volume at the sample plot level by means of image segmentation and Landsat TM imagery. *Remote Sensing of Environment*, Vol. 77, pp. 66-75

Mäkelä, H. & Pekkarinen, A. (2004). Estimation of forest stand volumes by Landsat TM imagery and stand-level field-inventory data. *Forest Ecology and Management*, Vol. 196, pp. 245-255

Mallinis, G.; Koutsias, N.; Makras, A. & Karteris, M. (2004). Forest parameters estimation in a European Mediterranean landscape using remotely sensed data. *Forest Science*, Vol. 50, pp. 450-460.

Maselli, F. & Chiesi, M. (2006). Evaluation of statistical methods to estimate forest volume in a Mediterranean Region. *IEEE Transactions on Geoscience and Remote Sensing*, Vol. 44, pp. 2239-2250

McRoberts, R.E. & Tomppo, E.O. (2007). Remote sensing support for national forest inventories. *Remote Sensing of Environment*, Vol. 110, pp. 412-419.

Meng, Q.; Cieszewski, C. & Madden, M. (2009). Large area forest inventory using Landsat ETM+: A geostatistical approach. *ISPRS Journal of Photogrammetry and Remote Sensing*, Vol. 64, pp. 27-36

MITYC (2010). *Plan de Energías Renovables 2011-2020. Evaluación ambiental estratégica. Documento de inicio.* Ministerio de Industria, Turismo y Comercio, Madrid, Spain, Available from

http://www.mma.es/portal/secciones/participacion_publica/eval_amb/pdf/2010_p_006_Documento_Inicio.pdf

MMA (1996). *Segundo Inventario Forestal Nacional (1986-1995): Aragón, Teruel*, Ministerio de Medio Ambiente, Madrid, Spain

Montero, G.; Ruiz-Peinado, R. & Muñoz, M. (2005). *Producción de biomasa y fijación de C02 por los bosques españoles*, Instituto Nacional de Investigación y Tecnología Agraria y Alimentaria - Ministerio de Ciencia y Tecnología, Madrid, Spain

Mourelatou, A. & Smith, I. (2004). *Energía y medio ambiente en la Unión Europea*, Ministerio de Medio Ambiente, Madrid, Spain

Muukkonen, P. & Heiskanen, J. (2005). Estimating biomass for boreal forests using ASTER satellite data combined with standwise forest inventory data. *Remote Sensing of Environment*, Vol. 99, pp. 434-447

Muukkonen, P. & Heiskanen, J. (2007). Biomass estimation over a large area based on standwise forest inventory data and ASTER and MODIS satellite data: A possibility to verify carbon inventories. *Remote Sensing of Environment*, Vol. 107, pp. 617-624

Pascual, J.; García-Galindo, D. & García-Martín, A. (2007). Optimum stands for forest residual biomass harvesting: development of a spatial index. *Proceedings of the 16th European Biomass Conference and Exhibition*, pp. 353-360, Berlin, Germany, May 7-11, 2007

Pekkarinen, A. (2002). Image segment-based spectral features in the estimation of timber volume. *Remote Sensing of Environment*, Vol. 82, pp. 349-359

Pilli, R.; Anfodillo, T. & Carrer, M. (2006). Towards a functional and simplified allometry for estimating forest biomass. *Forest Ecology and Management*, Vol. 237, pp. 583-593.

Powell, S.L.; Cohen, W.B.; Healey, S.P.; Kennedy, R.E.; Moisen, G.G.; Pierce, K.B. & Ohmann, J.L. (2010). Quantification of live aboveground forest biomass dynamics with Landsat time-series and field inventory data: A comparison of empirical modeling approaches. *Remote Sensing of Environment*, Vol. 114, pp. 1053-1068.

Proisy, C.; Couteron, P. & Fromard, F. (2007). Predicting and mapping mangrove biomass from canopy grain analysis using Fourier-based textural ordination of IKONOS images. *Remote Sensing of Environment*, Vol. 109, pp. 379-392

Raison, R.J. (2002). Environmental sustainability of forest energy production, In: *Bioenergy from Sustainable Forestry: Guiding Principles and Practice*, J. Richardson; R. Björheden; P. Hakkila, A.T. Lowe & C.T. Smith, (Eds.), 159-263, Kluwer Academic Publishers, Dordrecht, Netherlands

Roy, P.S. & Ravan, S.A. (1996). Biomass estimation using satellite remote sensing data. An investigation on possible approaches for natural forest. *Journal of Biosciences*, Vol. 21, pp. 535-561

Salvador, R. & Pons, X. (1998a). On the applicability of Landsat TM images to Mediterranean forest inventories. *Forest Ecology and Management*, Vol. 104, pp. 193-208.

Salvador, R. y Pons, X. (1998b). On the reliability of Landsat TM for estimating forest variables by regression techniques: a methodological analysis. *IEEE Transactions on Geosciences and Remote Sensing*, Vol. 36, pp. 1888-1897

Shoshany, M. (2000). Satellite remote sensing of natural Mediterranean vegetation: a review within an ecological context. *Progress in Physical Geography*, Vol. 24, pp. 153-158.

Steininger, M.K. (2000). Satellite estimation of tropical secondary forest above-ground biomass: data from Brazil and Bolivia. *International Journal of Remote Sensing*, Vol. 21, pp. 1139-1157.

Tangki, H. & Chappell, N.A. (2008). Biomass variation across selectively logged forest within a 225-km2 region of Borneo and its prediction by Landsat TM. *Forest Ecology and Management*, Vol. 256, pp. 1960-1970.

Tokola, T. & Heikkilä, J. (1997). Improving satellite image based forest inventory by using a priori site quality information. *Silva Fennica*, Vol. 1, pp. 67–78.

Tokola, T.; Pitkänen, J.; Partinen, S. & Muinonen, E. (1996). Point accuracy of a nonparametric method in estimation of forest characteristics with different satellite materials. *International Journal of Remote Sensing*, Vol. 17, pp. 2333– 2351

Vázquez de la Cueva, A. (2005). Variabilidad en la respuesta espectral de tres tipos de vegetación seleccionados de parcelas del IFN-3, In: *Teledetección. Avances en la observación de la Tierra,* Arbelo, M.; González, A. & Pérez, J.C., (Eds.), 113-116, Asociación Española de Teledetección, Tenerife, Spain

Velázquez, B. (2006). Situación de los sistemas de aprovechamiento de los residuos forestales para su utilización energética. *Ecosistemas,* Vol. 15, pp. 77-86.

Wang, C. (2006). Biomass allometric equations for 10 co-occurring tree species in Chinese temperate forests. *Forest Ecology and Management,* Vol. 222, pp. 9-16.

Wulder, M.A.; White, J.C.; Fournier, R.A.; Luther, J.E. & Magnussen, S. (2008). Spatially explicit large area biomass estimation: three approaches using forest inventory and remotely sensed imagery in a GIS. *Sensors,* Vol. 8, pp 529-560

Zianis, D. & Mencuccini, M. (2004). On simplifying allometric analyses of forest biomass. *Forest Ecology and Management,* Vol. 187, pp. 311–332.

Zheng, D.; Rademacher, J.; Chen, J.; Crow, T.; Bresee, M.; Le Moine, J. & Ryu, S.R. (2004). Estimating aboveground biomass using Landsat 7 ETM+ data across a managed landscape in northern Wisconsin, USA. *Remote Sensing of Environment,* Vol. 93, pp. 402-411

Zheng, G.; Chenb, J.M.; Tiana, Q.J.; Jub, W.M. & Xiaa, X.Q. (2007). Combining remote sensing imagery and forest age inventory for biomass mapping. *Journal of Environmental Management,* Vol. 85, pp 616-623

14

Geostatistical Estimation of Biomass Stock in Chilean Native Forests and Plantations

Jaime Hernández, Patricio Corvalán,
Xavier Emery, Karen Peña and Sergio Donoso
Universidad de Chile
Chile

1. Introduction

There are a variety of approaches to estimate above ground biomass (AGB), which can be classified according to the data source being used (Koch and Dees, 2008): field measurement, remotely sensed data or ancillary data used in GIS-based modeling. Field measurements are based on destructive sampling or direct measurement and the application of allometric equations (Madgwick, 1994). Recently, remotely sensed data, from both passive and active sensors, have become an important data source for AGB estimation. In this chapter we will focus on the use of optical multispectral data such as TM/ETM+ to estimate AGB. Generally, biomass is either estimated via a direct relationship between spectral response and biomass using multiple regression, k-nearest neighbor, neural networks, inverse canopy models or through indirect relationships, whereby attributes estimated from the remotely sensed data, such as leaf area index (LAI), structure (crown closure and height) or shadow fraction are used in equations to estimate biomass (Wulder, 1998). Here, we discuss the use of remote sensing data of moderate spatial resolution as input to estimate AGB. Research has demonstrated that it is more effective to generate relationships between field measurements and moderate spatial resolution remotely sensed data (e.g., LANDSAT), and then extrapolate these relationships over larger areas using comparable spectral properties from coarser spatial resolution imagery (e.g., MODIS) (Steininger, 2000; Lu, 2005; Phua and Saito, 2003; Foddy el al., 2003; Fazakas et al., 1999; Roy and Ravan, 1996; Zheng et al., 2004; Mickler et al., 2002). In general terms, LANDSAT TM and ETM+ data are the most widely used data of remotely sensed imagery for forest biomass estimation, but data from other moderate spatial resolution sensors have also been used, including ASTER and HYPERION data. In this chapter we present approaches that are currently being developed in Chile. Specifically, we introduce methods for the estimation of AGB using medium spatial resolution satellite imagery and digital elevation models. The main objective is to create, calibrate and validate such methods for applications. We developed an alternative approach in the estimation of AGB using LANDSAT ETM + imagery and SRTM digital elevation models as covariates for geostatistical modeling. From the spatial perspective, AGB data correspond to an array of points in space (x, y), while covariates correspond to a set of data that has a large number of samples in geographic space (extracted from each pixel), some of which having overlap with the available AGB data. The method can estimate the spatial variation of AGB at a stand or sub-stand level, and measure the uncertainty attached to the estimation, depending on local conditions. These results are promising and demonstrate the feasibility of using this approach in the evaluation and monitoring of stock biomass or communal farm scale. They are applicable to the actual landscape configuration of the country, open a series of new interest in research and development, and constitute a novel way to solve the problem of assessing biomass stocks.

2. Multispectral data for forest biomass estimation

Some remote sensing studies using LANDSAT TM/ETM + imagery have focused on estimating the forest attributes through correlation or regression analysis with the spectral response obtained from the images. The estimated attributes may include basal area, AGB, canopy closure, diameter at breast height (DBH), tree height, stand density, and leaf area index (LAI) (Franklin, 1986; Horler and Ahern, 1986; Peterson et al., 1986 and 1987; Lathrop and Pierce, 1991; Ardo, 1992; Curran et al., 1992; Cohen and Spies, 1992; Gemmell, 1995; Kimes et al., 1996; Trotter et al., 1997; Turner et al., 1999; Eklundh et al., 2001; Lu et al., 2004; Hall et al., 2006). One of the methods used to estimate attributes of forests from LANDSAT imagery is the direct correlation with radiometric values where regression relationships can be calculated between spectral bands, band ratios or vegetation indices (Franklin, 1986; Roy and Ravan, 1996; Jakubauskas and Price, 1997; Foody et al., 2003; Labrecque et al., 2006). The most used way to perform this kind of method is by applying stepwise forward variable selection. Table 1 summarizes some studies using this method in several countries.

Some of the advantages of using remotely sensed data are through the provision of a synoptic view of the study zone, capturing the spatial variability in the attributes of interest such as biomass, dominant height or crown closure (Zhen et al., 2007). Remotely sensed data can also be used to fill spatial and temporal gaps in forest inventory attributional data, improving estimates of forest biomass and carbon stocks, which can done for large areas as they are not limited by the extend of forest inventories (Birdsay, 2004; Fournier et al., 2003). However, the complicated vegetation stand structures in mature forest and in advanced successional forests often make the results less accurate given similar TM reflectance even if the above ground biomass varies significantly. This disadvantage is often called data saturation. For instance, in Manaus (Brazil) the canopy reflectance of tropical secondary successional forests saturated when AGB approached about 15 kg m-2 or vegetation age reached over 15 years (Steininger, 2000). Nelson et al. (2000) analyzed secondary forest age and AGB estimation using Landsat TM data and found that AGB cannot be reliably estimated without the inclusion of secondary forest age. The main approach to overcome this problem is to include further co-variables independent to reflectance values.

The k-NN method has become popular for forest inventory mapping and applications of estimation over the last years (McRoberts et al., 2007; McRoberts and Tomppo, 2007; Bafetta et al., 2009; Tomppo et al., 2009; Maselli et al., 2005; Tomppo and Halme, 2004; Thessler et al., 2008). The basic idea of this method is to estimate a target attribute of an object, i.e. AGB, using its similarity to objects with known attributes. Its application has focused on estimating continuous variables such as size, age, height, basal area, mean DBH, standing volume, and leaf area (Tomppo et al., 2009; McRoberts et al., 2007; Bertini et al, 2007; Tomppo and Halme, 2004).

Sensor	Study area	Estimation method	References
TM 5	Manaos, Brazil	Lineal and exponential regression	Steininger (2000)
TM 5	Pará state and Rondônia, Brazil	Multiple regression	Lu (2005)
TM 5	Sabah, Malaysia	Exponential regression	Phua and Saito (2003)
TM 4 and 5	Manaos, Brazil; Danum Valley, Malaysia; Khun Kong, Thailand	Multiple regression and neural network analysis	Foddy el al.(2003)
TM 5	Sweden	K nearest neighbor (k-NN)	Fazakas et al. (1999)
TM 5	Madhav National Park, India	Multiple regression	Roy and Ravan (1996)
ETM+	Wisconsin, USA	Multiple regression	Zheng et al. (2004)
ETM+	Southwest USA	PnET productivity model	Mickler et al. (2002)

Table 1. Examples of methods using LANDSAT data for estimation of forest biomass.

3. Estimation of biomass in Chilean forests

Estimation studies of forest biomass in Chile began to appear in the late 80's, primary for plantation of *Pinus radiata* (Caldentey, 1989) and, subsequently, for some local native species (Caldentey, 1995; Garfias, 1994; Herrera and Waisberg, 2002; Schlegel, 2001; Schmidt et al., 2009). In native forest, the background data is limited. The estimation method to be applied depends to the forest composition, structure and site variability. Natural forests are highly variable in the these attributes (Donoso, 1993; Gajardo, 1994; Luebert and Pliscoff, 2006) while plantations have less variation because they are monospecific and grow under intensive management regimes, designed to standardize the size and the quality of all trees (Lewis and Ferguson, 1993; Lavery, 1986; Gerding, 1991; Toro and Gessel, 1999). Secondary native forests, especially those dominated by the genus *Nothofagus*, have an intermediate degree of variation and heterogeneity (Donoso, 1981; Donoso, 1993; FIA, 2001). Traditionally, AGB estimation methods are based on sampling methods designed to assess standing timber (Husch et al., 1993; Anuchin, 1960; Bitterlich, 1984, Avery and Burkhart, 1994; Loetsch et al., 1973; Prodan et al., 1997). There is no reason for a different design because the volume/biomass ratio is relatively constant mainly depending on wood density. For the same reason, existing inventory plots can be used to estimate AGB directly. The AGB estimation method, which is usually performed for trees larger than 5 cm in diameter at breast height (DBH) and the understory is not included, should be done taking the following steps:

a. Estimate AGB at individual tree level. Given the high cost of measuring the biomass into its components (stem, bark, branches and leaves) it is preferable to use existing allometric equations for biomass by species and component. These equations depend on easy-to-measure state variables (i.e. DBH and height (H)) and allow estimating AGB in similar trees within the stand (Keith et al., 2000; Wang, 2006; Baker et al., 1984; Ares and Braener, 2005; Zewdie et al., 2009; Ketterings et al., 2001). The biomass components are estimated based on basic density samples (dry weight / green volume) multiplied by the total volume of the component (Keith et al., 2000; Steininger, 2000; Ishii and Tatedo, 2004; Hall et al., 2006). All these functions have the following generic form:

$$B_i^c = f(DBH_i, H_i, S_i) + \xi_i$$

where:

B_i^c	is the biomass of component c in the ith tree
DBH_i	is the diameter at breast height of the ith tree
H_i	is the total height for the ith tree
S_i	is an estimator of the sound/dead biomass proportion of the ith tree
ξ_i	is the regression error of the ith tree

(1)

b. Estimation of AGB for areal units. To this end, one has to use sampling plots within the target units and then their statistics as the estimates. In order for this to work properly, it is necessary to measure DBH, species and some health variable of the tree (usually subjective). In natural forests, due to the high variability for trees having the same DBH range, it is better make measurements of H and S in order to improve the precision of the results. In plantations (more homogeneous forests) the same can be made by applying a subsample for H and then to estimate the rest using a regression model for H as a function of the DBH.

The estimated AGB at plot level has the following generic form:

$$B_j^c = \sum_{i=1}^{n_j} B_{ij}^c \times F_j$$

where:

B_j^c	is the biomass of the component c for the jth plot (ton/ha)
n_j	is the number of trees in the jth plot
F_j	is the hectare expansion factor for the jth plot

(2)

c. Estimation of biomass at the stand, local, regional or national levels. The scales of interest for estimating AGB ranges from stand up to national levels according to the scale of the project. From stand level estimations the other aggregation levels can easily be archived by simply adding other stands estimation into the calculation. The biomass estimate at stand level has the following generic form:

$$B_r^c = a \sum_{k=1}^{m} B_k^c \times \frac{1}{m}$$

where:

B_r^c	is the biomass of the component c for the rth stand (ton)
a	is the total area of the stand (ha)
m	is the number of plots within a the stand

(3)

The use of optical satellite and topography data as auxiliary variables (covariables) allow the accuracy of AGB estimations to improve because they are based on their spatial covariation to field data by applying geostatistical interpolation. Using topographic data, the AGB variation is scaled to the actual values for that area, and then, AGB can be obtained by overlaying any available administrative division (stands, sites or districts).

| Survey data | | Survey data
+ Georefence (x,y) | | Survey data
+ Georefence (x,y)
+ Co-variates (TM/ETM+DEM) |

| Traditional statistics
Global estimators
(mean for each stand) | | Kriging
Local estimators
(sub-stands) | | Cokriging
Local estimators
(sub-stand) |

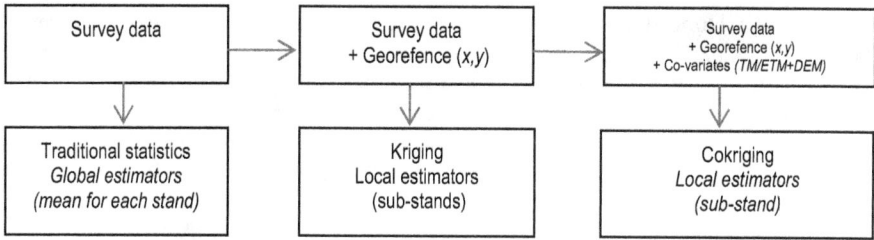

Fig. 1. Alternative AGB estimation based on the type of the available data.

According to the available data one can use different methods to estimate AGB (Figure 1).If data are traditional forest inventory plots, estimates should be made by using traditional statistical parameters for total and average quantities, which don't give any explicit information about the spatial variability of forest attributes. If data are georeferenced, distributed in such a way that they represent the whole territory of interest, then a spatial interpolation method, such as inverse distance interpolation or kriging, can be applied. The former approach has the advantage of being an unbiased probabilistic method where the degree of confidence (accuracy) can be calculated. The basic assumption behind kriging interpolation is that the spatial dependence models described in variograms assume the behavior of a regionalized variable, which is not necessarily true in reality and should be proved. On the other hand, if it is possible to have other variables (covariates), which besides being cheaper to obtained, are spatially well correlated to the variable of interest (AGB), then they can used to somehow correct the weakness of kriging by accounting for spatial discontinuities or irregularities found in nature (Coulston, 2008).

3.1 Native forest description
Chilean forests cover an area of 15.6 million hectares (20.7% of the national territory), of which 13.4 million hectares are natural forests (85.9% of the forested area). Currently 3.6 million hectares of forest are secondary forests (CONAF et al., 1999). Mixed forest of *Nothofagus obliqua*, *Nothofagus alpina* and *Nothofagus dombeyi* are one of the most important forest types in the country and cover 1.5 million hectares (12.2% of the total native forest). The genus *Nothofagus* has ten species that have a high economic value because of the quality of their wood. These *Nothofagus* forests area concentrated between 36° 30' S and 40° 30' S and between 100 and 1,000 m a.m.s.l. in Central Chile. They are present in both mountain ranges, costal and the Andes, where *N. obliqua* occupies the lowest areas (between 100 and 600 m, approximately), *N. alpina* intermediate ones (between 600 and 900 m) and *N. dombeyi* the highest (between 900 and 1,000 m), resulting in overlap ecotones with pure and mixed formations (CONAF et al., 1999; Donoso, 1981; Gajardo 1994). The main secondary species in these mixed *Nothogafus* second growth forest are *Aextoxicon punctatum*, *Genuine avellana*, *Laurelia sempervirens*, *Persea lingue* and *Eucryphia cordifolia* (Donoso, 1981). Today, a major part of these forests exhibit some state of degradation or have some form of human perturbation (Donoso, 1981; Gajardo, 1994). Nevertheless, they have a high productive potential and they need to be managed to improve their current condition. Usually, the quantification of these resources is done by applying volume tables or biomass functions, but these biomass functions rarely exist for native species and have only local applications

(Caldentey, 1995; Donoso et al., 2010; Garfias, 1994; Gayoso et al., 2002; Herrera and Waisberg, 2002; Sáez, 1991; Schlegel, 2001, Schmidt et al., 2009) (see Table 2). For native species, the biomass estimation methods most used in Chile are regression or allometry for the average tree (see section 3.0). In recent years, further attention has been paid in less expensive methods to assess AGB in native forests, based on remote sensing data and spatially explicit methods (Peña, 2007; Valdez et al., 2006).

3.2 Plantation description

In Chile, forest plantations are the main supply source of industrial raw material, with a total of near 40 million cubic meters. *Pinus radiata* D. Don is the main commercial species behind these massive commercial activities, with 80% of participation, followed by eucalyptus with 20% participation (CORMA, 2011). The plantations are concentrated in the central coastal area of Chile, between latitude 32 ° S and 42 ° S (Figure 2), covering a wide range of soils types (Schlatter, 1977; Schlatter and Gerding, 1984), environmental conditions (Fuenzalida, 1965; Madgwick, 1994) and silvicultural management regimes (Lewis and Ferguson, 1993; Lavery, 1986; Gerding, 1991). The land ownership structure is highly concentrated in two large forestry companies, which together owned 53% of the total planted area (CNE, GTZ / INFOR, 2008; Leyton, 2009), and the problem of quantifying wood supply has been primarily addressed by the private sector. Therefore, little public information exists on the area, location, age, species and management regime of plantations (CORMA, 2011). In contrast, the private sector has a large network of forest inventory information and has built growth simulators for the main species for different areas (i.e. *Radiata* and *Eucasim* models), types of site and management schemes and bucking (Fundación Chile, 2005; Morales et al., 1979). Some studies on the availability of wood for the forest industry (CORMA, 2011), carbon sequestration by plantations (Gilabert et al., 2010), AGB stocks for energy projects (INFOR, 2010, CNE / GTZ / INFOR, 2008) have been made by combining regional forest inventory data and these simulation models.

Chile Region	Forest Type	AGB (Mg/ha)	Author
VIII	*N. alpina* second growth forest	104	Garfias, 1994
X	Evergreen	194 to 663	Schlegel, 2001
X	Mixed *N. obliqua*, *N. alpina* and *N. dombeyi*	285	Herrera and Waisberg, 2002
XII	*Nothofagus pumilio*	380 to 447	Caldentey, 1995

Table 2. Above ground biomass of different types of forest and locality.

4. Remote sensing based biomass estimation

4.1 Description of the study areas and field data collection

We consider four study areas, two predominantly plantations (pine and eucalyptus) and two covered with mostly second growth Chilean oaks: *Nothofagus obliqua*, *Nothofagus alpina* and *Nothofagus dombeyi*. In both cases, one of the areas has a farm spatial scale and the other a municipality level (Table 3). Figure 2 shows the geographic location of the four areas of study.

Forest type	Study area	Ha	Description
Plantations	Pantanillos	400	Mainly *Pinus radiata* and few *Eucaliptus globulus* stands, both at different ages with intensive management.
	Quivolgo	20 000	
Native, secondary forest	Monte Oscuro	1 600	Mainly *N. obliqua* with several secondary tolerant species and informal management.
	Curacautín	20 000	Mixed of *N. obliqua, N.Alpina, N. dombeyi* at different ages and without management.

Table 3. Description of the study areas

We used a random sampling design and concentric circular sampling units (8m radius). All the data were collected in summer 2010 and the total size of the sampling was 348 plots, with an average sampling intensity of one sampling units every 63.2 hectares. Each selected tree was measured for DBH and one out of three for total height (H). For young plantations where understory foliage did not allow seeing trees at DBH we used circular plots with a radius of 5.65 m. In plantations stands where individuals grew in clusters - coppice type - we counted the number of individuals per unit and DBH and H was measured for the smaller and bigger trees. Biomass calculations were performed using the methods presented in Section 3.

4.2 Methods
4.2.1 ETM+ and SRTM data acquisition and preprocessing
LANDSAT ETM + images were obtained from the Earth Explorer Web site of the United States Geological Survey (USGS). Additionally, the SRTM digital elevation model (90 m) was freely downloaded from the site Earth Explorer. Bands ETM+ 1, 2, 3, 4, 5 and 7 were grouped into a single file and then projected to WGS84 UTM 18 South or 19 South, according to the area to which they correspond. Subsequently, there were a series of preprocessing steps as detailed below:

a. SLC-off Correction: The images acquired after 2003 have missing data due to malfunction of an instrument called Scan Line Corrector (Figure 3). This so-called SLC-off error was corrected using ENVI[RM] software application, which fits the flaw by using two images with the error in different areas, i.e. non-overlapping. Interpolation is performed considering the local histogram of the two images.

b. Geometric Correction: The images were rectified using Chilean regular cartography 1:50.000 with at least 30 control points per image and a root mean square error less than 30 m was achieved.

c. Radiometric corrections: Standard radiometric corrections were applied on all images to reduce the atmospheric effect following the method proposed by Chavez (1996), and for the topographic the one proposed by Riaño et al. (2003).

Once ETM+ and SRTM were co-registered, several variables were then derived:

a. Normalized Difference Vegetation Index or NDVI (Tucker, 1979).

b. Tasseled Cap bands: Brightness (TC1), Greenness (TC2) and Wetness (TC3). The other three components (TC4, TC5 and TC6) do not have a direct biophysical interpretation but were also calculated to take into account the complete data variation.

c. Slope, Aspect and Altitude were directly calculated from SRTM data.

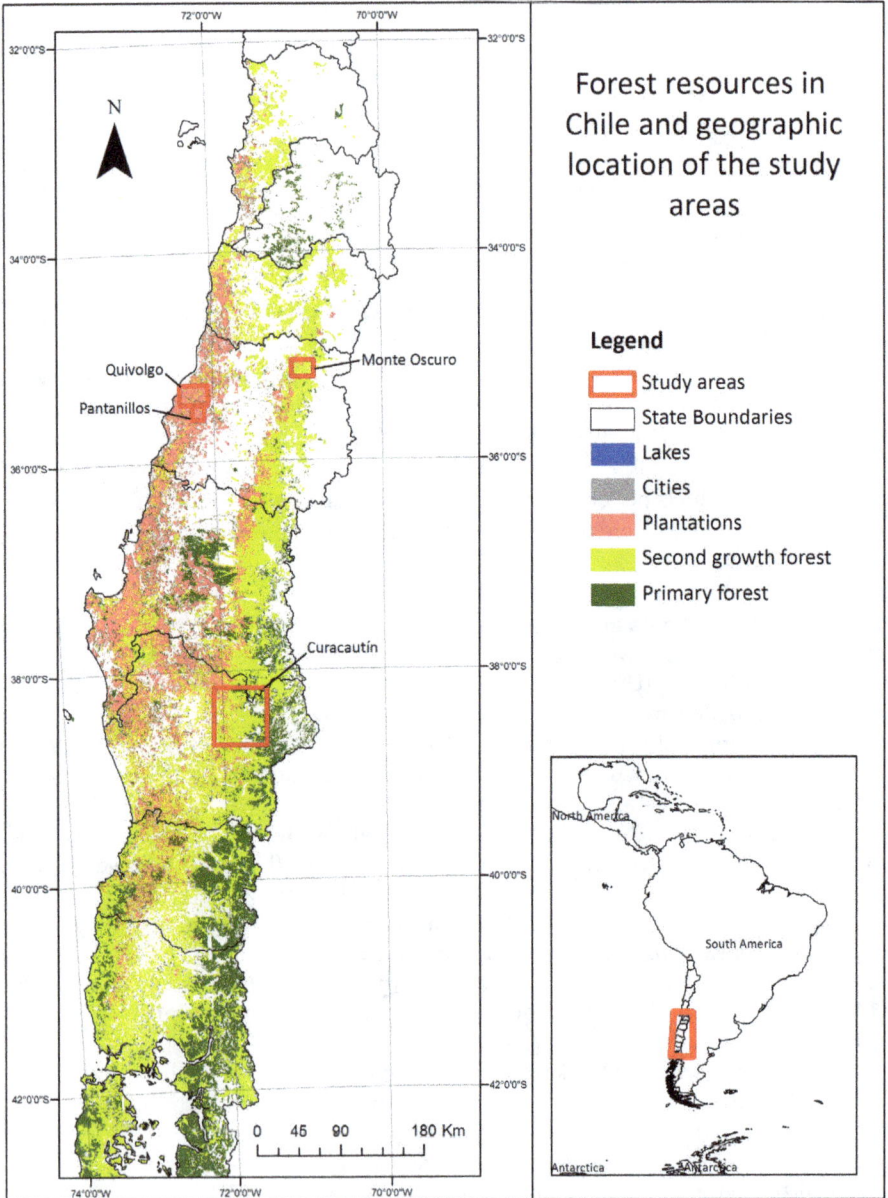

Fig. 2. Forest resources in Chile and geographical location of the study areas.

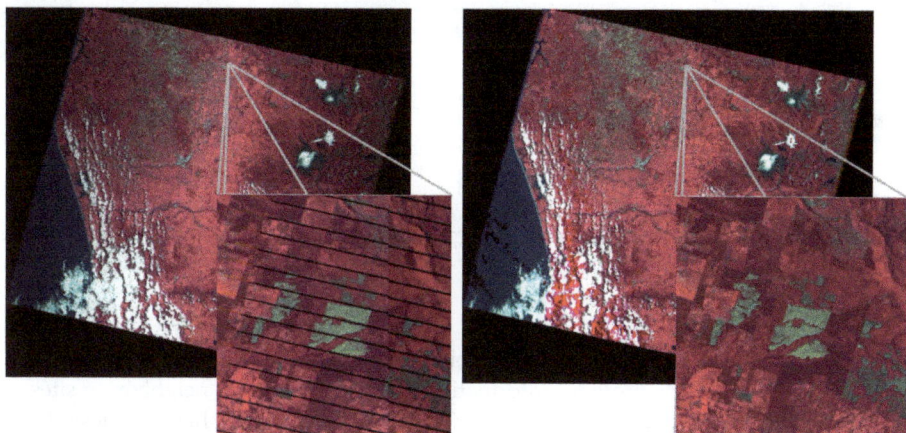

Fig. 3. LANDSAT ETM+ SLC-off Correction. Left: before, with the uncorrected data, right: after with the multidate correction.

4.2.2 Data integration

The integration of all layers of information was carried out in GIS environment (ArcGIS 9.3RM) and grouped into four classes:

a. Field data, which contain the observed AGB (target variable) and its coordinates.
b. ETM spectral bands, excluding the thermal band, plus their derivatives: NDVI and Tasseled Cap components (covariates).
c. Topographic information derived from SRTM data: elevation, slope, and orientation (covariates).
d. Vector files with property boundaries, stands, and base cartography support, which are useful in the stratification of the results and administrative aggregation.

For each of the four study areas a database was generated, which was used subsequently in the geostatistical analysis. Data were treated as point processes or count variables in the bi-dimensional space. Each set of field data is associated with a specific geographical area, for which all values to a resolution of 16 or 30 meters were considered in a comprehensive manner. Figure 4 shows the general methodological approach and the generic structure of the databases for each area.

From here and for the remaining sections, we considered biomass and all covariates mentioned above as regionalized variables to be able to follow the formality of geostatistical analysis. First, we performed a variogram analysis with the aim of studying the spatial autocorrelation of biomass and its spatial dependence with covariates. The result is a set of spatial models called variograms for the variable of interest (biomass) and the covariates. In all cases, we analyzed the anisotropy of the models, incorporating it explicitly in subsequent interpolations whenever possible. The variograms were used in the spatial estimation of biomass via cokriging. The exploratory analysis and modeling process were performed using IsatisRM geostatistical software and MatlabRM environment.

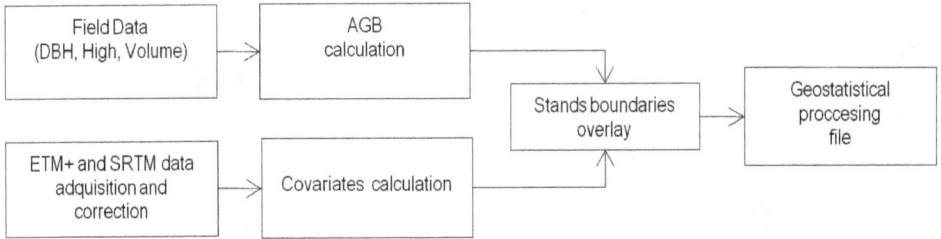

Fig. 4. Methodological framework for the integration of information and preparation of files for geostatistical analysis.

4.2.3 Variogram analysis

The values of a regionalized variable such as biomass are not independent, in the sense that the value observed in a site provides information about the values of neighboring sites (i.e. one is more likely to find a high value of biomass near other high values). In the geostatistical interpretation of the regionalized variable this intuitive notion of dependence is described by the formulation of random functions, which model the way the values observed in different sites relate to each other by a multivariate probability distribution.

The moment of order 1 of a random function (expected value or mean) involves a single spatial position and does not actually deliver information on spatial dependence. In contrast, the moments of order 2 (especially, the variogram) are defined with the help of two spatial positions, i.e. the smallest set that can be considered to describe the spatial "interaction" between values. These are the moments that give a basic description and operations of the spatial continuity of the regionalized variable.

Variogram analysis consists in calculating an experimental variogram from the available data, and subsequently fitting a theoretical variogram model. The experimental variogram measures the mean squared deviation between pairs of data values, as a function of the separation vector (distance and orientation) between the two data. The characteristics of this variogram are related to the characteristics of the regionalized variable (Isaaks and Srivastava, 1989; Chilès and Delfiner, 1999), in particular:

a.   The behavior of the variogram near the origin reflects the short-scale variability of the variable.

b.   The increase of the variogram reflects the loss of spatial correlation with the separation vector. It may depend on the direction of this vector, indicating the existence of directions with greater spatial continuity than others (anisotropy).

c.   The range of correlation, for which the variogram reaches a sill (plateau), indicates the distance at which two data do no longer have any correlation.

In practice, the anisotropy can be identified by comparing experimental variograms calculated along various directions of space, for example oriented 0°, 45°, 90° and 135° with respect to the x-axis. Often, this test is completed by drawing a "variogram map". When there is isotropy, the experimental variograms in different directions overlap and concentric circles are drawn in the variogram map. Otherwise, one may distinguish geometric anisotropies, in which the variogram map draws concentric ellipses; in such a case, the modeling rests upon the experimental variograms along the main anisotropy axes, which correspond to the axes of the ellipses.

Once calculated, the experimental variogram is fitted with a theoretical model, which usually consists of a set of basic nested structures with different shapes, sills and/or ranges (Gringarten and Deutsch, 2001). Examples of basic structures include the nugget effect (discontinuous component), as well as the spherical, exponential, cubic, Gaussian and power models (Chilès and Delfiner, 1999).

In the multivariate case, one has to calculate the experimental variogram of each variable (direct variogram), as well as the cross-variograms between each pair of variables, which measure the spatial correlation structure of the set of coregionalized variables. The fitting of a nested structure model is subject to mathematical constraints, reason for which one usually resorts to automatic or semi-automatic fitting algorithms (Wackernagel, 2003).

4.2.4 Interpolation by kriging

Kriging aims at estimating the value of the regionalized variable (here, AGB) at a position with no data, as accurately as possible, using the data available in other positions. It is structured according to the following steps:

a. Linearity constraint: The estimator must be a linear combination (i.e., a weighted average) of the data.

b. Unbiasedness constraint: This step expresses that the expected error is zero: if one considers a large number of identical kriging configurations, the average estimation error approaches zero. The absence of bias does not guarantee that errors are small, only that they cancel out on average.

c. Optimality constraint: according to the above steps, the estimator is subject to one or more constraints but is not fully specified. The last step consists in minimizing the estimation error variance, which measures the dispersion of the error around its zero mean, therefore the uncertainty in the true unknown value. Intuitively, this restriction means that if a large number of identical kriging configurations are considered, the statistical variance of the estimation errors is as small as possible. In addition to the estimated value, one can also calculate this minimal estimation error variance, known as the "kriging variance".

The weighting of the data depends on the spatial continuity of the regionalized variable, as modeled by the variogram, and on the spatial configuration of the data and location to estimate. In general, closer data receive larger weights than data located far from the target location, but this may be counterbalanced by other effects. For instance, data that are clustered in space tend to convey redundant information and may therefore receive small weights; data located along directions of little spatial continuity (with a small variogram range) may also be underweighted with respect to data located along directions of greater continuity.

There exist many variants of kriging, depending on the assumptions about the first-order moment or mean value (in particular, whether it is known or not, constant in space or not) and on the areal support of the value to estimate ("point" support identical to that of the data, or "block" support larger than that of the data) (Chilès and Delfiner, 1999). The multivariate extension of kriging is known as cokriging, which makes use not only of the data of the variable to estimate (primary variable), but also of covariates that are cross-correlated with this primary variable (Wackernagel et al., 2002). The determination of cokriging weights and of the error variance relies on the set of modeled direct and cross variograms (Wackernagel, 2003).

4.2.5 Kriging and cokriging neighborhood

Quite often, the amount of available data is large and using all of the available data is impractical and a selection of the most relevant data is required. One therefore defines a local window or "kriging neighborhood" in which to search for nearby data and to perform local estimation. The shape of the neighborhood should explicitly consider the detected anisotropies, by making the search radius dependent on the variogram range: the larger the range along a direction, the larger the search radius. Typically, the neighborhood is defined by placing an ellipse around the target location, and selecting a given number of data within this ellipse. The justification for such a practice is the so-termed screening effect, according to which the closest data screen out the influence of farthest data (David, 1976; Stein, 2002). Also, to avoid the selection of clustered data that convey redundant information, it is good practice to divide the neighborhood into several angular sectors (e.g., quadrants or octants) and to look for data within each sector (Isaaks and Srivastava, 1989).

Such a definition of the kriging neighborhood is, however, mainly heuristic and one is usually not aware of which data are really worth being included in the neighborhood, and which data can be discarded without deteriorating the estimation. The optimal neighborhood actually depends on the variogram of the variable of interest, as the screening effect does not occur with every variogram model. Some generic guidelines have been provided by Rivoirard (1987) to validate the choice of a given neighborhood in the univariate context.

In the multivariate case, the design of the neighborhood is more complex and critical. For instance, the data of a covariate may be screened out by the collocated data of the primary variable or, on the contrary, they may supplement the primary data and provide useful information to improve local estimation. As suggested by recent publications (Rivoirard, 2004; Subramanyam and Pandalai, 2008), the decision to include or not the covariate data should consider the correlation structure of the coregionalized variables and the sampling scheme (in particular, whether or not all the variables are measured at all the sampling locations). Some options include:

a.   Single search: the data points are selected according to their proximity to the target location, irrespective of which variable(s) is (are) informed at these data points.

b.   Two-part search: the first one is a selection of the nearby data of the primary variable, ignoring the information of covariates; the second one is a selection of the nearby data of the covariates, ignoring the information of the primary variable. The advantage of this strategy is to decouple the search of data according to the nature of the variable (primary or covariate). The disadvantage is that the search and the resolution of the cokriging equations must be carried out as many times as there are variables to estimate.

4.3 Results
4.3.1 Selection of variables

To estimate AGB we used ordinary cokriging (i.e., cokriging with unknown means that are assumed constant at the scale of the neighborhood) considering all the available covariates. Cokriging only works with linearly independent variables, for which there is no colinearity or "redundancy" of information. For this reason, the spectral bands TM1 to TM7 were excluded from the analysis since these variables can be deduced from that of the other variables such as Tasseled Cap components and NDVI. Also, the orientation variable corresponds to an angle between 0 and 360 degrees, so the value 0 represents the same information as the value 360. In order to prevent very different values to represent equal or nearly equal angles, we replaced the angles by their cosines and sines. Thus, in addition to the primary variable (AGB), we used the following 11 covariates in all the analyses: altitude (ALT), orientation cosine and sine (ASPECT), slope (SLOPE), normalized difference vegetation index (NDVI), and the six Tasseled Cap components (TC1, TC2, TC3, TC4, TC5 and TC6).

4.3.2 Variogram analysis

While there is much information from the covariates (tens of thousands to millions of records) from which experimental variograms can be calculated in a very detailed way, information is scarcer with the primary variable (AGB) that has only a few hundreds of positions with field data. Because of the limited data available for the country, the inference of the variograms of the primary variable and the cross variograms between this variable and all covariates is difficult. To determine the spatial correlation structure, we chose one of the following alternatives, depending on the case under study:

a.   Use of traditional experimental directional variograms, calculated along the identified main directions of anisotropy (Pantanillos case study).

b.   Use of omnidirectional variograms, when anisotropy was not detectable for the primary variable (AGB) due to the scarcity of data (Quivolgo and Curacautín case studies).

c.   Use of centered covariance as a substitute to the variogram (Chilès and Delfiner, 1999) in case the experimental variogram is too erratic (Monte Oscuro case study).

As an illustration, Figure 5 shows the experimental and modeled variograms for AGB, Brightness (TC1), Greenness (TC2) and Wetness (TC3), along the two identified main directions of anisotropy (north-south N0°E and east-west N90°E), for the Pantanillos area. For TC1, TC2 and TC3, the spatial variability appears to be greater along the north-south direction than along the east-west. These direct variograms, together with that of the other covariates and with the cross-variograms (a total of 78 variograms), have been jointly fitted thanks to the algorithm proposed by Goulard and Voltz (1992), using a nugget effect and nested exponential and power basic structures.

4.3.3 Cokriging neighborhood definition and application

To run cokriging, it is also necessary to define a neighborhood containing the data relevant to the local estimation. Given the very different number of data between the variables (AGB with very few data in comparison with covariates), we decided to use a two-part search:

a.   For AGB we considered the 15 closest available data.

b.   For each covariate, the 50 closest data were considered.

Cokriging was performed with an *ad hoc* MATLAB routine, since no known commercial software is able to perform cokriging with the above specifications and 11 covariates. The results are estimated values and error variances for AGB. The estimates were made for all the study areas, at the nodes of a grid with cells of 16m × 16m or 30m × 30m, depending on the case study, assuming unknown mean values for all the variables (ordinary cokriging).

Since not all the land in each area is covered by vegetation, we subsequently multiplied the estimates and error variances by the fraction of the cells located inside the identified stands, using vector digital layers of their boundaries. Figures 6 and 7 present the field data and identified stands, as well as the cokriging results for the Pantanillos area.

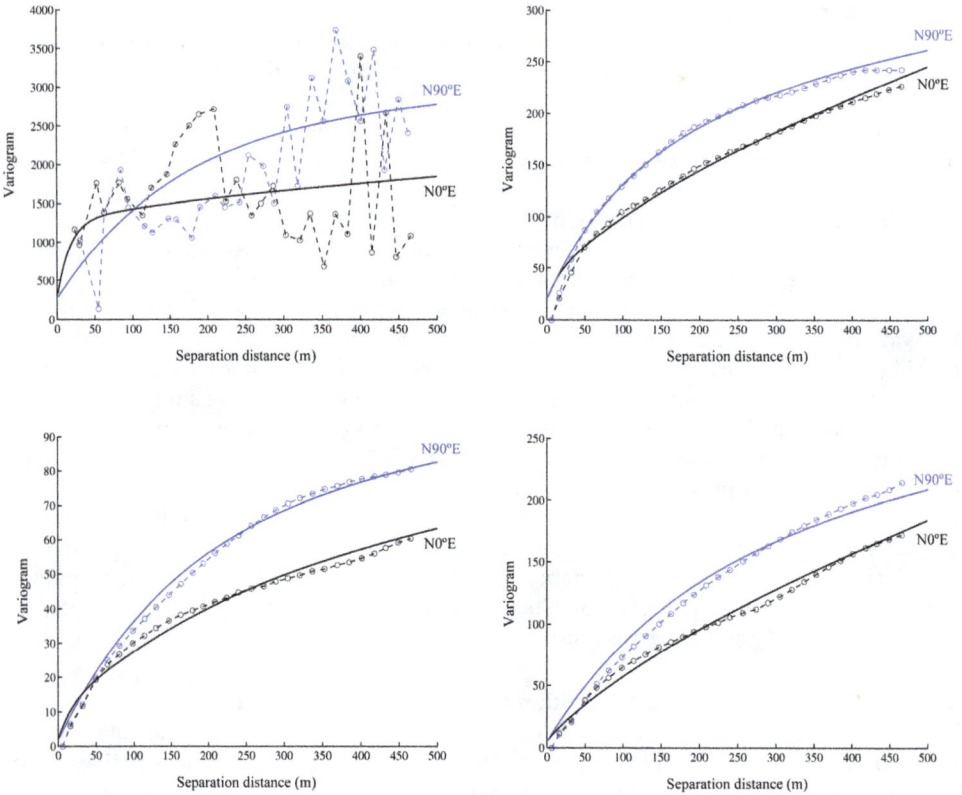

Fig. 5. Experimental (dots and dashed lines) and modeled (solid lines) variograms along the directions N0°E (black) and N90°E (blue) for AGB (top left), TC1 (top right), TC2 (bottom left) and TC3 (bottom right) in Pantanillos study area.

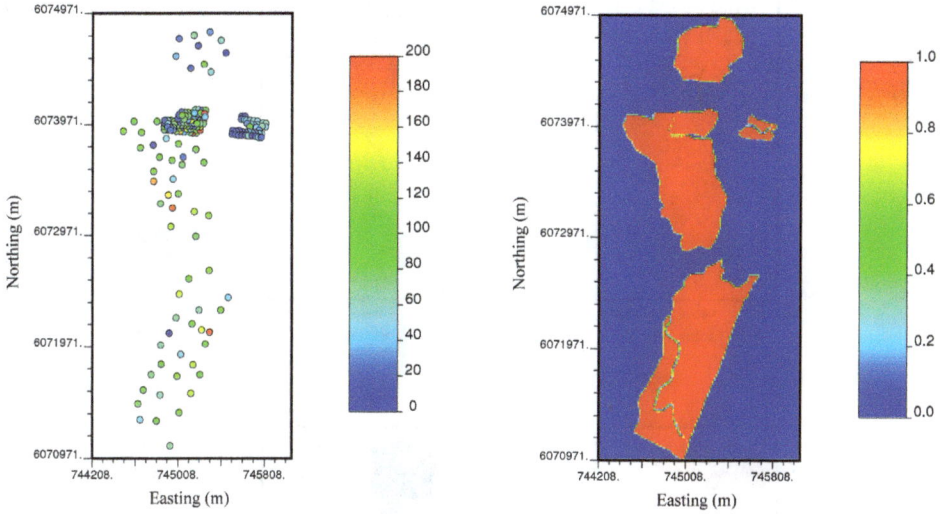

Fig. 6. Fi eld data for Pantanillos area (179 data), measuring above ground biomass (ton/ha) (left), and fraction of each grid cell (16m × 16m) contained in stands (right)

4.4 Conceptualizing and generalizing the method

The estimation approach we developed is a real alternative for assessing AGB at communal and regional scales. It has a relatively low implementation costs when compared to traditional methods (i.e. inventories of land) and can be used to assess and monitor the stock of biomass over time on a country-wide scale. Although it requires field data, it is possible to estimate other areas using the already gained knowledge (in particular, the spatial correlation structure, via the direct and cross variogram models). The idea is based on two hypotheses:

a. The amount of biomass available at any point in the geographical space depends, first, on the productivity of the site and, second, on the *status* of the vegetation that grows on it.

b. Using satellite data (reflectance in several bands of optical range of the electromagnetic spectrum) and topography attributes, we can estimate both the productivity of the site and the vegetation *status* in an integrated manner. Then we can estimate the spatial variation of biomass concentrations. Using terrain data, the variation is scaled to the actual values for specific areas and subsequently aggregated at any available administrative division (stands, sites or districts).

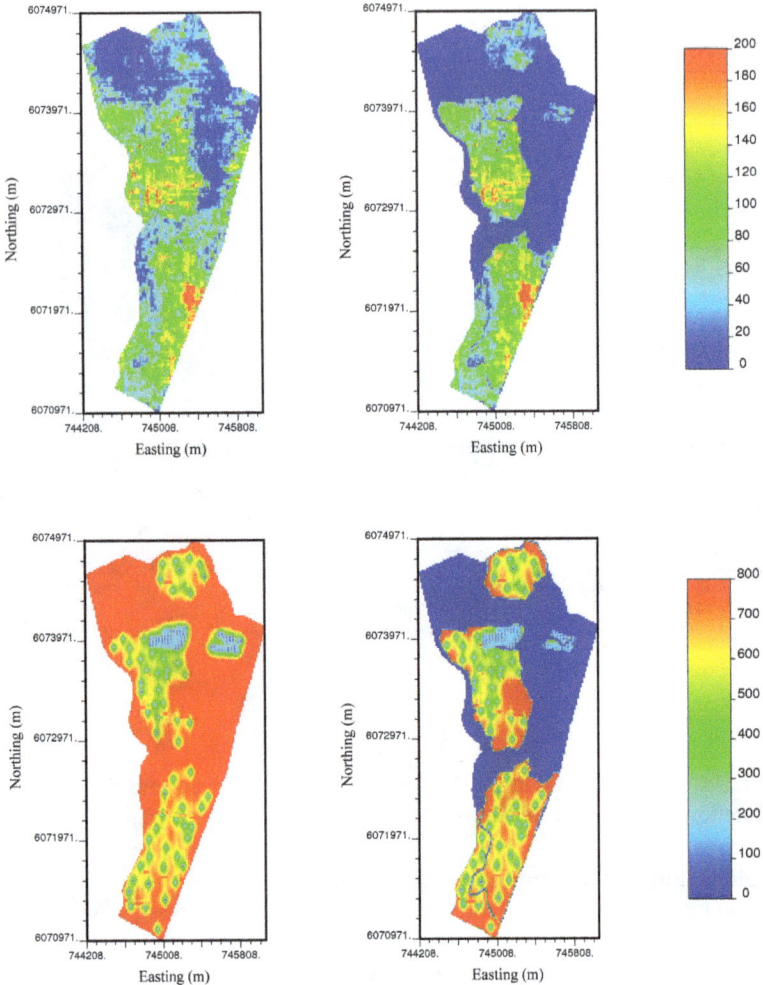

Fig. 7. Cokriging results for Pantanillos area. Above: Original estimate (left) and processed (right) biomass (ton/ha). Bottom: Variance of estimation error for AGB (ton^2/ha^2), original (left) and processed (right).

In order to apply this approach in the estimation of AGB in any area the following steps should be done:

a. Selection of the geographic area: It is first necessary to define the limits of the study area and the stands of interest. One may not be able to divide into stands if one does not have enough previous information and the results can be aggregated in the proper spatial division later (see step f).

b. Gathering and processing satellite images: The images used should be subject to the following treatments:

i.   Eliminate the presence of clouds and systematic errors. In the case of Landsat ETM + error must be corrected SLC-off explicitly.
ii.   Geometric correction for integration into GIS environment.
iii.   Radiometric correction to reduce topographic and atmospheric errors.
iv.   Calculation of vegetation indexes and Tasseled Cap components.
c.   Collection of field data: Field data can be collected especially for AGB estimation or can be available from previous inventories. Data should be expressed in biomass per unit area (i.e. hectare) and associated with a point in geographical formal space (i.e., UTM coordinates). Whatever the case, the field estimations of biomass can be made in any of the following cases:
i.   If data are available at the individual tree, apply allometric equations by species.
ii.   If data are aggregated for each plot, apply biomass functions.
d.   Data checking and cleaning: To perform the geostatistical analysis, it is necessary to create integrated data tables that incorporate both the biomass data, obtained from field data, plus all available covariates (satellite and topographical). The data should be reviewed, validated, and one should eliminate those displaying an aberrant behavior.
e.   Geostatistical modeling: Using the data structure obtained in step d, perform geostatistical analysis including the following sub-steps:
i.   Analysis of multivariate spatial correlation (variogram analysis) to obtain experimental direct and cross variograms and to fit variogram models required for interpolation.
ii.   Obtain local estimations of biomass and levels of uncertainty (estimation error variances) via cokriging throughout the study area.
iii.   Masking for estimating administrative units or stands of interest. To get accurate estimates, it is important to rule out areas with no vegetation.
f.   Report results and prepare associated maps.

In the previous paragraphs, one of the most important steps is the collection of field data. These data allow modeling the autocorrelation of biomass in space and its cross-correlation with satellite and topographic variables. When these data are available, the procedure for the estimation should be the one already described. However, sometimes no data are available or they cannot be collected on time or the costs are not reasonable. In this case, one may use known variogram models to make a "blind" estimation using a slightly alternative method: simple cokriging, which assumes that the average biomass is known. We propose that this average can be obtained using the forest growth simulators that are available in many countries, e.g., *Radiata* and *Eucasim* growing models and software in Chile.

5. Conclusions

The use of covariates extracted from SRTM elevation model and LANDSAT images allows for estimates with a greater level of detail than those obtained by using only data field. This can be corroborated by comparing the estimates by cokriging with those that would be obtained with univariate kriging. We hypothesize that the set of covariates accounts, through their spatial dependence, for two fundamental aspects to explain existing biomass:
a.   The quality of the forest site. Topographic variables (altitude, slope and orientation) inform in a synergistic way about the quality of the specific potential growing conditions (forest site).
b.   Vegetation health and vigor are jointly captured by the vegetation indices and Tasseled Cap components.

The approach to this scale of work, using medium spatial resolution sensors, is to estimate AGB independently of species or plant community types and then overlap stands boundaries, or any other administrative division, to obtain the associated total AGB stocks. Coregionalization models (direct and cross variograms) are simplifications of reality; in particular, they may not detect anisotropies when having a too small amount of field data, allowing only for an omnidirectional inference. The number of field data should be such that the inference of AGB variogram is feasible, which is usually achieved with more than 100 data in each stratum. In order for this number to be reduced, we recommend the sampling of the area (site) to be as regular as possible, i.e. avoiding samples clustered in space.

In the future, we suggest identifying other covariates correlated with AGB, for example, an indicator of the amount of sunlight available or a local indicator of the ground geometry (concave, convex or plane) that is related to the availability of water. These indicators could be calculated from a digital elevation model, already available at several spatial resolutions.

6. Acknowledgment

We would like to express our thanks to Biocomsa Consortium, and therefore INNOVA CORFO CHILE, for funding most of the field work, data acquisition and processing. Also, we thank Miss Paz Acuña and Miss Lissette Cortés from GEP Lab. at Universidad de Chile, for their support in reviewing the text and for providing the reference cartography. Also we thank Helen Grover for proof reading the text.

7. References

Anuchin, N.P. (1960). Forest Mensuration. Second edition. Goslesbumizdat. Moskova-Leningrado.454 pp

Ardo, J. (1992). Volume quantification of coniferous forest compartments using spectral radiance record by LANDSAT Thematic Mapper. *International Journal of Remote Sensing*, 13, 1779–1786

Ares & Braener (2005). Above ground biomass partitioning in loblolly pine silvopastoral stands: Spatial configuration and pruning effects. Forest Ecology and Management 219 (2005) 176-184

Avery, Th. & Burkhart, H. (1994). Forest Measurements. Fourth edition. McGraw-Hill. ISBN 0-07-002556-8. 408 pp

Baffetta, F.; Fattorini, L.; Franceschi, S. & Corona, P. (2009). Design-based approach to k-nearest neighbours technique for coupling field and remotely sensed data in forest surveys. *Remote Sensing of Environment*, 113(3):463-475

Baker, T.; Attiwill, P. & Stewart H. (1984). Biomass Equations for Pinus radiate in Gippsland, Victoria. New Zealand Journal of Forestry Science 14(1): 89-96

Bertini, R.; Chirici, G.; Corona, P. & Travaglini, D. (2007). Comparison between parametric and non-parametric methods for the spatialization of forest standing volume by integrating eld measures, remote sensing data and ancillary data. *Forest@*, 4110−117:1082−1089

Birdsey R. (2004). Data gaps for monitoring forest carbon in the United States: An inventory perspective. Environmental Management 33 supplement 1, S1-S8.

Bitterlich, W. (1984). The Relascope Idea. Relative Measurements in Forestry. Commonwealth Agricultural Bureaux. ISBN 0-85198-539-4. 241 pp

Caldentey, J. (1989). Beziehungen zwischen Klimaelementen und der Produktivitat von *Pinus radiata* Plantagen in Chile. Inaugural- Dissertation zur Erlangung der Doktorwurde der Forestwissenschaftlichen Fakultat der Ludwig-Maximilians-Universitat Munchen. 131 pp

Caldentey, J. (1995). Acumulación de biomasa en rodales naturales de *Nothofagus pumilio* en Tierra del Fuego, Chile. Investigación Agraria: Sistemas y Recursos Forestales 4(2): pp. 165-175

Chavez, P. (1996). Image-based atmospheric corrections - revisited and improved. *Photogrammetric Engineering & Remote Sensing*, Vol. 62, No. 9 (September 1996), pp. 1025-1036

Chilès, J.P. & Delfiner, P. (1999). *Geostatistics: Modeling Spatial Uncertainty*, Wiley, New York.

CNE/GTZ/INFOR. (2008). Disponibilidad de residuos madereros. Residuos de la industria primaria de la madera – Disponibilidad para uso energético. Santiago, octubre 2007. 54 pp

Cohen, W. & Spies, T. (1992). Estimating structural attributes of Douglas-fir/ western hemlock forest stands from LANDSAT and Spot imagery. *Remote Sensing of Environment*, 41 (1), 1–17

CONAF-CONAMA-BIRF. (1999). Catastro y evaluación de recursos vegetacionales nativos de Chile. Informe nacional con variables ambientales. Santiago. Chile. 90 pp

CORMA, (February 2011). Available from http://www.corma.cl/corma_info.asp

Coulston, J. (2008). Forest Inventory and Stratified Estimation: A Cautionary Note. Research Note SRS–16. Southern Research Station Forest Service. Department of Agriculture. United States. 6 pp

Curran, P.J.; Dungan, J. & Gholz, H. (1992). Seasonal LAI in slash pine estimated with LANDSAT TM. *Remote Sensing of Environment*, 39, 3–13

David, M. (1976). The Practice of Kriging, In: *Advanced Geostatistics in the Mining Industry*, M. Guarascio, M. David & C.J. Huijbregts, (Eds.), 31-48, Reidel, Dordrecht

Donoso, C. (1981). Tipos Forestales de los Bosques Nativos de Chile. Documento de Trabajo N°. 38. Investigación y Desarrollo Forestal (CONAF, PNUD-FAO) (Publicación FAO Chile) 70 pp

Donoso, C., (1993). Bosques templados de Chile y Argentina. Variación estructura y dinámica. Ecología forestal. Ed. Universitaria. Santiago, Chile 484 pp

Donoso, S.; Peña-Rojas, K.; Delgado-Flores, C.; Riquelme, A. & Paratori, A. (2010). Above-ground biomass accumulation and growth in marginal *Nothofagus macrocarpa* forest in central Chile. Interciencia, 35(11):65-69

Eklundh, L.; Harrie, L. & Kuusk, A. (2001). Investigating relationships between LANDSAT ETM+ sensor data and leaf area index in a boreal conifer forest. *Remote Sensing of Environment*, 78, 239–251

Fazakas, Z.; Nilsson, M. & Olsson, H. (1999). Regional forest biomass and wood volume estimation using satellite data and ancillary data. *Agricultural and Forest Meteorology*, 98–99, pp. 417–425

FIA. (2001). Bosque nativo en Chile: situación actual y perspectivas. Fundación para la Innovación Agraria. Ministerio de Agricultura. ISBN 956-7874-14-X, Santiago, Chile. 111pp

Foody, G.; Boyd, D. & Cutler, M. (2003). Predictive relations of tropical forest biomass from LANDSAT TM data and their transferability between regions. *Remote Sensing of Environment*, 85, 463–474

Fournier, R. A.; Luther, J. E.; Guindon, L.; Lambert, M.C.; Piercey, D.; Hall, R.J. & Wulder, M.A. 2003, Mapping above-ground tree biomass at the stand level from inventory information: test cases in Newfoundland and Québec. Canadian Journal of Forest Research 2003, 33, 1846-1863.

Franklin, J. (1986). Thematic Mapper analysis of coniferous forest structure and composition. *International Journal of Remote Sensing*, 7 (10), 1287–1301

Fuenzalida, H. (1965). Clima. Biogeografía. In: Geografía Económica de Chile, Texto refundido: pp. 98-152; 228-67. CORFO

Fundación Chile. (2005). Tablas Auxiliares de Producción. Simulador de Árbol Individual para Pino Radiata (*Pinus radiata* D. Don): Arquitectura de Copa y Calidad de Madera. 100pp

Gajardo, R. (1994). La Vegetación Natural de Chile. Clasificación y Distribución Geográfica. Editorial Universitaria, Santiago, Chile. 165 pp

Garfias, R. (1994). Crecimiento y biomasa en renoval raleado de *Nothofagus alpina* (Poepp, et Endl) Oerst, en la provincia de Bío-Bío, VIII Region. Tesis Ingeniería Forestal. Universidad Chile. Facultad de Ciencias Forestales y conservación de la naturaleza. Santiago, Chile. 64pp

Gayoso, J.; Guerra, J. & Alarcón, D. (2002). Contenido de carbono y funciones de biomasa en especies nativas y exóticas. Proyecto FONDEF D98I1076 Medición de la capacidad de captura de carbono en bosques de Chile y promoción en el mercado mundial. 157 pp

Gemmell, F. (1995). Effects of forest cover, terrain, and scale on timber volume estimation with Thematic Mapper data in the rocky mountain site. *Remote Sensing of Environment*, 51, 291–305

Gerding, V. (1991). Manejo de las plantaciones de Pinus radiata D. Don en Chile. BOSQUE 12(2):3-10

Gilabert, H.; Meza, F.; Cabello, H. & Aurtenenchea, M. (2010). Estimación del carbono capturado en las plantaciones de pino radiata y eucaliptos relacionados con el DL 701 de 1974. Informe final. Oficina de Estudios y Políticas Agrarias. ODEPA. Ministerio de Agricultura. Gobierno de Chile.

Goulard, M. & Voltz, M. (1992). Linear Coregionalization Model: Tools for Estimation and Choice of Cross-Variogram Matrix. *Mathematical Geology*, Vol.24,No.3, pp. 269-286

Gringarten, E. & Deutsch, C.V. (2001). Variogram Interpretation and Modeling, *Mathematical Geology*, Vol.33, No.4, pp. 507-534

Hall, R.; Skakun, R.; Arsenault, E. & Case, B. (2006). Modeling forest stand structure attributes using LANDSAT ETM+ data: Aplication to mapping of above ground biomass and stand volume. *Forest Ecology and Management*, 225, 378–390

Herrera, S.&Waisberg, R. (2002). Estimación de carbono almacenado en el Tipo Forestal Roble-Raulí–Coigüe (*Nothofagus obliqua-Nothofagus alpina–Nothofagus dombeyi*), para determinar los beneficios ambientales de someterlo a sumidero. Memoria de título Ingeniería de Ejecución en ambiente. Departamento de Ingeniería Geográfica, Facultad de Ingeniería, Universidad de Santiago de Chile.

Horler, D. & Ahern, F. (1986). Forestry information content of Thematic Mapper data. *International Journal of Remote Sensing*, 7, 405–428

Husch, B.; Miller, C. & Beers, T. (1993). Forest Mensuration. Krieger Publishing Company, Third Edition. ISBN 0-89464-821-7, Malabar, Florida. 402 pp

INFOR. (2010). Wisdom, Plataforma de Información de la Oferta de Dendrocombustibles en Chile, In: Seminario Centro de Energías Renovables "Energía sustentable de la biomasa: oportunidades y desafíos"

Isaaks, E.H. & Srivastava, R.M. (1989). *An Introduction to Applied Geostatistics*, Oxford University Press, New York

Ishii, T. & Tateda, Y. (2004). Leaf Area Index and Biomass Estimation for Mangrove Plantation in Thailand. In: Geoscience and Remote Sensing Symposium, 2004. IGARSS '04. Proceedings. 2004 IEEE International, 20-24 Sept. 2004. Vol 4: 2323 – 2326

Jakubauskas, M. & Price, K. (1997). Empirical relationships between structural and spectral factors of Yellowstone lodgepole pine forests. *Photogrammetric Engineering and Remote Sensing*, 63 (12), 1375–1381

Keith, H.; Barret, D. & Keenan, R. (2000). Review of Allometric Relationships for Estimating Woody Biomass for New South Wales, the Australian Capital Territory, Victoria, Tasmania and South Australia. National Carbon Accounting System Technical Report N° 5B. Australian Greenhouse Office 121 pp.

Ketterings, Q.; Coe, R.; van Noordwijk M.; Ambagau, Y. & Pal, Ch. A. (2001). Reducing uncertainty in the use of allometric biomass equations for predicting above-ground tree biomass in mixed secondary forests. Forest Ecology and Management 146: 199-2009

Kimes, D.; Holben, B.; Nickeson, J. & Mckee, W. (1996). Extracting forest ages in a pacific northwest forest from Thematic Mapper and topographic data. *Remote Sensing of Environment*, 56, 133–140

Koch, B. & Dees, M. (2008). Forestry applications. In: Advances in Photogrammetry, Remote Sensing and Spatial Information Sciences: 2008 ISPRS Congress Book, Chapter 32 - Li, Chen & Baltsavias (eds). Taylor & Francis Group, London. pp 439-465.

Labrecque, S.; Fournier, R.; Luther, J. & Piercy, D. (2006). A comparison of four methods to maps biomass from LANDSAT–TM and inventory data in western Newfoundland. *Forest Ecology and Management*, 226, 129–144

Lathrop, R. & Pierce, L. (1991). Ground-based canopy transmittance and satellite remotely sensed measurements for estimation of coniferous forest canopy structure. *Remote Sensing of Environment*, 36, 179–188

Lavery, P. (1986). Plantation Forestry with Pinus radiata. Review Paper N° 12. School of Forestry. University of Canterbury, Christchurch, New Zealand 255 pp

Lewis N. & Ferguson, I. (1993) in association with Sutton, W.; Donald, D. & Lisboa, H. Management of radiata Pine.ISBN 0-909605-79-3 Langley Editing, Melbourne, Australia 404 pp

Leyton, J. (2009). Tenencia Forestal en Chile. Available from http://www.fao.org/forestry/17192-0422df95bf58b971d853874bb7c5755f7.pdf

Loetsch, F., Zoehrer, F. & Haller, K.E. (1973). Forest Inventory. Vol. 2 B.L.V. Verlagsgesellschaft. München. 469 pp

Lu, D. (2005). Aboveground biomass estimation using LANDSAT TM data in the Brazilian Amazon Basin. *International Journal of Remote Sensing*, 26, 2509–2525

Lu, D.; Mausel, P.; Brondizio, E. & Moran, E. (2004). Relationships between forest stand parameters and LANDSAT Thematic Mapper spectral responses in the Brazilian Amazon basin. *Forest Ecology and Management*, 198, 149–167

Luebert, F.&Pliscoff, P. (2006). Sinopsis bioclimática y vegetacional de Chile. Santiago de Chile, Editorial Universitaria. ISBN 956-11-1832-7, 316 pp

Madgwick, H. (1994). *Pinus radiata* – Biomasa, Form and Growth. H.A.I. Madwick (Ed) ISBN 0-473-02375-X, Rotorua New Zealand. 428pp

Maselli, F.; Chirici, G.; Bottai, L.; Corona, P. & Marchetti, M. (2005). Estimation of Mediterranean forest attributes by the application of k-NN procedure to multitemporal LANDSAT ETM+ images. *International Journal of Remote Sensing*, 17:3781–3796

McRoberts R. & Tomppo, E. (2007). Remote sensing support for national forest inventories. *Remote Sensing of Environment*, 110:412–419

McRoberts R.; Tomppo, E.; Finley, A. & Heikkinen, J. (2007). Estimating areal means and variances of forest attributes using the k-Nearest Neighbors technique and satellite imagery. *Methods*, 111:466 - 480

Mickler, R.; Earnhardt, T. & Moore, J. (2002). Regional estimation of current and future forest biomass. *Environmental Pollution*, 116, S7–S16

Morales, R.; Weintraub, A.; Peters, R.&García, J. (1979). Modelos de simulación y manejo para plantaciones forestales. FO: DP/CHI/76/003. Documento de trabajo N° 36. Santiago de Chile. 155pp

Nelson, R.F., Kimes, D.S., Salas, W.A. & Routhier, M. (2000). Secondary forest age and tropical forest biomass estimation using Thematic Mapper imagery. Bioscience, 50, pp. 419–431.

Peña, M. (2007). Correcciones de una imagen satelital ASTER para estimar parámetros vegetacionales en la cuenca del río Mirta, Aysén. Bosque 28(2): 162-172p.

Peterson, D.; Spanner, M.; Running, S. & Teuber, K. (1987). Relationship of Thematic Mapper simulator data to leaf area index of temperate coniferous forests. *Remote Sensing of Environment*, 22, 323-331

Peterson, D.; Westman, W.; Stephenson, N.; Ambrosia, V.; Brass, J. & Spanner, M. (1986). Analysis of forest structure using Thematic Mapper simulator data. *IEEE Transactions in Geoscience and Remote Sensing*, 24, 113–121

Phua, M. & Saito, H. (2003). Estimation of biomass of a mountainous tropical forest using LANDSAT TM data. *Canadian Journal of Remote Sensing*, 29, 429–440

Prodan, M.; Peters, R.; Cox, F.&Real, P. (1997). Mensura Forestal. Serie Investigación y Educación en Desarrollo Sostenible. IICA BMZ/gtz

Riaño, D.; Chivieco, E.; Salas, J. & Aguado, I. 2003. Assessment of Different Topographic Corrections in Landsat-TM Data for Mapping Vegetation Types. *IEEE Transactions in Geoscience and Remote Sensing*, vol. 41(5): 1056:1061

Rivoirard, J. (1987). Two Key Parameters When Choosing the Kriging Neighborhood, *Mathematical Geology*, Vol.19, No.8, pp. 851-856

Rivoirard, J. (2004). On Some Simplifications of Cokriging Neighborhood, *Mathematical Geology*, Vol.36, No.8, pp. 899-915

Roy, P. & Ravan, S. (1996). Biomass estimation using satellite remote sensing data: an investigation on possible approaches for natural forest. *Journal of Bioscience*, 21 (4), 535–561

Saez, M. (1991). Biomasa y contenido de nutrientes e renovales no intervenidos de roble-raulí (*Nothofagus obliqua* (mirb) Oerst. – *Nothofagus alpina* (Poepp. et Ende.) Oerst.) en suelos volcánicos de la precordillera andina, IX Región. Memoria de título Ingeniería Forestal. Facultad de Ciencias Forestales. Universidad de Chile. 96 pp

Schlatter, J. & Gerding, V. (1984): Important site factors for Pinus radiata growth in Chile. In: Grey, D.C., Schonau, A.P.G., Schutz, C.J. & Van Laar, A. (Editors). Symposium on Site and Productivity of Fast Growing Plantations, Pretoria and Pietermaritzburg, South Africa 30 April 11- May 1984 pp 541-549

Schlatter, J. (1977). La Relación entre suelo y Plantaciones de *Pinus radiata* D.DON en Chile Central Análisis de la Situación Actual y Planteamientos para su Futuro Manejo. Bosque Vol. 2 N° 1 31 pp

Schlegel, B. (2001). Estimación de la biomasa y carbono en bosques del tipo forestal siempreverde. Simposio internacional medición y monitoreo de la captura de carbono en ecosistemas forestales. 18-20 Octubre 2001. Valdivia, Chile. 13 pp

Schmidt, A.; Poulain, M.; Klein, D.; Krause, K.; Peña-Rojas, K.; Schmidt, H. & Schulte, A. (2009). Allometric above-belowground biomass equations for *Nothofagus pumilio* natural regeneration in the Chilean Patagonia. Ann. For. Sci. 66 (5):315-323.

Stein, M.L. (2002). The Screening Effect in Kriging, *Annals of Statistics*, Vol.30, No.1, pp. 298-323

Steininger, M. (2000). Satellite estimation of tropical secondary forest above-ground biomass: data from Brazil and Bolivia. Int. J. Remote Sensing Vol 21 N° 6 & 7, 1139 – 1157

Subramanyam, A. & Pandalai, H.S. (2008). Data Configurations and the Cokriging System: Simplification by Screen Effects, *Mathematical Geosciences*, Vol.40, No.4, pp. 425-443

Thessler, S.; Sesnie, S. & Ramosbendana, Z. (2008).Using k-NN and discriminant analyses to classify rain forest types in a LANDSAT TM image over northern Costa Rica. *Remote Sensing of Environment*, 112(5):2485-2494

Tomppo, E. & Halme, M. 2004. Using coarse scale forest variables as ancillary information and weighting of variables in k-NN estimation: A genetic algorithm approach. *Remote Sensing of Environment*, 92:1–20

Tomppo, E.; Gagliano, C.; Denatale, F.; Katila, M. & McRoberts, R. (2009). Predicting categorical forest variables using an improved k-Nearest Neighbour estimator and LANDSAT imagery. *Remote Sensing of Environment*, 113(3):500-517

Toro, J. & Gessel, S. (1999). Radiata pine plantations in Chile. New Forests 18: 33–44

Trotter, C.; Dymond, J. & Goulding, C. (1997). Estimation of timber volume in a coniferous plantation forest using LANDSAT TM. *International Journal of Remote Sensing*, 18, 2209–2223

Tucker, C. (1979). Red and photographic infrared linear combinations for monitoring vegetation. *Remote Sensing of Environment*, 8: 127-150

Turner, D.; Cohen, W.; Kennedy, R.; Fassnacht, K. & Briggs, J. (1999). Relationships between leaf area index and LANDSAT TM spectral vegetation indices across three temperate zone sites. *Remote Sensing of Environment*, 70, 52–68

Valdez R, M.; González&De Los Santos, H. M. (2006). Estimación de cobertura arbórea mediante imágenes Satelitales multiespectrales de alta resolución. Agrociencia 40(3): 383-394 p. ISSN 1405-3195.

Wackernagel, H. (2003). Multivariate *Geostatistics: an Introduction With Applications*, Third Edition, Springer, Berlin

Wackernagel, H.; Bertino, L.; Sierra, J.P. & González del Río, J. (2002). Multivariate Kriging for Interpolating With Data From Different Sources, In: *Quantitative Methods for Current Environmental Issues*, C.W. Anderson, V. Barnett, P. Chatwin & A.H. El Shaarawi, (Eds.), 57-75, Springer-Verlag, London

Wang, Ch. (2006). Biomass allometric equations for 10 co-occurring tree species in Chinese temperate forests. Forest Ecology and Management 222 (2006) 9–16

Wulder, M. (1998). Optical remote-sensing techniques for the assessment of forest inventory and biophysical parameters. Progress in Physical Geography, 22, pp. 449–476.

Zewdie, M.; Olsson, M. & Verwijst, Th. (2009). Above-ground biomass production and allometric relations of Eucalyptus globulus Labill. coppice plantations along a chronosequence in the central highlands of Ethiopia. Biomass and Bioenergy 33 (2009) 421–428

Zheng, D.; Heath, L.S. & Ducey, M.J. (2007). Forest biomass estimated from MODIS and FIA data in the Lake States: MN, WI, and MI, USA. Forestry 2007, 80, 265-278.

Zheng, D.; Rademacher, J.; Chen, J.; Crow, T.; Bresee, M.; Le Moine, J. & Ryu, S. (2004). Estimating aboveground biomass using LANDSAT 7 ETM+ data across a managed landscape in northern Wisconsin, USA. *Remote Sensing of Environment*, 93, 402–411

Permissions

The contributors of this book come from diverse backgrounds, making this book a truly international effort. This book will bring forth new frontiers with its revolutionizing research information and detailed analysis of the nascent developments around the world.

We would like to thank Temilola Fatoyinbo, for lending her expertise to make the book truly unique. She has played a crucial role in the development of this book. Without her invaluable contribution this book wouldn't have been possible. She has made vital efforts to compile up to date information on the varied aspects of this subject to make this book a valuable addition to the collection of many professionals and students.

This book was conceptualized with the vision of imparting up-to-date information and advanced data in this field. To ensure the same, a matchless editorial board was set up. Every individual on the board went through rigorous rounds of assessment to prove their worth. After which they invested a large part of their time researching and compiling the most relevant data for our readers. Conferences and sessions were held from time to time between the editorial board and the contributing authors to present the data in the most comprehensible form. The editorial team has worked tirelessly to provide valuable and valid information to help people across the globe.

Every chapter published in this book has been scrutinized by our experts. Their significance has been extensively debated. The topics covered herein carry significant findings which will fuel the growth of the discipline. They may even be implemented as practical applications or may be referred to as a beginning point for another development. Chapters in this book were first published by InTech; hereby published with permission under the Creative Commons Attribution License or equivalent.

The editorial board has been involved in producing this book since its inception. They have spent rigorous hours researching and exploring the diverse topics which have resulted in the successful publishing of this book. They have passed on their knowledge of decades through this book. To expedite this challenging task, the publisher supported the team at every step. A small team of assistant editors was also appointed to further simplify the editing procedure and attain best results for the readers.

Our editorial team has been hand-picked from every corner of the world. Their multi-ethnicity adds dynamic inputs to the discussions which result in innovative outcomes. These outcomes are then further discussed with the researchers and contributors who give their valuable feedback and opinion regarding the same. The feedback is then collaborated with the researches and they are edited in a comprehensive manner to aid the understanding of the subject.

Apart from the editorial board, the designing team has also invested a significant amount of their time in understanding the subject and creating the most relevant covers. They scrutinized every image to scout for the most suitable representation of the subject and create an appropriate cover for the book.

The publishing team has been involved in this book since its early stages. They were actively engaged in every process, be it collecting the data, connecting with the contributors or procuring relevant information. The team has been an ardent support to the editorial, designing and production team. Their endless efforts to recruit the best for this project, has resulted in the accomplishment of this book. They are a veteran in the field of academics and their pool of knowledge is as vast as their experience in printing. Their expertise and guidance has proved useful at every step. Their uncompromising quality standards have made this book an exceptional effort. Their encouragement from time to time has been an inspiration for everyone.

The publisher and the editorial board hope that this book will prove to be a valuable piece of knowledge for researchers, students, practitioners and scholars across the globe.

List of Contributors

Sietse Los and Peter North
Swansea University, United Kingdom

Ross Nelson and Bruce Cook
NASA Goddard Space Flight Center, USA

Jacqueline Rosette
Swansea University, United Kingdom
NASA Goddard Space Flight Center, USA
University of Maryland College Park, USA

Juan Suárez
Forest Research, United Kingdom
University of Maryland College Park, USA

Christophe Proisy, Nicolas Barbier, Raphael Pélissier and Pierre Couteron
Institut de Recherche pour le Développement (IRD), UMR AMAP, France

Michael Guéroult
Institut National de la Recherche Agronomique (INRA), UMR AMAP, France

Jean-Philippe Gastellu-Etchegorry and Eloi Grau
Université Paul Sabatier, UMR CESBIO, France

Natasha Ribeiro, Micas Cumbana, Faruk Mamugy and Aniceto Chaúque
Faculty of Agronomy and Forestry, Eduardo Mondlane University, Mozambique

Stefano Tebaldini
Politecnico di Milano, Italy

Antoni Jordi and Gotzon Basterretxea
Institut Mediterrani d'Estudis Avançats, IMEDEA (UIB-CSIC), Spain

Tiffany A.H. Moisan
NASA Goddard Space Flight Center, Wallops Island, USA

Shubha Sathyendranath
Plymouth Marine Laboratory, Bedford Institute of Oceanography, Prospect Place, the Hoe
Plymouth, Canada

Heather A. Bouman
Department of Earth Sciences, University of Oxford, South Parks Road, Oxford, UK

Ioannis Gitas and Anastasia Polychronaki
Laboratory of Forest Management and Remote Sensing, Aristotle University of Thessaloniki, Thessaloniki, Greece

George Mitri
Biodiversity Program, Institute of the Environment, University of Balamand and Department of Environmental Sciences, Faculty of Science, University of Balamand, Lebanon

Sander Veraverbeke
Department of Geography, Ghent University, Ghent, Belgium
Jet Propulsion Laboratory, California Institute of Technology, Pasadena, CA, USA

Evan Ellicott and Eric Vermote
University of Maryland, USA

Jacquelyn Kremper Shuman and Herman Henry Shugart
University of Virginia, Department of Environmental Sciences, USA

Meng Zhen Kang and Jing Hua
State Key Laboratory of CompSys, Institute of Automation, CAS, China
LIAMA, Institute of Automation, CAS, China

Thomas Corpetti
LIAMA, Institute of Automation, CAS, China
CNRS, National Center for Scientific Research, France

Philippe de Reffye
Cirad-Amis, Montpellier Cedex 5, France

Kishore C. Swain
Department of Agricultural Engineering, Triguna Sen School of Technology, Assam University, India

Qamar Uz Zaman
Engineering Department, Nova Scotia Agricultural College, Nova Scotia, Canada

Wenjun Chen, Weirong Chen, Junhua Li, Yu Zhang, Robert Fraser, Ian Olthof, Sylvain G. Leblanc and Zhaohua Chen
Canada Centre for Remote Sensing, Natural Resources Canada, Canada

A. García-Martín, J. de la Riva, F. Pérez-Cabello and R. Montorio
Department of Geography and Spatial Management, University of Zaragoza, Zaragoza, Spain

Jaime Hernández, Patricio Corvalán, Xavier Emery, Karen Peña and Sergio Donoso
Universidad de Chile, Chile